Practical
Pulmonary Pathology

Practical
Pulmonary Pathology

—————————————— Edited by ——————————————

MARY N SHEPPARD MD, MRCPath

Senior Lecturer and Honorary Consultant in Histopathology
Royal Brompton National Heart and Lung Hospital
London, UK

Edward Arnold
A member of the Hodder Headline Group
LONDON BOSTON SYDNEY AUCKLAND

First published in Great Britain 1995 by
Edward Arnold, a division of Hodder Headline PLC,
338 Euston Road, London NW1 3BH

Distributed in the Americas by
Little, Brown and Company
34 Beacon Street, Boston, MA 02108

Whilst the advice and information in this book is believed to be true and
accurate at the date of going to press, neither the author nor the publisher
can accept any legal responsibility or liability for any errors or omissions
that may be made. In particular (but without limiting the generality of the
preceding disclaimer) every effort has been made to check drug dosages;
however it is still possible that errors have been missed. Furthermore,
dosage schedules are constantly being revised and new side effects
recognized. For these reasons the reader is strongly urged to consult the
drug companies' printed instructions before administering any of the drugs
recommended in this book.

British Library Cataloguing in Publication Data
A catalogue record for this book is available from the British Library

Library of Congress-in-publication Data
Available upon request

ISBN 0 340 57318 X

1 2 3 4 5   95 96 97 98 99

Typeset in Times and Souvenir by
Preset Graphics, Derby
Printed and bound in Great Britain by
The Bath Press, Bath, Avon

Contents

Contributors

R M du Bois Interstitial Lung Disease Unit, Royal Brompton National Heart and Lung Hospital, London, UK

A R Gibbs Consultant Histopathologist, Llandough Hospital, Penarth, Wales, UK

Samuel P Hammar Director, Diagnostic Specialities Laboratory Inc., Bremerton, Washington, USA

D M Hansell Department of Radiology, Royal Brompton National Heart and Lung Hospital, London, UK

P G Isaacson Professor, Department of Histopathology, University College London, London UK

A Kennedy Department of Histopathology, Northern General Hospital, Sheffield, UK

James Linder Department of Pathology and Microbiology, University of Nebraska Medical Center, Omaha, Nebraska, USA

Bruce Mackay Professor, Department of Pathology, Texas Medical Center, Houston, Texas, USA

Marian Malone Department of Histopathology, Hospital for Sick Children, Great Ormond Street, London, UK

Alberto M Marchevsky Director, Division of Anatomic Pathology, Cedars-Sinai Medical Center, Los Angeles, California, USA

G D Phillips Interstitial Lung Disease Unit, Royal Brompton National Heart and Lung Hospital, London, UK

Stanley J Radio Associate Professor, Department of Pathology and Microbiology, University of Nebraska Medical Center, Omaha, Nebraska, USA

Mary N Sheppard Senior Lecturer and Honorary Consultant Histopathologist, Royal Brompton National Heart and Lung Hospital, London, UK

Susan Stewart Consultant Pathologist, Papworth Hospital, Cambridge, UK

C A Wagenvoort Professor, Department of Pathology, Erasmus University Medical School, Rotterdam, The Netherlands

Preface

My decision to edit this book was based upon the recognition that pathologists dealing regularly with pulmonary material needed a concise and practical benchbook. Therefore, I have invited pathologists from both the UK and USA to contribute in their area of expertise in pulmonary disease and asked them particularly to emphasize the practical problems faced by the pathologist working at the microscope. This I hope has made the book practical and accessible, helping pathologists to answer immediate problems faced in the course of their work.

In my time at the Royal Brompton Hospital, I have come to understand the difficulties that are associated with the interpretation of bronchial, transbronchial and open lung biopsies and have therefore decided to devote two chapters to this topic. Pathologists also have to deal with resection and autopsy lung specimens and a chapter is also given to this important subject. Many pathologists have limited exposure to interstitial lung disease and the interpretation of open lung biopsies. The chapter on this subject written with our clinical colleagues, emphasizes the clinical and radiological information which is essential to combine with the histology in interpreting the changes seen.

Recent advances have been made concerning the pulmonary complications of AIDS and also in lung transplantation which has expanded considerably since transbronchial biopsies became an important part of the monitoring of these patients. Lesser known aspects of pulmonary lung disease including vascular disease and paediatric lung disease are also dealt with, since the general pathologist may be called upon to give an opinion in such cases. There is a chapter on pneumoconiosis since these diseases are of particular importance especially in the medico-legal field and in coroner's work.

Pulmonary carcinoma is the most common fatal tumour in the Western world today and the most frequent reason for doing bronchial biopsies. We have devoted three chapters to pulmonary and mesothelial malignancy and also a chapter on cytology.

I wish to thank all the authors who kindly accepted my invitation to contribute to this book. All are well established figures in their field and deal on a daily basis with pulmonary problems. They have worked hard and were rapid in their response to deadlines, which was essential for success, especially with a multiauthor work such as this. We have tried to avoid repetition and I hope that this book will have the effect of increasing pathological understanding on either side of the Atlantic.

I also wish to thank the staff of the Histopathology Department and my medical and surgical colleagues at the Royal Brompton Hospital, without whom this book could not have been completed. As Herbert Spencer, one of the founding fathers of pulmonary pathology so wisely said, 'you never really know about a subject unless you write about it'. I have learned a considerable amount during the production of this book from both my pathological and clinical colleagues which has given me and I hope, our readers, a better understanding of pulmonary disease.

My interest in pulmonary pathology began when I completed an MD thesis on the neuroendocrine system of the lung with Professor Julia Polak at the Royal Postgraduate Medical School and continued while I completed my training at University College/Middlesex Medical School with Professor

Peter Isaacson. Following my appointment at The Royal Brompton Hospital, a tertiary referral centre for pulmonary disease, I have been lucky enough to have learned from the large volume of material coming into the department both from the UK and abroad. I wish to thank those colleagues who have in the past sent me valuable referral material which

forms such a large part of our teaching and learning.

I wish to give a special mention to my family for their patience while this book was being produced and with excellent timing, my daughter Francesca was born just as we completed the final text!

Mary N Sheppard

1

Anatomy of the lungs

Mary N Sheppard

General

In this first chapter I wish to refer specifically to the detailed anatomy of the bronchi and lung with special reference to the interpretation of bronchial, transbronchial, open lung biopsies and resection specimens. More comprehensive anatomical textbooks deal much better with the gross anatomy of the lung and its relationship to the rest of the mediastimun and thorax.

Main airway anatomy

The trachea divides into two main bronchi at the sternal notch, one main bronchus supplying each lung. The main bronchi enter the lung at the hilum along with the pulmonary artery and veins. The right bronchus is shorter and straighter than the left, which explains the propensity of foreign objects to lodge on that side (Fig. 1.1a). The lobar bronchus to the upper right lobe branches off before entering the lung at the hilum. The right pulmonary artery often divides into two branches also and these lie in front of the two bronchial branches as they enter the lung on the right side (the right main bronchus is thus described as eparterial), while the pulmonary artery on the left lies above the bronchus (left main bronchus is hyparterial). These points, as well as lobar anatomy, are useful in identifying mirror imaging or isomerism of the lungs which occurs in congenital heart disease. Once within the lung, the pulmonary arterial branches hook around to lie behind and lateral to the bronchi. That is why one can dissect the pulmonary arterial tree from a lateral approach while

the bronchial system can be dissected medially from the hilum. The two pulmonary veins leave the lungs on either side anterior to and inferior to the bronchi and arteries. The aortic arch curves over the left hilum while the azygous vein curves over the right hilum to join the brachiocephalic vein which lies anteriorly. The bronchial arteries arise from the descending aorta and enter both hila behind the bronchi and travel with the bronchi into the lung parenchyma. They also give rise to the vasa vasorum of the pulmonary arteries.

Each lung is divided into an upper and lower lobe by the oblique fissure. In addition, on the right side there is a transverse fissure which creates a middle lobe so that there are three lobes on the right and two on the left. The right main bronchus, after giving off the upper lobar branch, continues through an intermediate part to reach the middle and lower lobes. The left main bronchus does not have an intermediate part and after giving off the upper lobe bronchus continues as the lower lobe bronchus (Fig. 1.1a). The main bronchi divide into five lobar bronchi. The lobar bronchi branch further to supply a segment within the lobe. This is surgically important since this is the smallest piece of lung that can be identified and removed by the surgeon, together with its airway and pulmonary artery. The branching is irregular and is described as being asymmetrical dichotomous in type. The right middle lobe is supplied by a separate branch from the intermediate bronchus which divides into medial and lateral segments. On the left side the lingula, which is the equivalent of the right middle lobe, is supplied by a branch of the left upper lobe bronchus; this divides into superior and inferior segments. There are 19 segments in total, three in

(a)

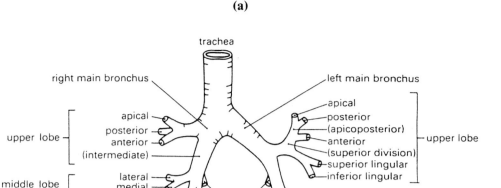

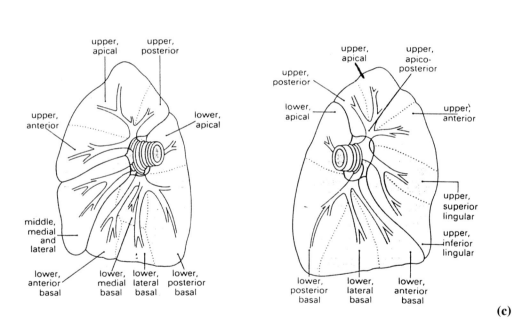

(b)   **(c)**

Fig. 1.1a   Shorter right main bronchus, longer left main bronchus, right intermediate bronchus, lobar and segmental branching of bronchi in both lungs.

Fig. 1.1b   Medial aspect of the right lung showing the segmental and subsegmental bronchi and the 10 lobar segments in the right lung.

Fig. 1.1c   Medial aspect of the left lung with segmental and subsegmental bronchi and the nine lobar segments.

the upper lobes, two in middle/lingual lobes and five in the right lower lobe (Fig. 1.1b). There are four segments in the left lower lobe, which does not have a medial basal segment since the heart occupies this area on the left side (Fig. 1.1c). The segments have been given internationally accepted titles (Figs 1.1 a–c) which were first agreed upon in 1949.[1] It is essential for the pathologist to become

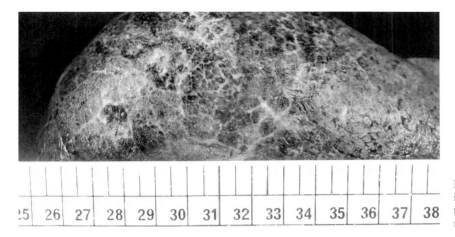

| ‖25| 26| 27| 28| 29| 30| 31| 32| 33| 34| 35| 36| 37| 38|

Fig. 1.2 Lung periphery with individual lobules outlined on the pleural surface with prominence of the interlobular septa.

familiar with these segments in order to localize accurately any lesion within the lung.

Beyond the segmental bronchi, the airways continue to branch. There can be up to 23 divisions before the air-exchanging area of the lung is reached at the periphery of the lung, while at the hilum this can occur after only eight divisions. The bronchi have cartilage in their walls which gradually disappears as branching continues and the airways become known as bronchioles. The earliest bronchiole has an internal diameter of 2 mm. The last of the purely conductive airways are known as terminal bronchioles. Around all the conduction airways is a wide sheath of connective tissue which also encloses the accompanying pulmonary artery and contains nerves and lymphatics, which disappear at the level of the respiratory bronchioles.

The working unit of the peripheral lung is the lobule. This is supplied by a central bronchiole, measures 1–2 cm across, and is delineated by incomplete fibrous septa that may be visible on the surface of the lung (Fig. 1.2). These septa are well developed on the lateral aspects of the lower lobes, where they can be seen as Kerley B lines when thickened, but are poorly developed medially and deep in the lungs. Each lobule contains 3–10 acini, each of which is supplied by a terminal bronchiole which arises from the central bronchiole (Fig. 1.3).

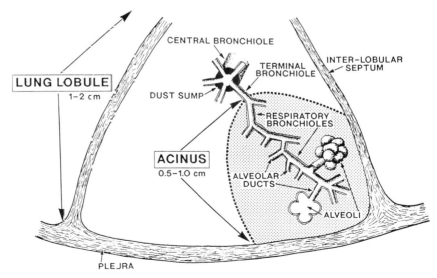

Fig. 1.3 A peripheral lung lobule and the individual acinus in diagrammatic form with central bronchiole, dust sump, terminal bronchioles, respiratory bronchioles, alveolar ducts and alveoli. (Courtesy of Dr Leon Gerlis, Royal Brompton National Heart and Lung Hospital.)

It has been estimated that there are about 20 000–30 000 terminal bronchioles, each supplying an acinus.[2] Each acinus is 0.5–1 cm long in the adult and contains up to 2000 alveoli.[3] Acini cannot be separated from each other and cannot be detected either macroscopically or microscopically. The centre of the acini can be seen macroscopically because carbon accumulates within the lymphatics in the adventitia around the central bronchioles, between the bronchiole and its accompanying artery in areas known as 'dust sumps' (Fig. 1.4).[4] As the airways divide, the cross-sectional area increases logarithmically so that there is minimal resistance to airflow once the periphery is reached. The terminal bronchioles branch to form respiratory bronchioles which,

because they have no alveoli opening from them, are known as membranous bronchioles. These are the last of the conductive airways with no gas exchange across them. The gas-exchanging zone comprises three generations of respiratory bronchioles and between two and nine generations of alveolar ducts which terminate in the alveolar sacs and alveoli. Alveoli open from all these branches (Figs 1.3 and 1.13) It is estimated that each lung contains 300 million alveoli.[5] The gas exchange occurs by diffusion across an area of some 70 m. Lambert's canals are tubules 30 µm in diameter which connect the terminal and respiratory bronchioles with adjacent peribronchiolar alveoli. These are early sites of dust accumulation in city dwellers and coalminers. In addition connections between segments, acini and respiratory bronchioles to have been described varying from 80 to 200 µm in diameter. Between alveoli are the pores of Kohn, 2–13 µm in diameter, which are pathways for the spread of infection or granulation/fibrous tissue in several disease processes. Thus an extensive interconnecting system is present which provides for the exchange of gases with the circulation.

Tracheal/bronchial wall histology

The tracheal/bronchial wall consists of a mucosal and submucosal layer and the outer advential layer of loose connective tissue. The mucosa contains ciliated pseudostratified columnar epithelium which lies upon a basement membrane (Figs 1.5–1.7). Beneath this is an ill-defined layer of connective tissue known as the lamina propria (Fig. 1.5) which contains blood vessels, occasional mononuclear cells, nerves, some collagen, fibroblasts and many elastic fibres. There is no clear boundary between the mucosa and the submucosa, but where cartilage, glands and muscle are, is considered to be submucosa (Figs 1.6 and 1.7). Ridges which are obvious in the mucosa represent thick longitudinal bundles of elastin fibres situated in the subepithelial mucosal lamina propria.[6] The cartilage forms horseshoe-shaped rings which are incomplete posteriorly in the trachea. They are joined together vertically by strong connective tissue and the tips of the horseshoes are joined together posteriorly by a transverse band of smooth muscle. All tissue outside the cartilage and muscle is considered adventitia and contains loose

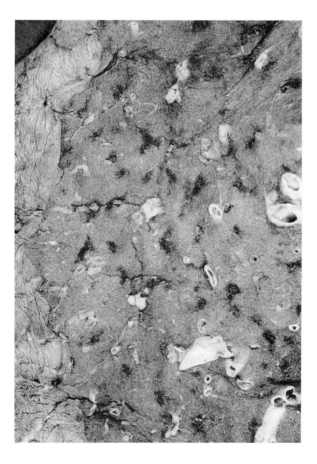

Fig. 1.4   Peripheral lung demonstrating the centres of acini with terminal bronchioles surrounded by a dark pigmented round spot containing carbon, the 'dust sump'.

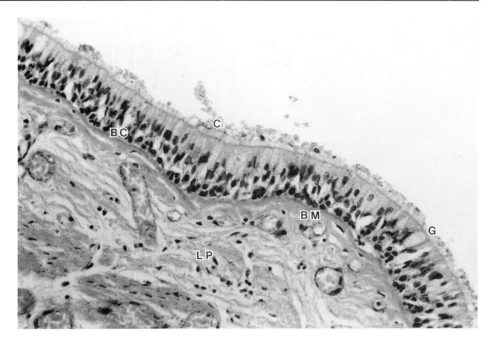

Fig. 1.5 Bronchial pseudostratified ciliated epithelium showing the ciliated columnar cells (C) with goblet cells (G) which appear vacuolated and basal cells with small hyperchromatic nuclei (BC). Note also the basement membrane (BM) and underlying lamina propria (LP).

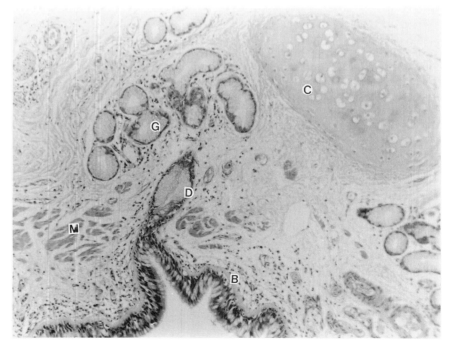

Fig. 1.6 Bronchial wall showing epithelium, basement membrane (B) secretory ducts (D), seromucous glands (G), smooth muscle (M) and irregular plates of cartilage (C).

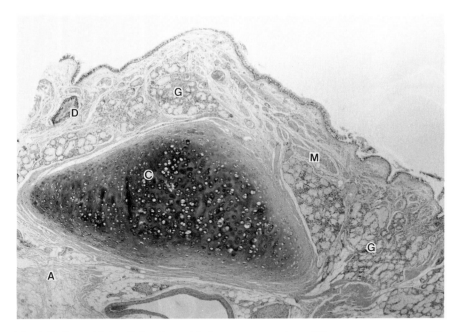

Fig. 1.7 Bronchial wall showing epithelium, secretory ducts (D), seromucous glands (G), smooth muscle (M) and irregular plates of cartilage (C). Outide the muscle and cartilage lies the adventitia (A).

connective tissue, ganglion cells, nerves and blood vessels (Fig. 1.7). In the main bronchi, the cartilage forms incomplete branching rings, so that in any cross-section it can appear as irregularly shaped plates (Figs 1.6 and 1.7). In the trachea and extra-pulmonary bronchi, smooth muscle lies dorsally in the intercartilaginous gap. In the intrapulmonary bronchi the muscle completely encircles the airways internal to the cartilage (Figs 1.6 and 1.7) The smooth muscle forms two bundles of fibres that wind around the bronchi as opposing spirals and appear incomplete in cross-section. When the respiratory bronchioles are reached the smooth muscle only appears as small knobs between the mouths of the numerous alveoli which open off the bronchiole (Fig. 1.8). These can at times be mistaken for interstitial thickening and are a normal feature, often seen prominently in the lungs of infants and children.

The epithelium of the respiratory tree is pseudo-stratified because all cells rest on the basement membrane, but not all reach the airway lumen (Fig. 1.5). Most of the epithelial cells are ciliated and columnar. At the ultrastructural level the ciliated cells possess numerous mitochondria and have short microvilli 0.4 µm in length and long slender cilia 6 µm in length (Fig. 1.9). There are 200–300 cilia per cell which beat at a frequency of 20 times per second, moving the surface mucus up into the oropharynx in a process known as the 'mucociliary escalator'. This is important in removing foreign antigens including bacteria from the respiratory tract. Other epithelial cells include mucous (goblet) cells, clear neuroendocrine cells and dark basal cells (Fig. 1.5). The airway epithelium decreases in height until it becomes simple cuboidal in the bronchioles. Mucous cells also decrease peripher-ally and are rarely found beyond bronchioles of less than 1 mm diameter. Two types of mucous cell are recognized. The most common mucous cell contains high-molecular-weight glycoproteins which are acidic and take up combined Alcian blue (AB) pH 2.5/periodic acid Schiff (PAS) or high iron diamine (HID)/AB stains. The acidic mucins stain blue while neutral mucins stain pink with AB/PAS. The combined HID/AB technique demonstrates the two main types of acidic glyco-proteins; those with sulphomucins stain brown and those with sialomucins stain blue. In adult man the majority of surface mucous cells contain both

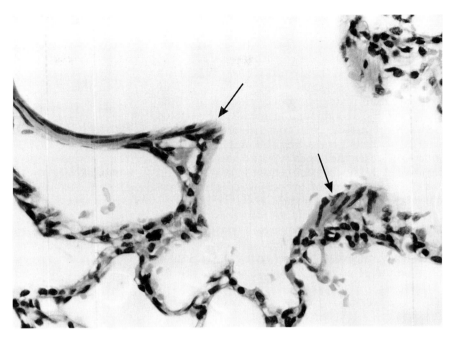

Fig. 1.8 Respiratory bronchiole with alveolar duct and alveoli opening from it with smooth muscle at the mouth (arrow).

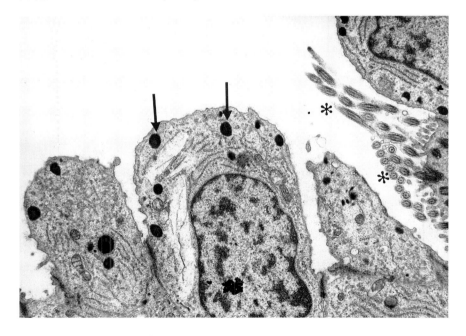

Fig. 1.9 Electron micrograph of human bronchiolar epithelium showing Clara cells with large electron-dense granules (arrows). Also note the ciliated cell with surface microvilli and cilia in both longitudinal and cross-section (asterisk) with internal microtubules just visible. × 10 800. (Courtesy of Ann Dewar, electron microscope unit, National Heart and Lung Institute.)

neutral and acidic glycoproteins with both sulphomucins and sialomucins being represented. Several studies indicate that a highly sulphated acidic secretion is associated with the surface.[7] The less common cell is the serous cell, which contains neutral mucins. Its secretion has less carbohydrate with no penultimate or terminal galactose and little sialic acid.[8] The submucosal glands are situated between the muscle coat and the cartilage plates with their ducts penetrating the muscle coat and mucosa to reach the lumen (Figs 1.6 and 1.7). They far exceed the surface epithelium in cell mass and are the main source of bronchial secretion. They are mixed seromucous glands.

The neuroendocrine cells are rarely recognized at light microscopy but possess intracytoplasmic dense-core granules 90–300 nm in diameter at the ultrastructural level. These cells are usually basal in position but often have a thin cytoplasmic projec-tion reaching the airway lumen which can be seen using immunocytochemical markers (Fig. 1.10). They are very rare in the adult respiratory tract but occur with greater frequency in the fetus and neonate, where they are most commonly found in bronchioles.[9]

Bronchiolar histology

The epithelium decreases in height distally, be-coming low cuboidal in the bronchioles (Fig. 1.11). The mucous cells also decrease in number, as do the basal cells. A cell unique to bronchioles called the Clara cell, which is considered to be the pro-genitor cell in this part of the airway, appears as a non-ciliated cell which protrudes above the surface of the ciliated cells and contains electron-dense granules of 500–600 nm diameter (Fig. 1.9). It is

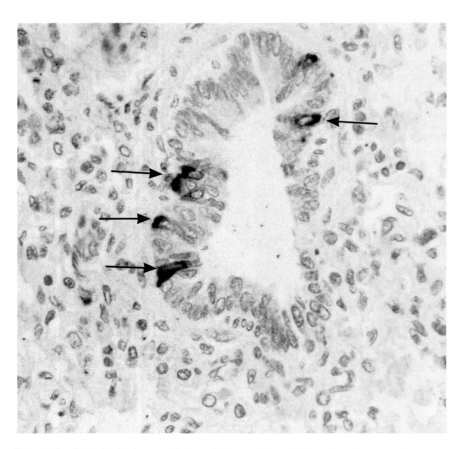

Fig. 1.10 Bronchiole from a fetus of 20 weeks' gestation showing several neuro-endocrine cells with long cytoplasmic processes extending towards the lumen (arrows). Immunostaining with neuron specific enolase antibody.

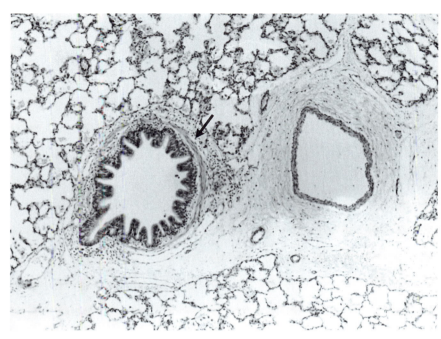

Fig. 1.11 Bronchiole with low cuboidal epithelium, no cartilage or glands and a complete ring of smooth muscle (arrow). Note the accompanying muscular pulmonary arterial branch with thin wall and wide lumen. Lymphatics and bronchial/pulmonary arterial branches are seen in the enveloping connective tissue sheath.

often impossible to recognize this cell at light microscopy but it can have a 'hobnail' appearance with the nucleus protruding above the surrounding basal nuclei of the columnar cells, particularly seen in bronchioloalveolar cell carcinomas with Clara cell differentiation. Beneath the epithelial basement membrane is the submucosa which contains only a thin layer of loose connective tissue. Both cartilage and submucosal glands disappear at the level of the bronchioles so that the mucosa and submucosa are completely surrounded by a band of circular smooth muscle (Fig. 1.11). Outside the muscle lies the adventitia containing loose connective tissue, nerves and blood vessels as well as lymphatics.

Peripheral lung histology

The periphery of the lung consist of the individual lobules with the overlying visceral pleura. Each lobule contains a central bronchicle with its accompanying pulmonary arterial branch (Figs 1.3 and 1.12a–b). The central bronchiole gives rise to

3–10 terminal bronchioles, which supply each acinus (Fig. 1.3); these are the last airways completely lined by ciliated columnar epithelium. The terminal bronchioles branch to give three generations of respiratory bronchioles, whose epithelium is interrupted by alveolar ducts with alveoli branching from them (Fig. 1.13). As the terminal and respiratory bronchioles penetrate into the lungs, the epithelium changes to become flat and squamoid in type in the alveolar ducts and alveoli (Fig. 1.13). The walls of the alveolar ducts consist of thin spiral bands of elastin and collagen, between which are the mouths of alveoli.[10] The alveoli are polyhedral and measure 250 μm in diameter when expanded, although gravitational forces result in them being bigger in the upper lobes. The alveolus is lined by type 1 and 2 pneumocytes (Figs 1.14 and 1.15). The epithelium is separated from the underlying connective tissue by a basement membrane. Type 2 cells are taller and twice as numerous as type 1 cells, but they cover only 7% of the alveolar surface. Type 2 cells are usually found in the corners of the alveolus. They contain lamellar bodies and are the source of

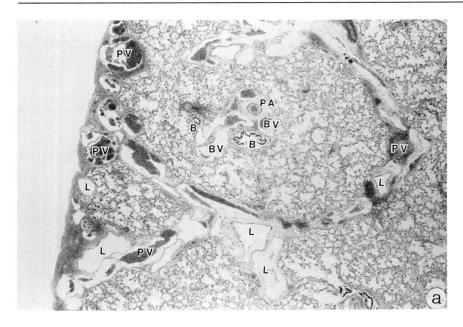

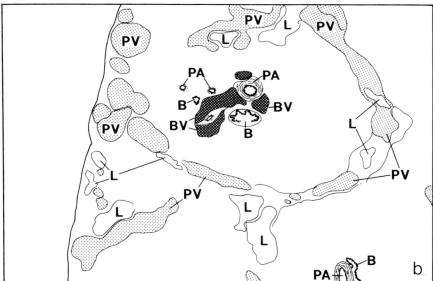

Fig. 1.12a,b Peripheral lung lobule with central bronchiole (B), accompanying pulmonary artery (PA) and accompanying bronchial veins (BV). Note that the pleura and interlobular septa are distended by dilated pulmonary veins (PV) and lymphatics (L). (Courtesy of Dr Leon Gerlis, Royal Brompton National Heart and Lung Hospital.)

surfactant, the surface tension lowering agent (Fig. 1.15). The type 2 cell acts as progenitor for type 1 cells when they become damaged and proliferate as a general response to injury. The type 1 cell has long cytoplasmic processes which enable it to cover most of the alveolar surface (Fig. 1.15). The alveolar interstitium contains collagen and elastin fibres with some fibroblasts, myofibroblasts, mast cells, macrophages and nerves. Capillaries are present (Figs 1.14 and 1.15) and the actual inter-

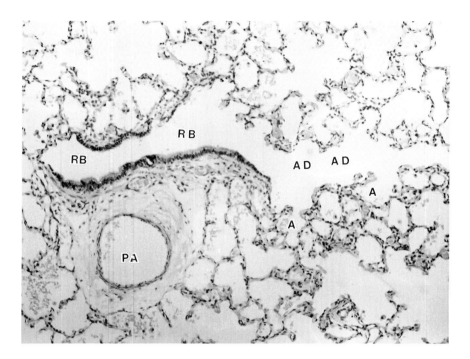

Fig. 1.13 Gas-exchanging airways of the peripheral lung with respiratory bronchioles (RB), alveolar ducts (AD) and alveoli (A) opening from it. Note the accompanying pulmonary artery (PA).

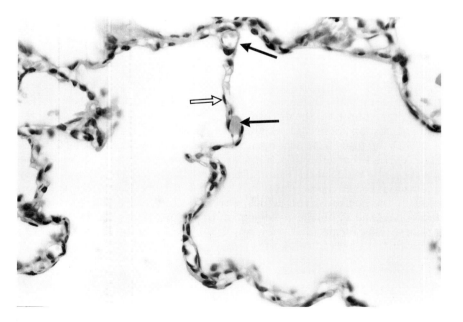

Fig. 1.14 Neighbouring alveoli with lining pneumocytes (open arrow) and capillaries between them (black arrows). Interstitium is inconspicuous.

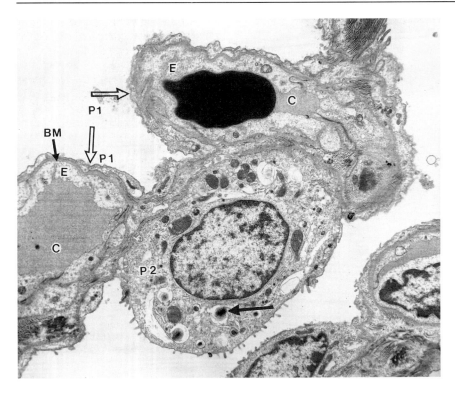

Fig. 1.15   Electron micrograph of human alveolar wall showing the type 2 pneumocyte (P2) with intracytoplasmic lamellar bodies (large black arrow). On the other side, thin cytoplasmic extensions of type 1 pneumocyte (P1) are seen (open arrows) which cover most of the alveolar surface and form the blood–gas barrier along with the basement membrane (BM) and the endothelial cell (E). Note the capillary lumen (C), one containing a red blood cell. × 5900. (Courtesy of Ann Dewar, electron microscope unit, National Heart and Lung Institute.)

stitium is bounded by the epithelial and endothelial basement membranes. The diffusion of gases takes place across alveolar and capillary endothelial walls with the fused basement membranes between (Fig. 1.15) so that any infiltration or thickening here results in reduced diffusion. The interstitium is continuous with the peribronchiolar connective tissue in the centre of the acinus. Free alveolar macrophages lie within the alveoli and mop up dust particles. This means that they are usually pigmented, particularly in smokers.

Blood supply

The pulmonary arteries enter the lungs at the hilum and accompany the bronchi and bronchioles down to the level of the respiratory bronchioles. Although the pulmonary veins also lie in loose connective tissue sheaths adjacent to the main bronchus at the hilum, once inside the lungs they follow Miller's dictum that the veins will generally be found as far away from the airways as possible. The elastic pulmonary arteries accompany the bronchi down to bronchiolar level. They usually exceed 1000 μm in external diameter. Their media consist of sheets of elastic fibres with intervening layers of smooth muscle, collagen and ground substance. The internal elastic membrane is not conspicuous because it is one of the many elastic layers in the wall. The muscular pulmonary arteries begin at the point where the airways lose their cartilage and become bronchioles. During this transition, the elastic laminae become disrupted

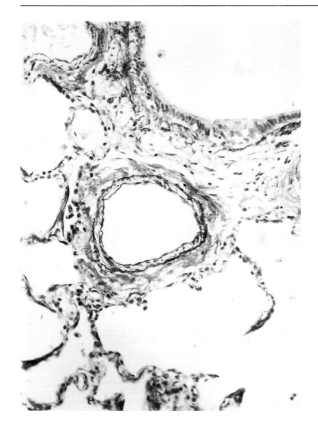

Fig. 1.16 Muscular pulmonary artery with internal and external elastic layers and muscle in between. Elastic van Gieson stain.

and incomplete and diminish in number. Muscle increasingly makes up most of the media, with the formation of prominent internal and external elastic lamina (Fig. 1.16). Most arteries below 500 μm in diameter are completely muscular. The muscular arteries have thin walls and a large lumen. These vessels are the chief site of resistance in the pulmonary circulation. They branch with the bronchial tree and lie close to the bronchioles, respiratory bronchioles and alveolar ducts Usually the bronchi/bronchioles and pulmonary arterial branches are similar in size (Figs 1.11 and 1.13) and lie side by side for easy comparison. However if the lungs have been inflation fixed, the bronchioles will be bigger. When arterial and bronchial divisions occur at different planes, a single bronchus may be accompanied by two arteries of much smaller calibre, or vice versa. The average thickness of the media when measured between internal

and external elastic lamina, is approximately 5% of the external diameter, with a range of 3–7%. The external diameter is measured between the external elastic lamina on either side. There can be individual variation between vessels and one needs to sample extensively to detect any changes as seen in hypertension. The pulmonary arterial wall is very thin compared with that of systemic arteries, reflecting the different pressures in the two systems. In the first few weeks of life, the muscular pulmonary arteries have a much thicker media with narrow lumen which gradually attenuates until the medial thickness assumes the adult ratio at approximately 1 year.

The muscle of the media gradually thins out as the arteries branch and become smaller. In vessels between 30 and 100 μm in diameter, the muscle layer consists of a spiral, so that in cross-section the muscle is visible only in part of the circumference of the vessel. Vessels with diameters below 30 μm have no muscle in their walls, which consist only of fibroelastic tissue. Therefore trying to separate arterioles from venules is impossible at this level and all the vessels are commonly referred to as arterioles. The term arteriole is reserved, therefore, for vessels less than 100 μm in diameter or where they lose their contact with the bronchioles and become intra-acinar. The walls of such vessels consist of an endothelial lining, a single elastic lamina and, in the normal state, no media. The adventitia is thin and ill-defined. It is impossible to distinguish arterioles from venules unless one traces the parent vessel by serial sectioning. The pulmonary arterioles divide to supply the pulmonary capillaries which form a network situated in the alveolar walls (Figs 1.14 and 1.15), regrouping at the periphery of the acinus as venules which pass into the interlobular septa as pulmonary veins (Figs 1.12a,b). The veins are 60–100 μm in diameter entering the septa. The pulmonary veins are thus separate from the pulmonary artery and airway in the centre of the lobule and when distended will outline beautifully the borders of the pulmonary lobule (Figs 1.12a,b). The walls of the larger veins contain many collagen and elastic fibres with fewer smooth muscle cells than arteries and do not form internal and external elastic laminae as the muscular pulmonary arteries do. The largest veins possess an internal elastic lamina. Beyond the level of the lobule they take up a position alongside the artery and bronchus. With increased venous pressure more muscle may be found in the venous walls so that in disease states, the structure of the vessel wall may not be a totally reliable criterion for separating arteries from veins. The location of the vessels is thus of utmost importance in distinguishing arteries from veins, with arteries being present in a centriacinar position and accompanying the airway whereas the veins are found in the periphery of the lobule and enter the interlobular septa (Figs 1.12a,b).

The bronchial circulation consists of one or more arteries branching from the aorta or upper intercostals at the level of the hilum. These arteries travel into each lung with the main bronchi as far as the terminal bronchioles. The bronchial arteries supply the walls of the airways, while the trachea is supplied by branches of the inferior thyroid artery which anastomose with the bronchial arteries. The bronchial arteries also supply the pleura and interlobular septa. Bronchial arteries have a thicker media than pulmonary arteries and lack an outer elastic membrane. In addition they are much smaller in calibre. They travel in the loose connective sheath which surrounds the airways and branch repeatedly, giving rise to a capillary network in the submucosa and adventitia. Beyond the bronchi the identification of these vessels is difficult because both bronchial and pulmonary arteries anastomose at the precapillary, capillary and venous levels. The bronchial veins are closely associated with the bronchial arteries but are difficult to distinguish from pulmonary veins with which they anastomose. Up to 75% of the bronchial venous flow drains into the pulmonary venous system. Venous return is by small bronchial veins which drain into the azygous or hemizygous veins. In long-standing inflammatory and proliferative diseases (cystic fibrosis, bronchiectasis, carcinoma) bronchial blood supply may be greatly increased. All scar tissue and tumours receive their blood supply from the bronchial circulation.

There are no lymphatics in the alveolar wall. They commence in the centre of the acini at the level of the terminal bronchioles and in the interlobular septa (Figs 1.12a,b). Fluid drains from the alveoli into the loose connective tissue of the adventitia which surrounds the terminal bronchioles and pulmonary artery and into the lymphatics. The lymphatics accompany the airways and pulmonary arteries to the hilum where they are joined by branches from a subpleural plexus on the visceral pleura. Lymph from the upper lung lobes drains into the tracheobronchial lymph nodes on either side of the trachea, while both lower lobes drain

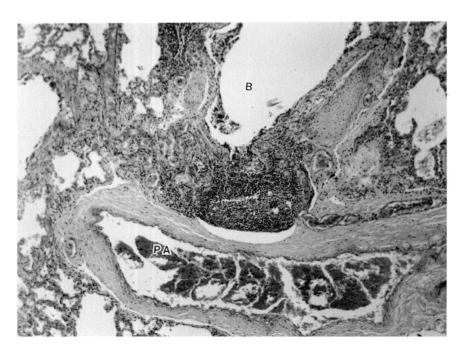

Fig. 1.17 Bronchus and accompanying pulmonary artery between which there is bronchial associated lymphoid tissue (BALT).

into the infratracheal groups of lymph nodes. The lymphatics are very wide in relation to their thickness.

Encapsulated lymph nodes can be found within the lungs, usually related to large bronchi, but can also be found in the peripheral lung and even in the visceral pleura. Lymphoreticular aggregates consist of small aggregates of lymphocytes with a few plasma cells and eosinophils and often lie in the connective tissue of the bronchi, in the interlobular septa and in the pleura. They lack the architecture of a normal lymph node. They are particularly prominent between terminal and respiratory bronchioles and their accompanying arteries. Some lymphocytes are arranged into lymphoid follicles which are closely related to the overlying epithelium. These are known as bronchial associated lymphoid tissue (BALT)[11] and are more prominent in smokers. These can be seen particularly between the bronchus/bronchiole and pulmonary artery (Fig. 1.17). The overlying epithelium is flattened, non-ciliated and contains lymphocytes which are derived from the outer mantle region of the underlying lymphoid follicle. Lymphatics are not seen in these follicles but they contain high endothelial postcapillary venules in the parafollicular region, through which lymphocytes traffic through the lung. Occasional lymphocytes, plasma cells, mast cells and macrophages can be found in the walls of bronchi and can extend down into the alveoli.

Innervation

Both components of the autonomic nervous system enter the lung at the hilum and there are ganglia and nerve bundles in the walls of the bronchi and bronchioles. After entering the lung, the nerves divide into peribronchial and perivascular plexuses. The peribronchial plexus gives rise to two zones, one lying outside the cartilage plates in the adventitia while the other lies in the mucosa beneath the surface epithelium. Using immunohistochemical techniques and general neural markers, a rich innervation within the bronchial wall with both ganglion cells supported by Schwann cells and large nerve bundles are identified in the airways down to bronchiolar level.[12] The adrenergic innervation which causes bronchodilatation with vasodilatation and stimulation of glandular secretion can be delineated using antibodies to thyrosine-hydroxylase (Fig. 1.18),[12] while adrenoceptors can be delineated using *in situ* hybridization.[13] The presumed cholinergic innervation which mediates bronchoconstriction and vasocon-

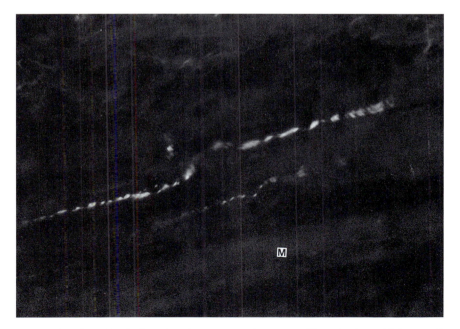

Fig. 1.18 Thyrosine-hydroxylase-immunoreactive nerve fibres in bronchial smooth muscle (M). Indirect immunofluorescence.

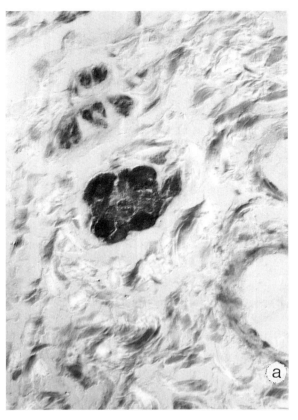

Fig. 1.19a   Group of cholinesterase-positive ganglion cells in the submucosa of human bronchus. Cholinesterase technique.

striction with decreased glandular secretion can be demonstrated by the acetylcholinesterase enzymatic reaction. This shows a dense innervation throughout the airways where the majority of ganglion cells and nerve fibres (Figs 1.19a,b) are acetylcholinesterase-positive. Antibodies to neuron-specific enolase are excellent general neuronal markers that work in formalin-fixed paraffin-embedded tissues and delineate much more than the other techniques. They indicate that in addition to the other two components, there is a rich nerve complex, known as the peptidergic nervous system,[14] which includes vasoactive intestinal polypeptide-immunoreactive ganglion cells and nerves (Figs 1.20a,b). The peptidergic nervous system is believed to mediate the non-cholinergic, nonadrenergic muscle relaxation seen in the mammalian respiratory tract. In addition sensory nerve fibres containing substance P can be detected in the lining epithelium of the bronchi and bronchioles.

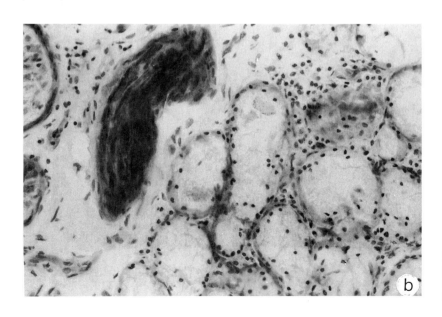

Fig. 1.19b   Large cholinesterase-positive nerve bundle close to seromucous glands in the bronchial submucosa. Cholinesterase technique.

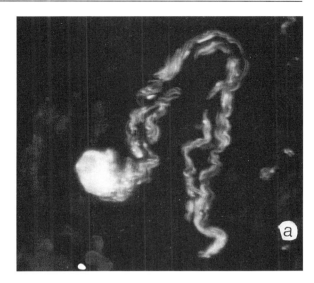

Fig. 1.20a Vasoactive-intestinal polypeptide-immuno-reactive ganglion cell in the submucosa of human bronchus. Indirect immunofluorescence.

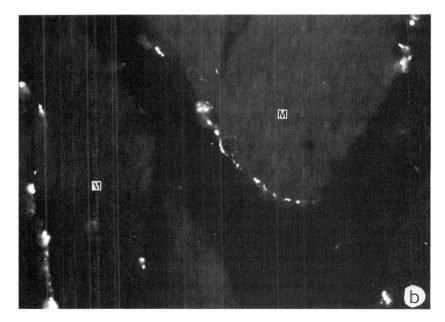

Fig. 1.20b Vasoactive-intestinal polypeptide-immunoreactive nerve fibres in the bronchial smooth muscle (M). Indirect immuno-fluorescence.

Pleura

In an open lung biopsy the specimen is usually from the lung periphery and will include pleura. The pleura consists of loose connective tissue covered by a layer of mesothelium. On the pleural surface the loose connective tissue continues into the interlobular septa. The loose connective tissue contains lymphatics, bronchial arteries, both pulmonary and bronchial veins (Figs. 1.12a, b) as well as nerves.

Acknowledgement

I wish to thank Dr Leon Gerlis for the artistic diagrams provided for this chapter and Mr Richard Florio and Ms Ann Dewar for the photographs.

References

1. Sealy WC, Connally SR, Dalton ML. Naming the bronchopulmonary segments and the development of pulmonary surgery. *Ann Thorac Surg* 1993; **55**, 184–188.
2. Horsefield K, Cummings G. Morphology of the bronchial tree in man. *J Appl Physiol* 1968; **24**, 373–378.
3. Haefeli-Bleuer B, Weibel ER. Morphometry of the human pulmonary acinus. *Anat Rec* 1988; **220**, 401–414.
4. Macklin CS. Pulmonary sumps, dust accumulations, alveolar fluid and lymph vessels. *Acta Anat* 1955; **23**, 1–33.
5. Dunnill MS. Postnatal growth of the lung. *Thorax* 1962; **17**, 329–333.
6. Monkhouse WS, Whimster WF. An account of the longitudinal mucosal corrugations of the human tracheobronchial tree, with observations on those of some animals. *J Anat* 1976; **122**, 681–695.
7. Spicer SS, Schulte BA, Chakrin LW. Ultrastructural and histochemical observations of respiratory epithelium and gland. *Lung Res* 1983; **4**, 137–156.
8. Jeffery PK, Gaillard D, Moret S. Human airway secretory cells during development and in mature airway epithelium. *Eur Respir J* 1992; **5**, 93–104.
9. Sheppard MN, Marangos PJ, Bloom SR, Polak JM. Neuron specific enolase: a marker of nerves and endocrine cells in the human lung. *Life Sci* 1984; **34**, 265–271.
10. Whimster WF. The microanatomy of the alveolar duct system. *Thorax* 1970; **25**, 141–149.
11. Gould SJ, Isaacson PI. Bronchus-associated lymphoid tissue (BALT) in human fetal and infant lung. *J Pathol* 1993; **169**, 229–234.
12. Sheppard MN, Kurian SS, Henzen-Logmans SC *et al*. Neuron-specific enolase and S-100: new markets for delineating the innervation of the respiratory tract in man and other mammals. *Thorax* 1983; **38**, 333–340.
13. Mak JCW, Hamid Q, Sheppard MN *et al*. Localization of beta-2 adenoreceptor mRNA in human and rat lung using *in situ* hybridization: correlation with receptor autoradiography. *Eur J Pharmacol* 1991; **206**, 133–138.
14. Sheppard MN, Polak JM. The localization of neuropeptides in the mammalian respiratory tract. In: *Asthma. Clinical Pharmacology and Therapeutic Progress*, Kay AB (ed.). London: Blackwell Scientific Publications, 1986, pp. 73–92.

2

Bronchial biopsy and its applications

Mary N Sheppard ─────────────────────────────

Introduction

The indications for bronchoscopy are numerous and arise mainly from abnormal radiographic appearances combined with respiratory symptoms. The procedure is generally well tolerated, with a low morbidity and few contraindications. Bronchoscopy is well established as a diagnostic tool for a large range of pulmonary disorders.[1,2] The most frequent indication is the diagnosis and assessment of bronchogenic carcinoma. The widespread acceptance of the flexible fibreoptic bronchoscope has to some extent eclipsed the use of the rigid bronchoscope. The rigid bronchoscope requires general anaesthesia while the flexible fibreoptic bronchoscope requires only topical anaesthesia and the range of visible airways is greater than with the rigid instrument. Today most biopsies are done with samples obtained by fibreoptic bronchoscopy. The bronchoscopist can view down to segmental bronchial level but can push the bronchoscope beyond this to get to the periphery of the lung. Bronchial biopsy is performed for endobronchial or bronchial wall lesions while the techniques of bronchoalveolar lavage and transbronchial biopsy are widely used for more diffuse parenchymal lung disease.[3] The rigid bronchoscope continues to be used because the wide calibre allows the better removal of blood, secretions and tumour tissue with more control over manipulations, i.e. for bleeding, laser therapy, diathermy, removal of foreign bodies, etc.

Handling and processing of biopsies

The biopsy pieces are about 2 mm in diameter. They are put immediately into formal saline. For rapid processing biopsies are put into a small nylon bag. Rapid processing requires a higher temperature than normal processing and biopsies are ready within 3 hours for cutting and staining. If the biopsies go on an overnight schedule, they can be embedded in agar to prevent loss of fragments in processing. The samples are levelled at 30–50 μm intervals to give three sections on each of three glass slides. Three spare unstained sections are taken at each of the three levels for further special stains, if required. In our laboratory we normally do a periodic acid Schiff stain with diastase (PASD) if neoplasia is suspected. Samples delivered to our laboratory by 11 a.m. can be reported on by 4 p.m. with rapid processing.

Another method used is serial sections when approximately 80 sections are placed on a total of three slides with the paraffin strips being placed across the narrow dimension of the glass slide. The pathologist samples these by reading the mid-point portion of tissue across the longer dimension of the slide. The fragments on either side may be then referred to if abnormalities are detected. Between every 20–30 sections, several unstained sections are placed on each of two slides and kept for further investigation. A total of approximately two-thirds of the paraffin block is cut through with one-third held for future reference. This method is often applied to transplant biopsies.

Special investigations

The following may be carried out:

1. Periodic acid Schiff with diastase (PASD) for neutral mucins in adenocarcinoma, pulmonary alveolar lipoproteinosis and *Pneumocystis*.
2. Alcian blue pH 2.5 (AB) stains acidic mucins, which combined with PASD will demonstrate both neutral and acidic mucins in the respiratory epithelium.
3. Alcian blue pH 2.5/PASD for adenoid cystic carcinoma to demonstrate basement membrane material as well as neutral mucin.
4. Periodic acid Schiff (PAS)/Grocott silver stains for fungi and *Pneumocystis*.
5. Ziehl–Neelsen/Auramine-O for mycobacteria.
6. Reticulin Elastic van Gieson for fibrosis.
7. Argyrophilic Grimelius silver stain for neuro-endocrine tumours.
8. Immunohistochemistry for tumour markers, *Pneumocystis* antibody, viral antibodies.
9. Congo red for amyloid.
10. Perls Prussian blue reaction for iron.
11. Von Kossa for calcium.
12. Polarizing for foreign birefringent material.

Reporting bronchial biopsies

The adequacy of the biopsy must first be assessed by looking at each individual layer. Is there epithelium, lamina propria, submucosa, bronchial glands and cartilage present within the biopsy? Is lung parenchyma included in the specimen? In addition the number of specimens should be noted. It is well established that the more specimens that are taken, the higher the diagnostic yield.[4] It should be remembered that the contents of the bronchial lumen may lie quite separate from the biopsy fragment so bacteria, neutrophils, necrotic tissue, fungi, tumour (Fig. 2.1), desquamated epithelial cells, eosinophils, Charcot–Leyden crystals and viral inclusions in cells can be detected by looking at these separate fragments. In addition all levels must be closely examined so that focal lesions such as granulomas will not be missed.

Incidental lipochrome is occasionally observed in bronchial mucosal cells with no link to disease. It is usually seen as variously sized tan-to-golden round-to-oval inclusions between the nuclei and borders of the ciliated respiratory cells. Oncocytic changes can be seen in the bronchial seromucous

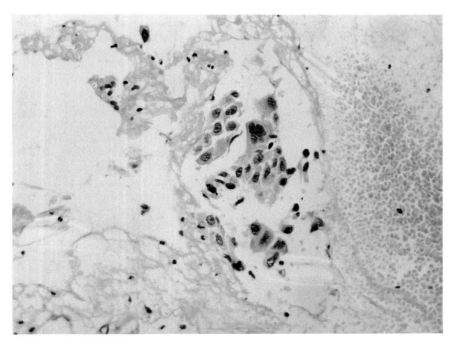

Fig. 2.1 Section of bronchial biopsy where a separate fragment shows carcinoma, non-small cell in type.

glands and become more common with age. No pathological significance is attached to this. The mucosal basement membrane also thickens with age and the elastic of the lamina propria becomes more prominent. The bronchial cartilage becomes calcified and in elderly patients there may even be bone with marrow formation within the cartilage.

Artefacts

One must be aware of artefacts that can occur in bronchial biopsies and trap the unwary. Squeeze artefact by the biopsy forceps causes distortion and destruction of the bronchial wall and surrounding parenchyma if included. This can give the erroneous impression of thickening of the bronchial wall and lung parenchyma (Fig. 2.2). In addition it can disrupt tumour cells as in small cell carcinoma where the cells are very friable and may make it impossible to come to a definite diagnosis (Fig. 2.3). Herniation of lung parenchyma into the submucosa of the bronchial wall can occur with the accumulation of macrophages in the alveolar lumen which can give a false impression of desquamative interstitial pneumonia. In addition, herniation of cartilage into the bronchial wall (Fig.

2.4) may give an erroneous impression of tracheobronchopathia osteochondroplastica; the clinical history is important in order to eliminate this confusion. Desquamation of epithelium can occur with poor fixation and is not unique to asthma (Fig. 2.5). There can be small numbers of chronic inflammatory cells in the bronchial wall of normal individuals, particularly plasma cells and lymphocytes with the occasional eosinophil. In addition there can be accumulation of lymphocytes at the adventitial/parenchymal junction in the normal bronchus (Figs 2.4 and 2.5). Bronchial associated lymphoid tissue with flattening and infiltration of the overlying epithelium by lymphocytes can be seen in smokers and in infants who have had repeated infections.[5,6] These lymphocytes are liable to crush artefact (Fig. 2.4) and can be mistaken for small cell carcinoma. The smooth muscle of the bronchial wall can be mistaken for collagen (Fig. 2.5).

General changes seen in inflammation

Desquamation of the surface epithelium is seen in asthma. At times clumped epithelial cells in which

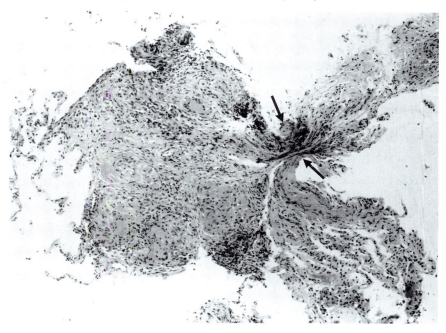

Fig. 2.2 Bronchial biopsy in which the wall and surrounding parenchyma is squeezed by the forceps giving the impression of thickening. Note three well-defined non-necrotizing granulomas in the bronchial wall extending out into the parenchyma.

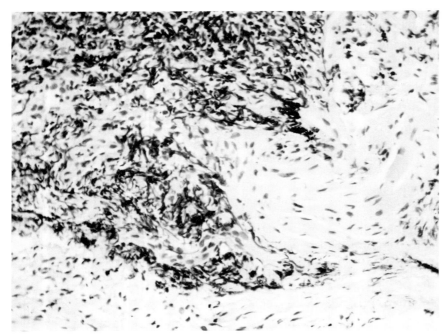

Fig. 2.3   Tumour cells from a small cell carcinoma in which the cellular detail is obscured by crush artefact, making a definitive diagnosis impossible in this area.

the nuclei vary in shape and size can resemble adenocarcinoma; great care must be taken in interpretation. Acute and chronic inflammation with ulceration is seen in allergic reactions, viral and bacterial infection, intubation, tumour infiltration and sites of previous biopsy. The respiratory epithelium gives a stereotyped response to different types of injury. Basal cell/mucous cell hyperplasia is frequent. The basal cells rest on the basement membrane and have a high nuclear cytoplasmic ratio with dark hyperchromatic nuclei. It has been assumed in the past that all cells in the basal or suprabasal layers were basal cells, but this is not true. Many of them are in fact mucous cells on electron microscopy, so the term basal cell hyperplasia is incorrect. Mucous cell hyperplasia is the main established response to many forms of injury. Stratification can occur, consisting of multiple layers of mucous cells overlying one or more layers of basal cells. Animal experiments have shown that both basal cells and mucous cells dividing in response to diverse injury, give rise to cells expressing both epidermoid and mucus-secreting characteristics, resulting in eventual squamous metaplasia with keratinization on the surface. Squamous metaplasia is common in smokers and is seen in many forms of insult to the respiratory epithelium.

The epithelial basement membrane is $5.0\,\mu m$ in thickness and lies subjacent to the basal layer of epithelium. It becomes thickened in any inflammatory process affecting the bronchi, particularly in chronic bronchitis and asthma, and it also thickens as a normal feature of ageing. Beneath the basement membrane lies the loose connective tissue of the lamina propria, which is bordered by a fine sheet of elastic where it abuts onto the muscle coat. The elastic layer can increase, particularly in chronic inflammatory conditions such as chronic bronchitis, asthma and bronchiectasis.

Assessment of airway inflammation in chronic bronchitis

Chronic bronchitis is a clinical entity defined by the production of excessive mucus in the bronchial tree with resultant cough. There are usually increased numbers of mucous 'goblet' cells in the surface epithelium and the submucosal glands. The submucosal glands of patients with chronic bronchitis contain a relatively larger proportion of acini

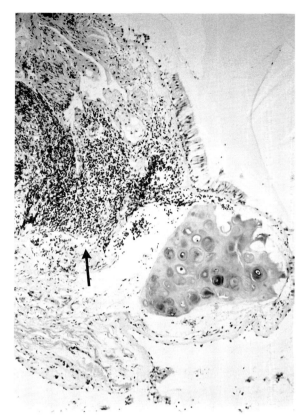

Fig. 2.4 High power view of bronchial wall showing herniation of cartilage beneath the mucosa. There is also a chronic mainly lymphocyte infiltrate at the adventitial parenchymal junction (arrow).

which are mucous in type and secrete higher quantities of sulphamucins than is normal.[7] The Reid index of gland thickness to full wall thickness has been used to assess the severity of chronic bronchitis.[8] But it appears that quantification of these changes seen in biopsy specimens has no relation to the severity of symptoms and biopsy is not used to diagnose, grade or follow chronic bronchitis clinically. Recent studies have demonstrated that airway inflammation is central to the clinical changes associated with chronic bronchitis and this has been correlated with obstruction.[9,10] It is possible that in the future the use of biopsy with more sophisticated immunological markers may prove useful in assessing chronic obstructive airways disease.

Bronchiectasis

This is also a clinical diagnosis with the production of copious amounts of purulent sputum from dilated, infected bronchi. Biopsy can show ulceration, squamous metaplasia, epithelial hyperplasia with papillary proliferation, increase in goblet cells (Fig. 2.6), and acute and chronic inflammation with basement membrane thickening. Bronchiectasis is a prominent feature of cystic fibrosis. Look carefully for fragmented fungal hyphae which can be seen in allergic bronchopulmonary aspergillosis; this is common in bronchiectasis (Fig. 2.7). However, the features are non-specific

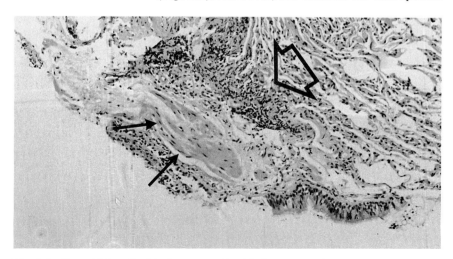

Fig. 2.5 Bronchial wall with desquamated epithelium, smooth muscle (arrows) and a mild chronic inflammatory infiltrate at the adventitial parenchymal junction (open arrow).

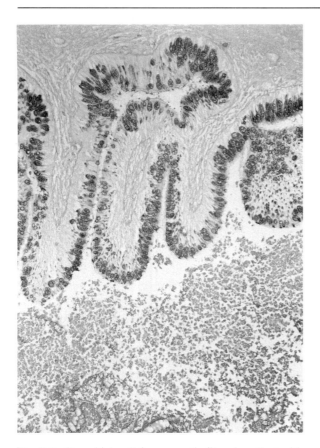

Fig. 2.6 Bronchial wall from a cystic fibrosis patient with bronchiectasis showing an increase in epithelial goblet cells and mucus in the lumen. PASD stain.

and can be seen in any inflammation of the bronchus. Bronchoscopy is very useful in assessing that there is no local lesion, i.e. tumour, causing the bronchiectasis. Reactive lymphoid follicles can be seen in the bronchial wall in follicular bronchiectasis with bronchial-associated lymphoid cells infiltrating the overlying epithelium.

Asthma

This is a clinical diagnosis manifested by an increased resistance to airflow and is difficult to characterize histologically. Classically asthma has been associated with mucous plugging, bronchial smooth muscle hyperplasia, mucous gland hyperplasia, thickening of the epithelial basement membrane, eosinophilic and mast cell infiltration and

epithelial desquamation. However these features are not seen in all asthmatics and the changes are not related to the severity of the disease. In addition there is an overlap with chronic bronchitis. Like chronic bronchitis, airway inflammation is a prominent feature even in mild asthmatics.

Bacterial infection

Diagnosis of bacterial infection is rarely made on bronchial biopsy. Large colonies of bacteria are occasionally seen in necrotic debris adjacent to malignancy. The finding of occasional bacteria in the inflammatory exudate in bronchial biopsies is not generally acceptable as a diagnostic finding since contamination or colonization is so common. However organisms may be seen on bronchial biopsy which may help the pathologist make a specific diagnosis occasionally.

Actinomyces israeli

Actinomyces israeli can present with endobronchial infection[11] and the presence of acute inflammation with sulphur granules and surrounding Gram-positive branching filamentous organisms with outer club-shaped projections is highly suggestive. They can also be well demonstrated with silver stains. Severe gingivitis, chronic obstructive airway disease and other debilitating conditions predispose to infection. However infection is more likely in the pulmonary parenchyma and bronchoscopy is rarely diagnostic.

Nocardia asteroides

Nocardia asteroides is also a Gram-positive filamentous organism but differs from *Actinomyces israeli* by not forming sulphur granules and can be acid-fast on staining with Ziehl-Neelsen, but this may not succeed and a modified Wade–Fite stain may be required. It can be associated with abscess formation which may rupture into the bronchial wall and present as an endobronchial mass.[12] Specific antibodies for immunocytochemical detection can be applied to detect these organisms.

Mycobacteria

Mycobacteria can be detected by use of the acid-fast Ziehl-Neelsen stain, which shows the typical curved banded staining of the bacillus. The fluor-

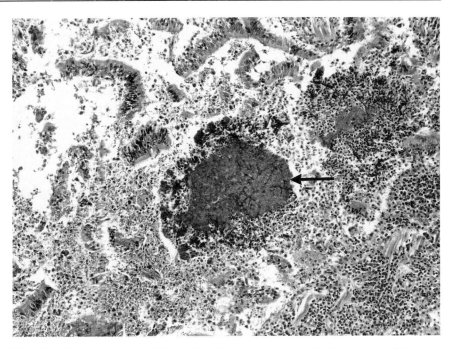

Fig. 2.7 Sections of bronchial wall which show mucus in the lumen with many eosinophils in the lumen and bronchial wall. In the centre a small colony of fungal hyphae (arrow) is present from a case of allergic bronchopulmonary aspergillosis.

escent Auramine-O stain can also be used and is claimed to be more sensitive. The finding of even one organism is significant but care must be taken to avoid false-positive staining from saprophytic mycobacteria found in tap water used in staining. It is advisable to screen a negative as well as a positive control to detect this possibility. Mycobacteria can also contaminate the fibreoptic bronchoscope.[13]

Granulomatous reactions in bronchi

Second to the finding of neoplasm, the detection of granulomas in bronchial biopsies is a very significant observation (Fig. 2.2). They should be described as non-necrotizing, necrotizing or with minimal central necrosis. The number of granulomas should also be mentioned and their location. Granulomas in bronchial biopsies should put into operation a protocol of analyses to determine if possible the specific aetiology. This includes a detailed clinical and occupational history. It requires the use of special stains to identify microorganisms including silver and PAS stains for

fungi, Ziehl–Neelsen or Auramine-O stains for acid-fast bacilli and the Gram stain. If a fresh specimen is received a direct smear may be prepared and stained for microorganisms. Polarizing filters to detect birefringent particles and possibly specialized equipment such as electron microprobe and X-ray analysis to detect and identify material not seen at light microscopy. A large study of surgical material showing granulomatous inflammation revealed that the most common causes were sarcoidosis, mycobacterial infection, foreign particulate material, fungal infection and rheumatoid arthritis.[14]

Tuberculosis

The incidence of tuberculosis (TB) has decreased to 10.2 per 100,000 in 1988 in the UK but it is much higher in the Asian and West Indian ethnic subgroups who have been emigrating to the UK since the late 1950s.[15] In addition the number of notifications has risen since then, which may be related to the incidence of AIDS. In the USA the rate of decline lessened in 1986 and there have been 9226 more cases of TB than expected; this increase is related to the AIDS epidemic. TB is the classic delayed hypersensitivity granulomatous reaction. The granulomas are usually well formed and typically confluent, with easily recognized epithelioid cells, Langhans giant cells and necrotic centres. Lymphocytes are usually present in the adjacent tissue. The presence of necrosis within an epithelioid granuloma makes mycobacterial infection the most likely cause.[14] The finding of acid-fast bacilli confirms the diagnosis but it can be rare to detect organisms since the formation of granulomas indicate a strong immune response. Most mycobacteria range up to 10 µm in length so they can be seen at $\times$ 400 magnification. Removal of the blue filter from the microscope may increase contrast and aid detection. The Auramine-O fluorescent technique is believed to give more positive results than Ziehl-Neelsen acid-fast staining.[14] In biopsies the histological pattern of tuberculosis can be variable, as might be expected owing to the variety of types of mycobacteria as well as the minimal amount of tissue sampled. Even in proven cases of tuberculosis, the findings can be of non-specific chronic inflammation. In addition the granulomas may be non-necrotizing so that distinguishing them from sarcoidosis may be impossible.

Atypical mycobacteria can evoke a granulomatous reaction which tends not to be necrotic. Atypical varieties include *Mycobacterium avium-intracellulare*, *M. kansasii*, *M. terrae* and *M. malnoense*. These are found particularly in immunosuppressed patients. In atypical mycobacterial infection, such as *M. avium-intracellulare* associated with AIDS, the reaction can resemble lepromatous leprosy with many macrophages distended by numerous mycobacteria with little or no granuloma formation – a histiocytic reaction. It may be possible to assess the type of mycobacterium species only with *M. kansasii*, where the bacilli are larger, more curved and banded in staining than the other species, which are impossible to tell from each other on purely morphological grounds. Culture is considered more sensitive for detecting mycobacteria than microscopy but a time interval of 3–8 weeks must be allowed for. Microscopic demonstration in sputum requires 5000–10 000 organisms per ml, whereas culture will be positive with 10–100 organisms per ml. Bronchoscopy will confirm a diagnosis of tuberculosis in about 67% of patients with radiographic evidence of active disease and in whom the sputum is negative for acid-fast bacilli.[16] In addition, cultures obtained at fibreoptic biopsy have been the only positive cultures in 33–50% of such patients.[17]

It is important to remember that all types of fungi as well as *Actinomyces*, *Nocardia* and even *Pneumocystis carinii*, can induce a necrotizing granulomatous reaction in the bronchial wall, so sections must be carefully examined for these. In contrast to mycobacterial infection, these agents are usually visible in tissue sections.

Immunosuppressed patients

Bronchial biopsy can be useful in diagnosing infections in immunosuppressed patients but transbronchial biopsy combined with lavage specimens are more useful. Obviously tuberculosis, fungal disease, and viruses can be detected on bronchial biopsy. However one specific entity should be looked for carefully – cryptosporidiosis – which lies on the surface of the bronchial epithelium. The small round bodies can be detected by acid-fast or Giemsa stains but other more specific stains may also be used.[18] The patients usually have gastrointestinal infection as well. Poorly controlled diabetics are liable to widespread endobronchial infection by mucormycosis (Fig. 2.8).

Allergic bronchopulmonary aspergillosis (ABPA)

This represents a hypersensitivity reaction to inhaled aspergillus within the airways and the patients usually are atopic and asthmatic. Patients suffering from cystic fibrosis are also at risk of developing this complication.[19] It is important to understand that this disease has a wide variety of histological reactions with overlapping being frequent.[20] The appearances in the bronchi can take many forms. There are increased numbers of eosinophils within the bronchial wall and mucoid plugs are seen in the bronchial lumen with distention of the bronchial/bronchiolar wall. These plugs have a typical appearance with layers of mucus being interspersed with eosinophils. Hyphae of aspergillus are frequently found in sections of these plugs but can be few and distorted, so a careful search has to be made for them (Fig. 2.7). If the inflammatory reaction is severe the bronchial wall may be destroyed with resultant bronchiectasis which is often proximal and hilar in

Fig. 2.8 Distorted hyphae of mucormycosis resembling 'twisted ribbon' with irregular thick walls and no septae; these features distinguish it from other fungi. Grocott stain.

distribution. Scattered granulomas may be seen in a peribronchial pattern.

Bronchocentric granulomatosis

Bronchocentric granulomatosis represents a more severe hypersensitivity reaction in the spectrum of ABPA. There is marked granulomatous inflammation affecting the bronchial wall with extensive necrosis leading to obstruction of the airway and eventual obliterative bronchiolitis.[2] There are usually many eosinophils as well as epithelioid cells present. Rarely bronchocentric granulomatosis can present as a single lung nodule and mimic tumour.

Bronchocentric granulomatosis can also occur in the setting of infection, particularly tuberculosis and fungal disease. It can be seen as a non-specific reaction in bronchiectasis, often with polymorphonuclear cells more than eosinophils as seen in ABPA. A search for causative organisms must be thoroughly checked.

Bronchocentric granulomatosis may also occur in the context of inhalation of stomach contents with a florid giant cell reaction. Vegetable matter

will be detected in the inflammatory tissue. Bronchographic material can also induce a granulomatous response in the bronchial wall as well as any other foreign material and endogenous products such as keratin produced by a squamous cell carcinoma (Fig. 2.9). Thus the differential diagnosis is wide and bronchocentric granulomatosis really covers a spectrum of conditions and is not a specific disease entity.

Granulomatous inflammation with eosinophils

Granulomas associated with parasites are accompanied by numerous eosinophils. In paragonimiasis, the fluke can become sexually mature in the lung, form abscesses and rupture into the bronchial wall with ova escaping. Ascariasis, strongyloidiasis (seen particularly in AIDS) and filariasis can be present in the airways and elicit an eosinophilic reaction in the bronchial wall as well as eosinophilic pneumonia. Hypersensitivity reactions to drugs can result in an eosinophilic arteritis with granulomas which may extend to the bronchial wall. The Churg–Strauss syndrome also

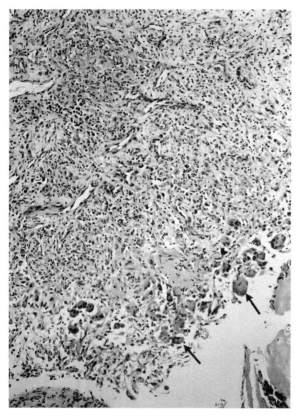

Fig. 2.9 Part of an inflammatory polyp which is overlying a squamous cell carcinoma with multinucleate giant cells (arrows) which are a response to the keratin produced by the tumour.

gives an eosinophilic necrotizing vasculitis with involvement of the bronchial wall, but these are rare conditions and the clinical findings are of paramount importance. Bronchial biopsy is rarely diagnostic but may aid in the clinical picture by displaying some, if not all, of the features mentioned, particularly the eosinophils, granulomas and necrosis. However both transbronchial and open-lung biopsy are more diagnostic for these conditions.

Foreign body reaction

Many foreign materials causing granulomatous reactions in bronchial specimens results from intravenous injection of drugs meant for oral use. The fillers used are most commonly cellulose, talc and corn starch, each of which can evoke a granulo-

matous reaction often associated with blood vessels which may rarely involve the bronchi. Cellulose is a 10–25 μm pale-grey PAS diastase resistant crystal which is birefringent. Talc is a plate-like crystal which is also birefringent. The granulomas are dominated by multinucleate giant cells which ingest the crystals. Foreign particulate material causing granulomatous reactions in the bronchial wall may also arrive by inhalation of dusts such as talc or embolization from implanted surgical prosthetic material. In addition previous surgical procedures in the bronchial tree or lung can result in the deposition of surgical talc, so a history of previous surgery is always important. A florid giant cell reaction can follow inhalation of stomach contents and bronchographic material or keratin from a squamous cell carcinoma which may mask the underlying tumour (Fig. 2.9). All these possibilities must be considered and careful examination made of the bronchial biopsy with polarizing light filters.

Sarcoidosis

Bronchial involvement with narrowing is common in sarcoidosis. Different studies have shown varying degrees of histopathological changes supporting the diagnosis (19–69%) in bronchial biopsies.[22] The widespread variation may be due to patient selection and the variable number of biopsies performed in each patient. Recently a study has shown that swelling and narrowing of the visible bronchus is more frequent in biopsy-positive patients. In addition correlation with bronchoalveolar lavage shows a significant increase in lymphocytes and mast cells in bronchial biopsy-positive patients, indicating more intense inflammatory activity in patients with bronchial sarcoidosis compared with those without. This appears to influence the clinical course with a more severe impairment of lung function and a greater use of systemic steroids.[23]

Sarcoidosis typically produces well-defined non-necrotizing granulomas in the bronchial wall (Figs 2.2 and 2.10). Rarely, the granulomas can have necrotic centres, but extensive areas of necrosis as seen in tuberculosis are absent. The granulomas are well defined and separate and do not become confluent as in tuberculosis. Granulomas may become detached from the wall during the procedure and form isolated granulomas in the biopsy, so care must be taken to examine every fragment and

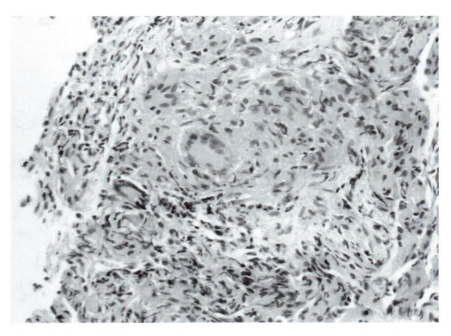

Fig. 2.10 Bronchial biopsy showing a non-necrotizing granuloma with multinucleate Langhans'-type giant cells within it; this is typical of sarcoidosis.

level. The examination of all levels cannot be overstressed in my experience. Epithelioid cells and multinucleate giant cells, similar to the Langhans' giant cells of tuberculosis, are found in the centres of granulomas. The giant cells can contain irregular clumps of birefringent material which bears no relation to pathological inhalation of foreign material but is just a general reaction of these giant cells which will mop up any passing debris which enters the lung. Asteroid and Schaumann bodies are common in the cytoplasm of the giant cells. A well-defined halo of lymphocytes usually surrounds the granulomas. In old healed sarcoid only Schaumann bodies may remain in the wall to indicate the previous existence of granulomas. However Schaumann bodies are not pathognomonic of sarcoidosis but can be seen in granulomas due to any cause. The granulomas can become fibrosed, leaving rounded scars in the bronchial wall. Active granulomas can co-exist with healed scars, indicating an ongoing process. Scarring can be present from early in the clinical disease. so that its presence gives no indication of the clinical status of the patient. It must be remembered that the presence of granulomas alone in no way indicates ongoing activity or progression in sarcoidosis. Worsening respiratory symptoms and deteriora-

tion in lung function are more useful indications.[24]

It can be impossible to separate sarcoid granulomas from infectious granulomas on a purely morphological basis and it is stressed that specific aetiological agents must be exhausted and the clinical picture analysed before reaching the conclusion that a granulomatous lesion is due to sarcoidosis. It is inappropriate for the pathologist to make a diagnosis of sarcoidosis based upon biopsy alone.

Berylliosis

Berylliosis is a far less common cause of granulomatous pulmonary reaction and the granulomas are identical to sarcoid. A careful history of exposure to beryllium with direct analysis of tissue are the only means of making the diagnosis.

Granulomatous vasculitis

Wegener's granulomatosis

Bronchial biopsy can show variable features and the three essential criteria for the diagnosis vasculitis, necrosis and granulomatous inflammation are unlikely to be present all together in a bronchial biopsy. One must look out for necrosis with a full range of both neutrophilic and granulomatous inflammation as pointers. Macrophages, epithelioid cells and multinucleated giant cells are usually scattered throughout the biopsy and are not as organized as in the reactions already discussed. They are easily missed as a result and bronchial biopsy rarely gives a specific answer. The epithelioid cells often palisade around arteries and veins and around foci of necrosis. An elastic stain such as the elastic van Gieson is essential to delineate the vessels in the biopsy and demonstrate the elastic lamina disruption seen in vasculitis. Capillaries and bronchial arterioles/venules are the only vessels usually seen in bronchial biopsy and they may be impossible to identify in a sheet of inflammation.

Histologically the differential diagnosis should include Churg–Strauss syndrome, in which the predominant cell is the eosinophil; hypersensitivity angiitis, in which both eosinophils and neutrophil predominates and capillaries are mainly affected; necrotizing sarcoid granulomatosis, in which the granulomas are well defined; and lymphomatoid granulomatosis with an atypical lymphoid infiltrate. Infections can also give rise to granulomatous vasculitis, particularly mycobacteria, fungi and parasites. Bacterial pneumonias of the Gram-negative type as well as *Pneumocystis carinii* can show vascular involvement. Viral and rickettsial infections characteristically involve capillaries, arteries and veins. It is often impossible to make a definitive diagnosis on bronchial biopsy because the tissue is so small and tests are limited.

Systemic diseases affecting the bronchi

These are rare but a few can affect bronchial tissues, the principal one being sarcoidosis, which has already been mentioned. The others are even rarer and usually known to the clinician. Bronchoscopy is often undertaken for suspected secondary disease. Nevertheless biopsy material may reveal unsuspected systemic disease to the observant pathologist.

Amyloidosis

Amyloid may be deposited as part of a generalized process in the respiratory tract, but this is rare. The amyloid can be found in the bronchial wall as well as the lung parenchyma. In the bronchial wall the amyloid diffusely involves the bronchial blood vessels of the lamina propria and submucosa and is usually of the protein A (AA) type with disappearance of staining on application of potassium permanganate or trypsin.

Amyloid confined to the respiratory tract is more likely to be primary with no underlying disease and usually has a nodular pattern.[25] In those cases where amyloid is associated with local disease, such as tuberculosis or tumour, it can have a nodular or diffuse pattern. The aetiology of the majority of localized nodular deposits in the respiratory tract is unknown. It usually occurs in middle-aged individuals with no sex predominance. Tracheobronchial amyloidosis takes the form of multifocal submucosal plaques or masses giving an irregular appearance to the bronchial tree simulating neoplasia. This results in progressive distortion and obstruction of the larger airways. The amyloid forms multiple submucosal nodules below the epithelium with spreading around the seromucous glands. There may be a chronic inflammatory reaction with plasma cells and multinucleate giant cells around the amyloid (Fig. 2.11). This inflammatory reaction is usually not seen in generalized amyloidosis. The eosinophilic material stains with Congo red and shows apple-green birefringence with polarized light. The staining with Congo red is not abolished by pretreatment with potassium permanganate, indicative of its non-AA amyloid nature. It appears that tracheobronchial amyloid is of the amyloid light chain (AL) type derived from immunoglobulin light chain.[25] Calcification with ossification can occur in the lamina propria, making the lesions histologically similar to tracheobronchopathia osteochondroplastica. Solitary polypoid endobronchial masses are uncommon.

Histiocytosis X

The characteristic lesions with Langerhans' cells, large numbers of eosinophils and a mixture of other inflammatory cells mainly involve the lung and are rarely sampled on bronchial biopsy. On such a small fragment of tissue it is difficult if not impossible to make a definitive diagnosis and

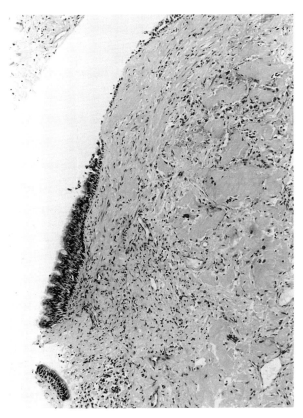

Fig. 2.11 Several nodular deposits of amyloid within the bronchial submucosa with plasma cells and multinucleate giant cells around the nodules.

immunocytochemistry is limited because Langerhans' cells can be found in the normal respiratory epithelium and in the underlying interstitium in inflammatory and fibrotic conditions.

Cartilage

Tracheobronchopathia osteochondroplastica

Prominence and irregularity of the cartilage may indicate tracheobronchopathia osteochondroplastica.[26] This is a rare condition characterized by submucosal nodules confined to the trachea and main bronchi. Bronchoscopy reveals multiple mucosal nodules but an important point is that the membraneous portion of the trachea is spared. These nodules microscopically consist of cartilage which, like normal cartilage, can calcify and

ossify. They are located between the normal cartilage and the surface epithelium and can cause narrowing of the airways. The nodules are due to development of new cartilage and lamellar bone between the cartilaginous rings and the mucosa. They are believed to be forms of exostosis from the cartilage rings and are not associated with systemic disease. The diagnosis relies on clinical information since the cartilage and bone can easily be dismissed by the pathologist as normal and the cartilage only differs from the normal in its location.

Beware the calcification and ossification seen in damaged tissue, amyloid and in inflammatory pseudotumours. One must also be aware of cartilage present in hamartomas and the metaplastic calcification and ossification in bronchial carcinoid tumours[27] which may be confused with the above. In addition the rare pleomorphic adenoma of bronchial gland can have chondroid stroma.

Relapsing polychondritis

This is an inflammatory disease of cartilage throughout the body, the nose and pinna being particularly affected. Laryngotracheal involvement is well known. Isolated laryngotracheal as well as bronchial involvement is described which can be fatal.[28] On biopsy the airway cartilage is surrounded by inflammatory cells, including giant cells with destruction of the cartilage. The cartilage loses its smooth outline and has an irregular border. End-stage disease shows irregular cartilage with surrounding fibrosis of the perichondrium and bands of fibrous tissue penetrating into the cartilage plates. This inflammation and fibrosis leads to severe stenosis and stricture. The disease is episodic but generally progressive. It often starts with auricular involvement and the respiratory tract is involved later. The trachea, main, lobar and segmental bronchi are affected while the smaller bronchi show non-specific inflammation. It is one of the causes of tracheomalacia because of the destruction of the cartilage.

Tracheomalacia/bronchomalacia

A congenital form of this occurs with cartilage plate deficiency and in association with chondrodystrophies. In the bronchi it often affects the left main-stem bronchus. There is softening with collapse of the airway with distal obstruction and infection. More distal involvement of the bronchi can occur from the fourth to the eighth generations in the Williams–Campbell syndrome. Bronchoscopy will show the anomaly with a decrease in size of the larger airway due to collapse. Histologically the usual cartilage plates are replaced by scattered islands of immature cartilage with non-specific inflammation more distally. In the subsegmental bronchi the airways are often dilated with a reduction in the number and size of the cartilage plates. The muscular wall is present but is often attenuated. Chronic inflammation of submucosal connective tissue is present in varying degrees. This inflammation has been suggested as the primary lesion with secondary cartilage destruction as a result of respiratory infection. The clinical information is most important because cartilage is often missing from a biopsy and the findings may be non-contributory unless a deep biopsy is done to obtain cartilage.

Acquired tracheomalacia is a more frequent condition than the congenital form. It commonly occurs in neonates treated with prolonged nasotracheal intubation for respiratory distress. The airways are dilated. Microscopically squamous metaplasia with focal ulceration and necrosis is prominent with acute and chronic inflammation. The cartilage rings are usually normal.

Tracheobronchomegaly

This is also known as Mounier–Kuhn syndrome and affects males in the 30–40 year age group. Familial cases have been described, suggesting an autosomal recessive type of inheritance. There is outpouching of the tracheal lumen between the cartilage rings producing multiple diverticula. It has also been seen in association with connective tissue disorders such as cutis laxa and Ehlers–Danlos syndrome. Again bronchoscopy is diagnostic with non-specific chronic inflammation associated with the diverticula. The disorder is believed to result from a defect in the development of connective tissue elements within the trachea and mainstem bronchi.

Broncholithiasis

This is the presence of calcified masses within the bronchial lumen. They can arise from the calcification of aspirated food or foreign body. A second mechanism is protrusion into the lumen of calcified bronchial cartilage due to necrosis, as in bronchiectasis or relapsing polychondritis. A third

and most common mechanism results from the erosion of a calcified granuloma from lymph nodes adjacent to the bronchi such as occurs in histoplasmosis or tuberculosis.[29] Complications include ulceration, fistulas and inflammatory papillomas. These can be removed either bronchoscopically or surgically.

Inflammatory pseudotumour

Inflammatory pseudotumours can present as endobronchial masses in up to 12% of cases.[30] They are large nodules composed of a mixture of inflammatory cells that infiltrate the bronchial wall. The cells include mononuclear histiocytes, foamy macrophages (xanthoma) cells, multinucleated Touton-type giant cells, plump spindle fibroblasts with associated collagen, lymphocytes and plasma cells. The lesions can be predominantly inflammatory or fibrous. In fibrous lesions the spindle cells are arrayed in interlacing fascicles or in a whorled (storiform) pattern. While the fibroblasts and histiocytes can be bizarre in shape and size, mitoses are claimed to be rare. However in my experience they can be frequent. Foci of calcification and osteoid metaplasia may be present. It is impossible to reach a firm diagnosis of such a lesion on bronchial biopsy alone. It is more frequent in childhood and in non-smokers which may point to the diagnosis. Many cases can resolve spontaneously.[31] In addition similar histological changes may be seen in sclerosing mediastinitis in which the bronchial wall can be involved.[32]

Bronchial carcinoma

The majority of bronchial carcinomas arise from the bronchial epithelium. When the tumour is visible bronchoscopically, the histological diagnosis can be made in up to 90% of cases, but this falls to 45% in cases not visible at bronchoscopy. Obviously the experience of the clinician doing the procedure is an important factor in diagnostic yield. For peripheral lung lesions the positive yield is partly dependent on size. Lesions less than 4 cm in size yield a rate of 49% or less but for larger lesions this rises to 65%.[33]

The histological classification of lung cancer is considered relevant to planning of treatment and to prognosis. The World Health Organization classification has been most commonly used but has undergone revision which highlights the problems inherent in this classification.[34] Cell typing may be difficult and the criteria used not always uniform. Because of the complex morphology and heterogeneity of lung cancer[35] interobserver and intraobserver discrepancies may occur, a problem which is compounded by the use of fibreoptic bronchoscopy. This procedure yields specimens about 2 mm in diameter, which frequently show tumour cells without any recognizable morphological pattern. Although the diagnosis of malignancy by fibreoptic biopsy has proven extremely reliable, the specific diagnosis of cell type often differs from the resection specimen. In one study 41 of 107 (37%) bronchial biopsies differed in cell type from their resection specimen[36] and it has been suggested that the term 'lung carcinoma, non-small cell or small cell in type' should be used for reporting those biopsies in which the cell type can be identified, but clear-cut morphological criteria are lacking (Fig. 2.12).[36] This is in agreement with clinical opinion, in which it is the cell type, small cell versus non-small cell, as well as staging-which determines treatment.[37] Thus the existing classification fails to allow for the smallness of the

specimen which may not permit specific categorization. Provision of a 'non-small cell carcinoma, not otherwise specified' category allows the tumour to be classified as far as the clinician is concerned, with the view that the diagnosis may be refined further should more diagnostic material become available. In a more recent study of biopsy versus resection specimens by a panel of pathologists with an interest in pulmonary pathology, two categories were introduced:

1. non-small carcinoma of specified type, i.e. giant cell, adenosquamous, carcinoid;
2. non-small cell carcinoma, not otherwise specified.

In addition they used a panel of specific criteria for the diagnosis of squamous, adenocarcinoma and small cell carcinoma. and included mucin stains, Alcian blue/periodic acid Schiff with diastase (AB/PASD) in the diagnosis of adenocarcinoma. They also stated that undifferentiated large cell carcinomas are not expected to be recognized in small biopsy specimens.[38] Nevertheless, even though the panel were considered experts, 7% of resection small cell carcinomas were called non-small cell on the original biopsy and 3% of

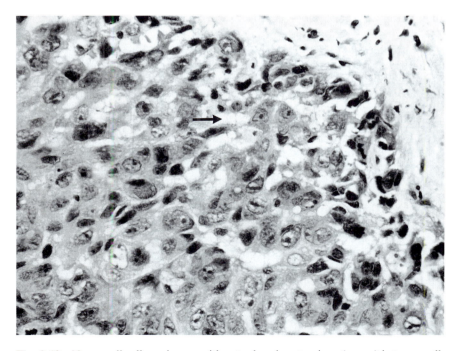

Fig. 2.12 Non-small cell carcinoma with cytoplasmic extensions (arrow) between cells but no well-defined prickles.

resection non-small cell carcinomas were called small cell carcinomas on the original biopsy. In particular carcinoid tumours were wrongly categorized in 32% of cases. The diagnosis of lung cancer itself is reliable in small biopsy specimens but the diagnosis of subtypes on which treatment is decided can be unreliable, a fact which must be borne in mind by all clinicians and pathologists working in this field.

Dysplasia, carcinoma *in situ* and squamous cell carcinoma

Several studies stress the development of invasive squamous carcinoma from areas of squamous metaplasia with progressive increase in the nuclear/cellular atypia to carcinoma *in situ* and finally invasion.[39] Squamous metaplasia is followed by dysplasia of variable degree and ultimately by carcinoma *in situ*. In uranium miners who have been closely monitored, mild dysplasia required approximately 6 years to progress to carcinoma *in situ*. With continued exposure, invasive carcinoma developed in approximately 3.2 years.[40] The degrees of atypia are similar to those applied to cervical epithelium and have been labelled as bronchial intraepithelial neoplasia. Mild atypia is characterized by cells and nuclei varying slightly in shape and size, usually seen in the basal layers of the epithelium. Moderate atypia shows more pronounced variation in nuclei with hyperchromasia, mitoses and pleomorphism of cells. The changes affect the lower half of the stratified epithelium. Severe atypia shows cells and nuclei varying markedly in shape and size with hyperchromasia, prominence of nucleoli and many mitoses. The changes extend into the upper half of the stratified epithelium. Carcinoma *in situ* involves marked nuclear and cellular change throughout the whole epithelium with multiple mitoses. It is characterized by an intact basement membrane. Periodic acid Schiff, silver and trichrome stains are recommended in order to assess the basement membrane. The dividing line between marked atypia and carcinoma *in situ* is not clear. Usually it is an academic division since both are of significance in the pathogenesis of neoplasia and should be reported as severe atypia/carcinoma *in situ*. Many pathologists find these criteria difficult to apply to the respiratory tract in view of the common phenomenon of inflammation and mild epithelial changes which can be seen in many individuals, especially smokers, which would label many patients as bronchial intraepithelial neoplasia. A more simple classification divides into atypical squamous metaplasia in which nuclear atypia, hyperchromasia and increased nucleo-cytoplasmic ratio are present, but in which the atypical cells do not occupy the full thickness of the epithelial layer. If these changes are seen in the full thickness of the epithelium the changes are labelled carcinoma *in situ*.[41] Most squamous cell carcinomas arise from large and medium sized bronchi, especially in the segmental branches and to a lesser extent in the lobar divisions. Atypia and carcinoma *in situ* are most frequent in these locations and are usually multifocal. Recent screening of heavy smokers in a Japanese study, in which positive sputum samples led to investigation of clinically and radiologically occult disease, showed up 19 cases of carcinoma *in situ* of the bronchus[42] while 130 others had invasive early tumours. Bronchoscopically small polypoid protuberances, micronodular swelling, thickening of spur and mucosal granularity could be seen in 15 of the 19 cases but 4 had no mucosal abnormalities, highlighting the difficulties which can face the clinician having to deal with such disease. Maximum length or diameter was 12 mm while 13 of the 19 cases were less than 4 mm in size. None had nodal involvement on resection and all had a favourable prognosis with none dying of tumour up to 7 years after diagnosis.

Often a detailed examination of the airways with repeated brush cytology and biopsy may be needed to localize the lesion. Making a diagnosis of carcinoma *in situ* only by cytology is difficult. Cytological features have been described progressing from squamous metaplasia to dysplasia to carcinoma *in situ* to early invasive carcinoma.[39] Although cytology has been claimed to differentiate between carcinoma *in situ* and invasive squamous cell carcinoma, most pathologists rely on biopsy for determining whether the lesion is confined within the basement membrane or not.

Squamous metaplasia/dysplasia can be focal and occur at multiple sites throughout the respiratory tract with intervening normal areas. The changes can extend into gland ducts and the glands themselves so the pathologist must be careful not to mistake glandular involvement for invasion. An abrupt transition is often seen between carcinoma *in situ* and respiratory epithelium that appears normal. Ciliated epithelium can also cover an area of carcinoma *in situ*. *In situ* lesions can extend proximally and distally from the site of invasion and can be multifocal.

Reporting of dysplasia must be taken in the context of the clinical setting. Viral pneumonia and chemotherapy/radiation can give rise to significant changes in epithelial cells with degrees of atypia simulating malignancy, so a clinical history is essential in these situations. The reporting of severe atypia/carcinoma *in situ* in the absence of a bronchoscopically visible lesion must lead to multiple sampling to localize the area of change. Often because the lesions are multiple it is impossible to operate on them all. The patients are followed at regular intervals with bronchoscopy in order to detect any changes and initiate operative procedures if required. Conservative treatment with local bronchial resection and conservation of lung tissue or laser therapy, have been recommended.

Invasive squamous cell carcinoma

As already demonstrated, squamous cell carcinoma is often accompanied by metaplastic, dysplastic and carcinoma *in situ* changes in the surrounding epithelium. However, these changes may not be present and there can be an abrupt transition to invasive squamous cell carcinoma from normal epithelium. The criteria of invasion is destruction of the basement membrane with extension into the lamina propria and underlying mucosa. The infiltrating edge of the tumour is usually irregular with surrounding chronic inflammation and granulation tissue in areas of ulceration.

The criteria for the diagnosis of squamous cell carcinoma are the presence of keratinization and prickles. Large polygonal eosinophilic cells with large vesicular nuclei, prominent nucleoli or stratification are not sufficient by themselves and can be seen in other non-small cell carcinomas. One must be very precise about the definition of prickles and keratinization on bronchial biopsy. Prickles are fine lines between cells, spaced at very regular intervals, best seen in areas of intercellular oedema and with the condenser reduced down for increased contrast. Irregular cytoplasmic processes cannot be accepted as prickles (Fig. 2.12). Keratinization is defined as deeply eosinophilic change in the cytoplasm of nucleated cells with flattening of the cells and the formation of epithelial pearls complete with keratohyaline granules within the cytoplasm. Necrosis, particularly within nests of

tumour cells, can give rise to increased eosinophilia within the cytoplasm of dead cells with pyknotic nuclei and desquamation of the cells into the necrotic centre which can mimic keratinization. The nuclei of keratinized cells must be viable in order to accept this criterion in a biopsy.

Most squamous cell carcinomas show a significant degree of heterogeneity with well-differentiated areas intermingled with poorly differentiated areas. Basaloid areas can be seen in squamous cell carcinoma with small cells with hyperchromatic nuclei and little cytoplasm (Fig. 2.13) which are believed to derive from the basal cells of the respiratory epithelium. They will often be seen at the outer areas of the cell nests or beneath the areas of stratification. Care has to be taken not to confuse these areas with small cell carcinoma. Always look for more differentiated areas in the biopsy. In addition, extensive areas of dead cells with pyknotic nuclei and cells undergoing apoptotis may lead to a misdiagnosis of small cell carcinoma on small biopsies. Tumour giant cells can also be seen as well as a granulomatous reaction to the keratin produced by these tumour cells. Spindle cells areas, oncocytic and clear cell change due to the accumulation of glycogen can also be observed in these tumours.

Early central squamous cell carcinoma has been defined as tumour confined to the bronchial wall with no infiltration of surrounding lung parenchyma or nodal metastases.[43] However it is not recognized in the TNM classification and its definition has not gained worldwide acceptance. Obviously a bronchial biopsy will not be able to define this early lesion and resection is needed to establish that the tumour has not spread beyond the bronchial wall.

Adenocarcinomas

Adenocarcinoma of the lung is the most morphologically heterogeneous major type of lung cancer and has replaced squamous cell carcinoma as the most frequently diagnosed lung neoplasm in the USA. It is accepted that the majority of adenocarcinomas arise from peripheral bronchi, bronchioles and alveoli. Bronchial biopsy usually only detects those arising from small bronchi. The majority form glands/acini, are mucin-secreting and are presumed to arise from mucous cells. The histological criteria for diagnosing adenocarcinoma, include gland formation by tumour cells or

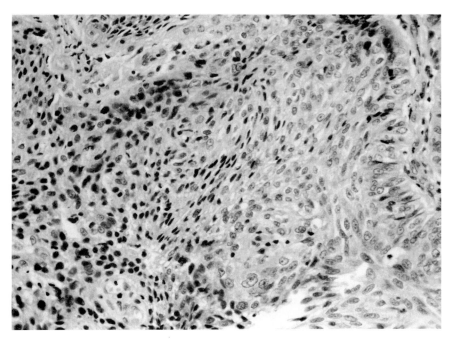

Fig. 2.13   Squamous cell carcinoma with basal cells at one edge of a nest of tumour cells showing hyperchromatic dense nuclei and less cytoplasm around the cells.

the demonstration of a mucosubstance within the cells. Mucin, both extracellular and more specifically intracellular, particularly in the form of intracytoplasmic lumina, should be looked for. Acinar, papillary and solid patterns can all be seen on biopsy. A PASD or Meyers mucicarmine stain should be done on all non-small cell carcinomas to detect mucin secretion. However, one must be aware that many lung tumours can have small foci of mucin production, even small cell carcinomas, so interpretation of extent of staining is important in the context of the cell type.

There is usually an abrupt transition from the normal epithelium to invasive tumour but squamous dysplasia/carcinoma *in situ* can be seen in the epithelium overlying adenocarcinomas. Progression from atypical glandular changes to invasive adenocarcinomas for bronchial adenocarcinoma has rarely been described. In dense inflammation of the bronchi, such as occurs in bronchiectasis, the seromucous glands can sometimes become larger and plump or may be broken up and separated by the inflammation. Great care must be taken to avoid misinterpretation of these changes as invasive adenocarcinoma.

Trying to distinguish primary adenocarcinomas

of the lung from metastases may be impossible in most cases, especially on small biopsies. Keratin and other epithelial markers give too much overlap with tumours in other parts of the body, to be able to separate them.[44] The use of surfactant apoprotein antibody has been put forward as a marker of primary adenocarcinomas while metastatic tumours are negative,[45] but this is usually positive in peripheral non-mucinous tumours and not in central mucous/goblet cell or bronchial gland adenocarcinomas, which are more likely to be detected on bronchial biopsy. Clear cell change can be seen in primary adenocarcinomas of the lung. The differential of these clear cell tumours lies with metastatic renal cell carcinoma and the rare benign clear cell tumour of lung which is usually peripheral. One cannot diagnose bronchioloalveolar cell carcinoma on bronchial biopsy.

Grading of tumours

It is impossible to grade a squamous cell carcinoma or adenocarcinoma on bronchial biopsy alone because it is only representative of a small piece of the whole tumour. We do not usually comment on grade but if there is another element with the tumour such as a definite small cell area or a giant cell area, it is important that this is commented on, since both are associated with a much poorer prognosis than a pure squamous or adenocarcinoma.

Adenosquamous cell carcinoma

Up to 7% of lung tumours have a mixed morphology according to light microscopic appearances. The most common is adenosquamous cell carcinoma. This tumour contains both malignant squamous and glandular elements. The WHO definition does not include specific criteria as to how much of each tumour element is needed to establish the diagnosis. One has to be very careful to avoid mistaking reactive benign glands commonly seen in bronchial biopsies for malignant glands in an infiltrating squamous cell carcinoma. Often malignant glandular differentiation can be seen in one region while squamous differentiation may be present in another. This makes it likely that only one element will be picked up on bronchial biopsy. Both glandular and squamous differentiation may be seen in the one nest of tumour cells, but they are clearly separate in contrast to the intimate admixture and transitional cells seen in mucoepidermoid carcinoma. This mixed type is more likely to be diagnosed on bronchial biopsy. It is a rare event, however, because the majority are peripheral.

Small cell carcinoma

This highly aggressive tumour can be difficult to diagnose on bronchial biopsy because of the extensive crush artefact that can occur with the biopsy forceps (Fig. 2.3). Diagnosis on the presence of a few cells can be dangerous; one needs definite cytological evidence of malignancy before a confident diagnosis can be made. Oval, round or fusiform cells with a high mitotic rate and the characteristic fine stippling 'salt and pepper' chromatin pattern, nuclear moulding, inconspicuous nucleoli, high nuclearcytoplasmic ratio and large numbers of apoptotic cells arranged in clusters, cords, sheets or trabeculae with absence of stromal desmoplasia – all these point to the diagnosis. Formation of rosettes and a fusiform pattern can be found. The tumour cells can infiltrate the overlying epithelium in a pagetoid

pattern (Fig. 2.14). The cells are usually larger than lymphocytes but a lymphoid infiltrate can be misdiagnosed as small cell carcinoma, especially if crush artefact is present. The use of lymphocyte common antigen immunostaining has eliminated the problem of differentiating these because small cell carcinoma will not stain with this antigen, while with keratin antibodies, small cell carcinoma cells often display a dot-like paranuclear positivity. Lymphatic, vascular and perineural invasion are prominent. In necrotic areas, basophilic nuclear material is deposited on the blood vessel walls and adjacent connective tissue (so-called nuclear encrustation). This can be a very useful feature in diagnosis, because it is rarely found with other tumour types. The tumour cells can aggregate around blood vessels, appearing as pseudorosettes, or are present in cords or trabeculae reminiscent of the growth pattern of carcinoid tumours. Occasionally the cells form round spaces, usually secondary to necrosis that could be misinterpreted as evidence of adenocarcinoma or carcinoid differentiation. Because of this variable pattern, mistakes in diagnosis can be made. Remember that the morphology of small cell carcinoma can change to a large cell type after chemotherapy.

Neuroendocrine lung tumours

Small cell carcinomas form the aggressive malignant end of the spectrum of neuroendocrine lung tumours, with carcinoid tumours forming the benign end.[46] All these tumours are characterized by the production of peptide hormones and the presence of dense-core neurosecretory granules at the ultrastructural level. There can be difficulties in the diagnosis of neuroendocrine lung tumours, especially on bronchial biopsy. The more recent classifications are difficult for pathologists to apply, particularly for the non-small cell types.[47] Classification requires basic light microscopy plus electron microscopy and the use of immunocytochemical markers, which are expensive and time-consuming. A second problem is that lung tumours show heterogeneity with both light microscopy and electron microscopy, which can lead to problems in interpreting neuroendocrine features. For definitive diagnosis large quantities of tissue may be required for special investigations, quantities which may not be available with bronchial biopsy, the only diagnostic technique used for many lung tumours as the majority of patients are inoperable.

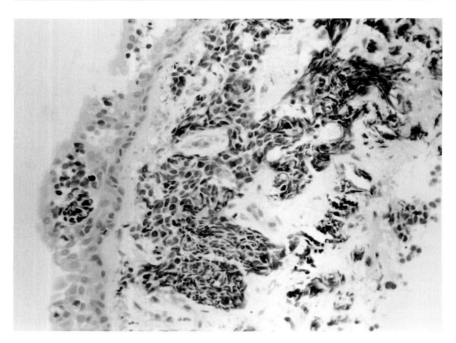

Fig. 2.14 Small cell carcinoma of bronchus with crush artefact and infiltration of tumour cells into the overlying epithelium.

Most neuroendocrine classifications rely heavily on the architecture of the cells within the tumour and this may not be seen easily in small biopsy specimens. Crush artefact, cellular distortion and degeneration may make it difficult at times to separate carcinoid tumours from small cell carcinomas, a fact of great importance for treatment and prognosis. Finally, the specificity of the general neuroendocrine markers has been questioned as these markers have been found in other lung tumours and in nonneuroendocrine tumours elsewhere such as breast or kidney. The use of only one marker for immunocytochemical diagnosis of neuroendocrine differentiation is open to criticism and a panel of markers including neuron-specific enolase, chromogranin, synaptophysin and neural cell adhesion molecule (NCAM) is now recommended.[47] In addition it is generally true that immunohistochemistry becomes less positive as the tumours become more malignant, so that the majority of small cell carcinomas do not express neuroendocrine markers immunocytochemically.

About 10% of non-small cell carcinomas express neuroendocrine features and the finding is most common in adenocarcinomas followed by large cell carcinomas and rarely in squamous cell carcinomas.[48] Obviously conventional light microscopy will not identify these tumours but immunocytochemistry may be positive. The clinical significance of this differentiation remains unproven and the routine use of immunocytochemical neuroendocrine markers is not recommended in bronchial biopsies.

Carcinoid tumours

This tumour usually arises in a main bronchus (85%) but may be peripheral (15%).[49] Carcinoids usually protrude into the bronchus as polypoid smooth masses and have a yellow/tan colour on sectioning. Often the tumour has an 'iceberg' appearance with much of the tumour infiltrating beneath the surface of the bronchus into the surrounding lung, so care must be taken to ensure complete resection. Carcinoid tumours contain uniform cells with small round-to-oval nuclei, few nucleoli and granular eosinophilic cytoplasm. The cells are most commonly in islands interweaving with one another in a mosaic pattern, as interconnecting ribbons of cells in a trabecular pattern, or in an adenopapillary pattern; mixed forms are also seen.[50] The stroma is usually very vascular, which explains their propensity to bleed on biopsy. Tumours with the typical patterns are usually easy to diagnose on bronchial biopsy but as already stated, problems can arise with diagnosis because of the less common patterns.[38] Carcinoid cells have been mistaken for plasma cell because of their dense eosinophilic cytoplasm. Spindle cells can be seen[51] as well as glandular differentiation and bone and cartilage formation can be extensive. In a typical carcinoid, nuclear pleomorphism and mitoses are rare, which is the most useful pointer to the diagnosis together with the dense pink granular cytoplasm and architecture. These tumours usually contain large numbers of secretory granules and argyrophilic stains, particularly the Grimelius stain, are positive in many of the tumours. However stains are capricious and immunocytochemistry is claimed to give more positive results.[52]

Most carcinoid tumours are strongly immunoreactive for general neuroendocrine markers and immunocytochemical methods and radio-immunoassay show that bronchial carcinoids produce a wide range of amines and peptide hormones. S-100 positive sustentacular cells have been reported in pulmonary carcinoids[53] and labelled as 'paragangloid' carcinoids. The feature that distinguishes these from the rare primary pulmonary paraganglioma is that paraganglioma rarely express keratin and are strongly immunoreactive for neurofilaments.[54] A minority of typical carcinoids spread to lymph nodes or other organs and histology cannot predict the behaviour. Deeply invasive tumours,[55] tumours greater than 3.0 cm in diameter, and lymph node metastases predispose to recurrence.[56]

A subgroup of tumours with a carcinoid architecture but with evidence of malignancy, characterized by nuclear pleomorphism, mitoses and necrosis are labelled as atypical carcinoids. They are also known as pleomorphic or malignant carcinoids. On bronchial biopsy it would be impossible to separate them from other non-small cell carcinomas and resection is required to reach a definitive diagnosis.[57]

Large cell carcinoma

This should never be diagnosed on bronchial biopsy alone since it is a diagnosis of exclusion. Any tumour with large cells, prominent nucleoli, well-defined cell borders without any features of differentiation is referred to as 'non-small cell carcinoma, not otherwise specified' (NOS).

Giant cell carcinoma

This tumour is classified as a variant of large cell carcinoma but can also be seen in association with adenocarcinomas or more rarely squamous cell carcinoma. The tumour cells are very pleomorphic and multinucleated with polymorphonuclear leukocytes in their cytoplasm. While one cannot make a specific diagnosis on bronchial biopsy, this histological appearance is important to mention because it carries a poor prognosis even in tumours showing differentiation.

Follow-up of patients

As already mentioned, small cell carcinoma may change to a large cell morphology after chemotherapy. In addition synchronous tumours with different histology can occur in the lung,[58] so the finding of a second pulmonary nodule may not necessarily indicate a metastases. In addition patients are also at high risk of developing metachronous lung tumours of different histology later, which can be as high as 10–15% in surviving patients. If the tumour is of similar histological type to the original primary tumour, an interval of 3 years must have elapsed before this is considered a new primary.[59]

Rarer tumours

Papillomas

In the presence of a polyp in the bronchial tree, the following lesions must be considered:

1. Inflammatory polyp which contains granulation tissue and may be covered by squamous epithelium. Sometimes infective organisms may be identified or foreign material (Fig. 2.9).
2. Both benign squamous cell and mucous cell papillomas can occur in the bronchi (Fig. 2.15). Often multiple fragments of the tumour are included in the biopsy and assessment of resection is impossible. Look for areas of dysplasia, carcinoma *in situ* and invasive carcinoma in any papillary lesion. Squamous cell

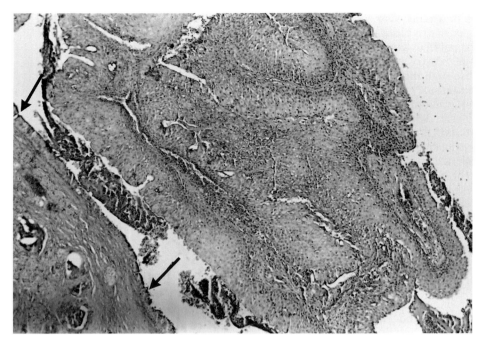

Fig. 2.15 Squamous cell papilloma of the bronchus. Note bronchial wall epithelial surface (arrows).

papillomas are rare lesions. Multiple benign squamous cell papillomas are usually associated with juvenile laryngeal papillomatosis in young people. They can occur in association with mucous gland adenomas.[60] Rare examples of transitional cell papillomas are also described. They are composed of a stroma lined by papillae covered by a stratified layer of transitional cells with central hyperchromatic nuclei and amphophilic cytoplasm.

3. Squamous cell carcinoma. This endobronchial polypoid tumour is a rare variant of squamous cell carcinoma which is not included in the WHO classification. It is important to recognize because it is nearly always an early tumour. Biopsy of the polypoid mass will reveal infiltration into the lamina propria, which should be carefully looked for.
4. Carcinoid.
5. Adenoid cystic carcinoma.
6. Pleomorphic adenoma.
7. Mucous gland adenoma.
8. Mucoepidermoid carcinoma.
9. Carcinosarcoma.

Tumours of bronchial gland origin are rare tumours and represent less than 0.5% of primary lung tumours. Adenoid cystic adenocarcinoma is the most common, followed by mucoepidermoid carcinoma.

Mucous gland adenoma

This is usually a soft well-circumscribed endobronchial polypoid tumour often covered by intact mucosa like a carcinoid. Histologically mucous gland adenomas are composed of multiple mucous-filled tubular cystic glands lined by a single layer of columnar goblet-type cells or cuboidal cells with eosinophilic cytoplasm. The mucous-filled glands can be calcified in the centre. Focal areas with oncocytic change may be seen. The tumour can be distinguished from adenocarcinoma by the lack of mitotic activity and the uniformity of the cells. The tumour is usually well defined. At times separation from a well-differentiated mucoepidermoid carcinoma can be difficult but these tumours do not generally contain squamous areas.

Oncocytoma

Oncocytic change can be seen in carcinoid tumours and mucoepidermoid carcinoma as well as mucous gland adenoma. Pure benign oncocytomas are extremely rare. In the bronchus they present as a well-circumscribed mass. They are composed of solid sheets of uniform cells with central nuclei and abundant granular eosinophilic cytoplasm. The cells do not infiltrate outside the bronchial wall and there is no lymphatic or vascular invasion. Differentiation from a carcinoid tumour can be difficult but oncocytomas are not argyrophilic and are not immunoreactive for neuroendocrine markers.

Acinic cell tumour

These present as endobronchial masses composed of sheets of clear cells in small nests or in a trabecular pattern admixed with eosinophilic cells. It is not surprising that they can be misdiagnosed as carcinoids. They have PASD-positive granules in their cytoplasm but contain no significant glycogen. This helps to distinguish them from benign clear cell tumours of lung (which is usually a peripheral intraparenchymal tumour) or carcinoid with clear cell areas which are argyrophilic and immunopositive for neuroendocrine markers.

Pleomorphic adenoma

Rare cases have been reported as mainly endo-

bronchial lesions. The histology shows myxoid, chondroid or hyalinized stroma with sheets of myoepithelial cells and acinar spaces lined by a double layer of epithelium with inner epithelial layer and outer myoepithelial layer. S-100 immunostaining can be useful in delineating the myoepithelial cells. On small biopsies the double layer and lack of cytological atypia and mitoses should point to the diagnosis but many cases have been misdiagnosed as adenocarcinoma. A rare case of pure myoepithelioma of the lung has been reported.

Adenoid cystic carcinoma

This is the most frequent malignant tumour of bronchial gland origin and can make up to 33% of carcinomas of the trachea. It can be polypoid or annular with either intact overlying mucosa or ulceration. Histologically the tumour is composed of regular glands that are lined by epithelial and myoepithelial cells and contain mucin. The cells are hyperchromatic with low nuclearcytoplasmic ratios. Occasional mitoses and necrosis can be seen. The glands frequently exhibit a cribriform or trabecular pattern which is typical and points to the diagnosis (Fig. 2.16). Solid tumour sheets may be more difficult to diagnose. Hyaline material

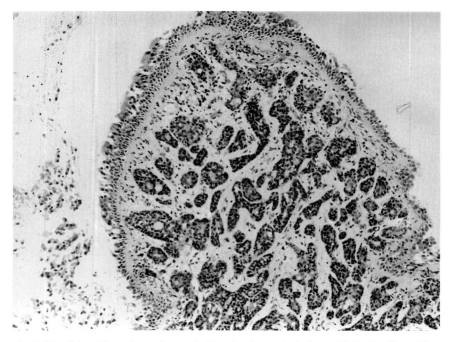

Fig. 2.16 Adenoid cystic carcinoma infiltrating beneath the bronchial epithelium. Note the double layer of cells and the cribriform pattern.

representing replicated basement membrane is seen which stains with Alcian blue while the mucin stains red/pink with PASD. Therefore AB/PASD stain is useful in this tumour. Tumour cells infiltrate widely into perineural spaces and blood vessels. The tumour can be mistaken for a straight-forward adenocarcinoma or even small cell carcinoma on biopsy but with adequate resection its prognosis is much better than these, so it is important to detect its architecture and double cell layer to come to a proper diagnosis.

Mucoepidermoid carcinoma

These can be either high-grade or low-grade. Histologically the low-grade tumours appear as endobronchial polypoid lesions that focally infiltrate the surrounding lung parenchyma. They consist of solid and cystic areas with multiple glands lined by mucin-rich goblet cells, admixed with basaloid cells, intermediate (transitional), squamous, oncocytic and clear cells (Fig 2.17). Focal areas of squamous differentiation recognized by the presence of intercellular bridges are seen but no keratin pearls or extensive keratinization is present which can help to differentiate it

from adenosquamous carcinoma. They have low mitotic activity (less than 1 per 20 high-powered fields), minimal pleomorphism and no necrosis.

High-grade tumours have a more variable appearance with nuclear pleomorphism, increased mitotic activity (4 per 10 high-powered field) and foci of necrosis.

The differential diagnosis includes mucous gland adenoma for the better-differentiated tumours but the latter lack intermediate cells and areas of squamous differentiation. The high-grade tumours can be mistaken for squamous cell carcinoma if the presence of intermediate and mucin-secreting cells is missed. Mucoepidermoid carcinomas are also difficult to separate from adenosquamous carcinomas. This distinction is important as adenosquamous carcinomas are more aggressive tumours. Points favouring mucoepidermoid carcinoma include (a) endobronchial polypoid appearance, (b) lack of dysplasia and carcinoma *in situ* changes in the overlying epithelium, (c) lack of pearl formation and extensive keratinization, (d) intermingled mixture of basal cells, intermediate cells, squamous cells and mucin-secreting cells, and finally (e) presence of better-differentiated mucoepidermoid areas.

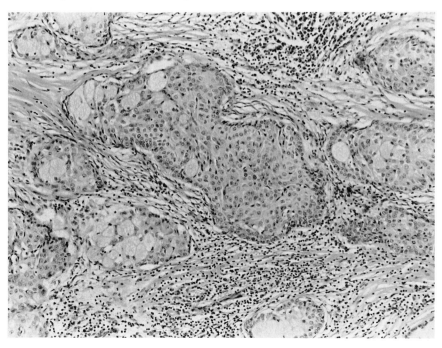

Fig. 2.17   Low-grade mucoepidermoid carcinoma of bronchus showing intimate mixture of mucous cells with vacuoles, intermediate cells and squamous cells. Note the lack of mitotic activity.

Hamartomas

Endobronchial or central hamartomas constitute 6–19% of all hamartomas and projects as a polypoid nodule within the lumen of a large bronchus.[61] Central hamartomas usually have a predominance of cartilage arranged in irregular nodules, fibromyxoid connective tissue, fat, bone and smooth muscle with ingrowing clefts of respiratory epithelium surrounding the cartilage. The chondrocytes can have an atypical appearance, being more cellular and hyperchromatic than normal, and chondrosarcoma, either primary or secondary, must be included in the differential diagnosis. The admixture of other mature mesenchymal elements will point to the diagnosis.

Connective tissue tumours

Other rare primary bronchial tumours include fibromas, leiomyomas[62] (Fig. 2.18), lipomas, granular cell tumours, glomus tumours and neural tumours. Sarcomas of all types are very rare but can occur. Obviously metastatic lesions must be eliminated[63] before a lesion can be labelled as a primary pulmonary sarcoma and an epithelial element carefully looked for in order to avoid missing a carcinosarcoma or spindle squamous cell carcinoma.

Carcinosarcoma

These are composed of intimately mixed epithelial and mesenchymal elements and can present as central endobronchial tumours in 25% of cases, often in elderly smokers. Microscopically the carcinomatous element is usually squamous, adenocarcinoma or large cell or mixtures of all three, while the sarcomatous element is most commonly fibrosarcoma. However, the ability of squamous cell carcinomas to spindle suggests that many of these are spindle squamous cell carcinomas, which can be confirmed by the finding of keratin in the spindle areas as well as the carcinoma areas. The merging of squamous areas into spindle cell areas has been put forward as a useful feature in spindle cell carcinomas, while carcinosarcomas have separate elements, but many pathologists find it hard to delineate this feature with confidence in most tumours.[64] For this reason the World Health Organization suggests that specific differentiation other than fibrosarcoma must be present before a

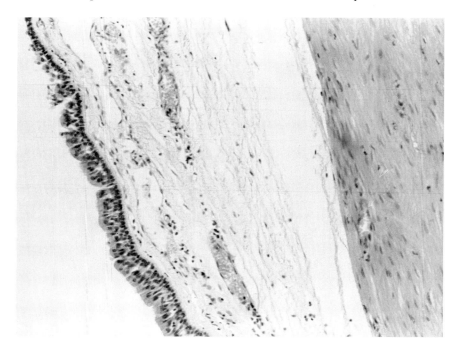

Fig. 2.18 Bronchial biopsy showing a well-defined spindle cell tumour mass in the submucosa which on immunostaining is positive for vimentin and smooth muscle actin, confirming that it is a leiomyoma.

lesion is considered a carcinosarcoma which is also the opinion of Humphrey *et al.*[65] The second most common stromal component is chondrosarcoma, while osteogenic sarcoma and rhabdomyosarcoma have also been reported. Bronchial biopsy may well miss the dual elements to these tumours and resection will reveal the true picture.

Immunocytochemistry in bronchial biopsies

Immunocytochemistry has limited value in diagnosing lung tumours. All carcinomas contain keratins and other epithelial markers such as carcinoembryonic antigen (CEA) and epithelial membrane antigen (EMA).[44] No unique marker can be used to separate squamous cell carcinoma from adenocarcinoma or large cell carcinoma. Unique keratin profiles have been suggested for squamous cell carcinoma and adenocarcinomas, but the spectrum of keratin antibodies is large and they only work on frozen material. Most importantly from a therapeutic point of view, no unique marker exists to distinguish small cell carcinoma from other types of carcinoma in small biopsies. The general neuronal markers like NSE and PGP9.5 are detected in all cell types and the more specific markers like chromogranin are not found in many small cell carcinomas.[47]

Immunocytochemistry can prove very useful in the biopsies where one has difficulty in separating small cell carcinomas from a dense lymphocytic infiltrate. The application of keratin antibodies such as CAM 5.2, elicits a positive response in small cell carcinomas, the keratin marker often appearing as a small dot-like blob near the nucleus, while lymphocytes are negative. This keratin-positive dot-like immunoreactivity in small cell carcinoma has been claimed as unique to small cell carcinoma, but it can also be seen in large cell carcinomas. Vimentin can be expressed in adenocarcinomas and large cell carcinomas so it is not useful in separating sarcomas from carcinomas. Surfactant has already been mentioned in the diagnosis of primary lung adenocarcinomas.

Metastatic tumours

Always remember that the lung is the most common place for blood-borne metastases and these can mimic any primary lung tumour. Metastases

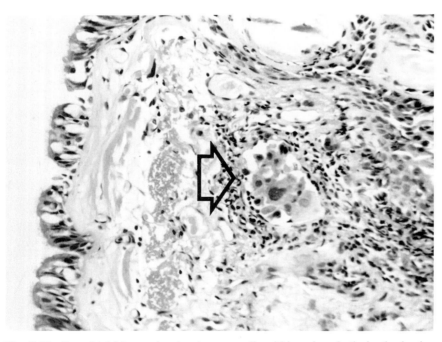

Fig. 2.19 Bronchial biopsy showing tumour cells within a lymphatic in the lamina propria. Lymphangitis carcinomatosa.

may be endobronchial, simulating a primary tumour on bronchoscopy, and a long latent period of several years may have elapsed since the original diagnosis was made.[66] It is impossible, especially with adenocarcinomas, to tell whether a tumour is primary or secondary. However the presence of tumour cells in lymphatics of the mucosa and submucosa indicates lymphangitis carcinomatosa which is easily diagnosed on bronchial biopsy (Fig. 2.19). The use of immunocytochemistry with thyroglobulin and prostate-specific antigen and prostatic acid phosphatase as well as germ cell markers can help in some cases, but this is the exception rather than the rule. Most epithelial markers are shared between a wide variety of tumours throughout the body. Others have suggested a panel of antibodies to predict the primary tumour site.[67]

Pulmonary lymphomas

In our experience bronchial biopsy can be used to diagnose primary lymphomas of the lung. The most common are B cell lymphomas, in which malignant cells can infiltrate the over lying mucosa forming lymphoepithelial lesions. Immunocytochemistry can be very valuable for the diagnosis of lymphomas on bronchial biopsy. Use of the leukocyte common antigen, light chain, B and T antibodies can be applied to routinely fixed tissue, as seen in Chapter 12.

Summary

Bronchial biopsy is a part of the everyday routine work of the practising pathologist and knowledge concerning the various entities, particularly tumours, is essential. As more sophisticated endoscopic techniques are introduced we are seeing less and less tissue and yet we are expected to make highly specific diagnoses on these tiny fragments. A pragmatic conservative approach is needed in many cases and the pathologist needs to be aware of the various pitfalls lying in wait for the unwary. It is essential to have a close working relationship with clinical colleagues in order to extract the most clinically useful information provided by the material itself.

References

1. Mathey RA, Moritz ED. Invasive procedures for diagnosing pulmonary infection. *Clin Chest Med* 1987; **2**, 3–18.
2. Fulkerson WJ. Fibreoptic bronchoscopy. *New Engl J Med* 1984; **311**, 511–515.
3. Mitchell DM, Emerson CJ, Collins JV, Stapleforth DE. Transbronchial lung biopsy with the fibreoptic bronchoscope: Analysis of results of 433 cases. *Br J Dis Chest* 1981; **75**, 258–262.
4. Fraire AE, Cooper SP, Greenberg SD, Rowland LP, Langston C. Transbronchial lung biopsy – histopathologic and morphometric assessment of diagnostic utility. *Chest* 1992; **102**, 748–752.
5. Gould SJ, Isaacson PI. Bronchus-associated lymphoid tissue (BALT) in human fetal and infant lung. *J Pathol* 1993; **169**, 229–234.
6. Richmond I, Pritchard GE, Ashcroft T, Avery A, Corris PA, Walters EH. Bronchus associated lymphoid tissue (BALT) in human lung – its distribution in smokers and non-smokers. *Thorax* 1993; **48**, 1130–1134.
7. Lopez-Vidriero MT, Reid L. Bronchial mucus in health and disease. *Br Med Bull* 1978; **34**, 63.
8. Reid L, Haller R. The bronchial mucous glands. Their hypertrophy and change in intracellular mucin. *Mod Probl Pediatr* 1967; **10**, 195–199.
9. Cosio MG, Ghezzo H, Hogg JC et al. The relations between structural changes in small airways and pulmonary function tests. *New Engl J Med* 1977; **298**, 1277–1281.
10. Mullen JBM, Wright JL, Wiggs BR, Pare PD, Hogg JC. Reassessment of inflammation of airways in chronic bronchitis. *Br Med J* 1985; **291**, 1235–1239.
11. Lau KY. Endobronchial actinomycosis mimicking pulmonary neoplasm. *Thorax* 1992; **47**, 664–665.
12. Mcneil KD, Johnson DW, Oliver WA. Endobronchial nocardial infection. *Thorax* 1993; **48**, 1281–1282.
13. Brown NM, Hellyar EA, Harvey JE, Reeves DS. Mycobacterial contamination of fibreoptic bronchoscopes. *Thorax* 1993; **48**, 1283–1285.
14. Woodard BH, Rosenberg SI, Farnham R, Adams DO. Incidence and nature of primary granulomatous inflammation in surgically removed material. *Am J Surg Pathol* 1982; **6**, 119–129.
15. Medical Research Council Cardiothoracic Epidemiology Group. National survey of notifications of tuberculosis in England and Wales in 1988. *Thorax* 1992; **47**, 770–775.
16. Anon. Fibreoptic bronchoscopy and sputum negative tuberculosis. *Lancet* 1983; **i**, 377–378.
17. Danek SJ, Bower JS. Diagnosis of pulmonary tuberculosis by flexible bronchoscopy. *Am Rev Respir Dis* 1979; **119**, 677–679.
18. Ma P, Villaneuva TG, Kaufman D, Gillooley JF. Respiratory cryptosporidiosis in the acquired im-

munodeficiency syndrome. Use of modified cold kinyoun and hemacolor stains for rapid diagnosis. *J Am Med Assoc* 1984; **252**, 1298–1301.

19. Katzenstein AL, Liebow AA, Friedman PJ. Bronchocentric granulomatosis mucoid impaction and hypersensitivity reactions to fungi. *Am Rev Respir Dis* 1975; **111**, 497–537.

20. Bosken CH, Myers JL, Greenberger PA, Katzenstein AL. Pathologic features of allergic bronchopulmonary aspergillosis. *Am J Surg Pathol* 1988; **12**, 216–222.

21. Koss MN, Robinson RG, Hochholzer L. Bronchocentric granulomatosis. *Hum Pathol* 1981; **12**, 632–638.

22. Koontz CH, Joyner LR, Nelson RA. Transbronchial lung biopsy via the fibroptic bronchoscope in sarcoidosis. *Ann Intern Med* 1976; **85**, 64–70.

23. Bjermer L, Thunell M, Rosenhall L, Stjernberg N. Endobronchial biopsy positive sarcoidosis: relation to bronchoalveolar lavage and course of disease. *Respir Med* 1991; **85**, 229–234.

24. World Association of Sarcoidosis and Other Granulomatous Disorders. Consensus conference: Activity of sarcoidosis. *Eur Respir J* 1994; **7**, 624–627.

25. Da Costa P, Corrin B. Amyloidosis localized to the lower respiratory tract: probable immunoamyloid nature of the tracheobronchial and nodular pulmonary forms. *Histopathol.* 1985; **9**, 703–710.

26. Pounder DJ, Pieterse AS. Tracheopathia osteoplastica. A study of the minimal lesion. *J. Pathol.* 1982; **138**, 235–239.

27. Vanmaele L, Noppen M, Frecourt N, Impens N, Welch B, Schandevijl W. Atypical ossification in bronchial carcinoid. *Eur Respir J* 1990; **3**, 927–929.

28. Sheffield E, Corrin B. Fatal bronchial stenosis due to isolated relapsing chondritis. *Histopathol* 1992; **20**, 442–443.

29. Galdermans D, Verhaert J, Van Meerbeeck *et al.* Bronchiolitiasis: present clinical spectrum. *Respir Med* 1990; **84**, 155–156.

30. Matsubara O, Tan-Liu NS, Kenny RM, Mark EJ. Inflammatory pseudotumour of the lung: Progression from organizing pneumonia to fibrous histiocytoma on to plasma cell granuloma in 32 cases. *Hum Pathol* 1988; **19**, 807–814.

31. Bush A, Sheppard MN, Wahn U, Warner JO. Spontaneous arrest of growth of a plasma cell granuloma. *Respir Med* 1992; **86**, 161–164.

32. Yacoub MH, Thompson VC. Chronic idiopathic pulmonary hilar fibrosis, a clinicopathological entity. *Thorax* 1971; **26**, 365–373.

33. Ellis JH. Transbronchial biopsy via the fibreoptic bronchoscope. *Chest* 1975; **68**, 524–533.

34. World Health Organization. The World Health Organization histological typing of lung tumours, 2nd edn. *Am J Clin Pathol* 1982; **77**, 123–136.

35. Roggli VL, Vollmer RT, Greenberg SD, McGavran MH, Spjut HJ, Yesner R. Lung cancer heterogeneity: a blinded and randomized study of 100 consecutive cases. *Hum Pathol* 1985; **16**, 569–579.

36. Chuang MT, Marchevsky A, Teirstein A, Kirschner PA, Kleinerman J. The diagnosis of lung cancer by fibreoptic bronchoscopy. 1 Problems in the histologic classification of non-small cell carcinomas. *Thorax* 1984; **39**, 175–178.

37. Mountain CF. A new international staging system for lung cancer. *Chest* 1986; **89**, 225S–233S.

38. Thomas J St J, Lamb D, Ashcroft T *et al.* How reliable is the diagnosis of lung cancer using small biopsy specimens? Report of a UKCCCR Lung Cancer Working Party. *Thorax* 1993; **48**, 1135–1139.

39. Saccomanno Y, Archer VE, Auerbach O. Development of carcinoma of the lung as reflected in exfoliated cells. *Cancer* 1974; **33**, 256–270.

40. Saccomanno G, Yale C, Dixon W *et al.* An epidemiological analysis of the relationship between exposure to Rn progeny, smoking and bronchogenic carcinoma in the U-mining population of the Colerado plateau – 1960–1980. *Health Phys* 1986; **50**, 605–620.

41. Yoneda K, Boucher LD. Bronchial epithelial changes associated with small cell carcinoma of the lung. *Hum Pathol* 1993; **24**, 1180–1183.

42. Nagamoto N, Saito Y, Sato M *et al.* Clinicopathological analysis of 19 cases of isolated carcinoma *in situ* of the bronchus. *Am J Surg Pathol* 1993; **17**, 1234–1243.

43. Ishida T, Inoue T, Sugio K *et al.* Early squamous lung cancer and longer survival rates. *Respiration* 1993; **60**, 359–365.

44. Sheppard MN, Hamid QA, Polak JM. Immunohistology of lung tumours. *Diagn Oncol* 1991; **1**, 252–261.

45. Mizutani Y, Nakajima T, Moringa S. Immunohistochemical localization of pulmonary surfactant apoproteins in various lung tumours. Special reference to non-mucus producing lung adenocarcinomas. *Cancer* 1988; **61**, 532–540.

46. Sheppard MN. Neuroendocrine differentiation in lung tumours. *Thorax* 1991; **46**, 843–850.

47. Sheppard MN. Neuroendocrine differentiation in lung tumours. In: *New Perspectives in Lung Cancer*, Thatcher N, Spiro S (eds). London: British Medical Association, 1994, p.50.

48. Linnoila RI, Mulshine JL, Steinberg SM *et al.* Neuroendocrine differentiation in endocrine and nonendocrine lung carcinomas. *Am J Clin Pathol* 1988; **90**, 641–652.

49. Oike N, Bernatz PE, Woolner, LB. Carcinoid tumours of the lung. *Ann Thorac Surg* 1976; **22**, 270–277.

50. Jones RA, Dawson IMP. Morphology and staining patterns of endocrine tumours in the gut, pancreas

and bronchus and their possible significance. *Histopathol* 1977; **1**, 137–150.

51. Nelson EL, Houghton DC. Concurrent spindle cell peripheral pulmonary carcinoid tumour and Merkel cell tumour of the skin. *Arch Pathol Lab Med* 1990; **114**, 420–423.

52. Sheppard MN, Corrin B, Bennett MH, Marangos PJ, Bloom SR, Polak JM. Immunocytochemical localisation of neuron-specific enolase in small cell carcinoma and carcinoid tumours of the lung. *Histopathol* 1984; **8**, 171–181.

53. Barbareschi M, Frigo B, Mosca L *et al*. Bronchial carcinoids with S-100 positive sustentacular cells. A comparative stiudy with gastrointestinal carcinoids, pheochromocytomas and paragangliomas. *Pathol Pract* 1990; **186**, 212–222.

54. Hoefler H, Denk H, Lackinger E, Helleis G, Polak JM, Heitz PU. Immunocytochemical demonstration of intermediate filament cytoskeletal proteins in human endocrine tissues and (neuro-) endocrine tumours. *Virchows Arch A Pathol Anat* 1985; **409**, 609–629.

55. Hajdu SI, Winawer SJ, Myers WPL. Carcinoid tumours: a study of 204 cases. *Am J Clin Pathol* 1974; **61**, 521–528.

56. McCaughan BC, Martini N, Bains MS. Bronchial carcinoids. *J Thorac Cardiovasc Surg* 1985; **89**, 8–17.

57. Valli M, Fabris GA, Dewar A, Hornall D, Sheppard MN. Atypical carcinoid tumour of the lung. A study of 33 cases with prognostic features. *Histopathol* 1994; **24**, 371–376.

58. Seo JW, Im JG, Kim YW, Kim JH, Sheppard MN. Synchronous double primary lung cancers of squamous and neuroendocrine type associated with cryptogenic fibrosing alveolitis. *Thorax* 1991; **46**, 857–858.

59. van Bodegom PC, Wagenaar SJ, Sc, Corrin E *et al*. Second primary lung cancer: importance of long term follow up. *Thorax* 1989; **44**, 788–793.

60. Brightman I, Morgan JA, Zwehl D, Sheppard MN. Cytological appearances of a solitary squamous cell papilloma with associated mucous gland adenoma in the lung. *Cytopathol* 1992; **3**, 253–257.

61. Butler C, Kleinerman J. Pulmonary hamartoma. *Arch Pathol* 1969; **8**; 584–592.

62. Douzinas M, Sheppard MN, Lennox SC. Leiomyoma of the trachea. An unusual tumour. *Thorac Cardiovasc Surg* 1989; **37**, 285–287.

63. Robinson MH, Sheppard MN, Moskovic E, Fisher C. Lung metastastectomy in patients with soft tissue sarcomas. *Br J Radiol* 1994; **67**, 129–135.

64. Sheppard MN. Spindle cell variant of pulmonary adenocarcinoma – Commentary. *Pathol Res Pract* 1993; **189**, 591–593.

65. Humphrey PA, Scroggs MW, Roggli VL, Shelburne JD. Pulmonary carcinomas with a sarcomatoid element; an immunocytochemical and ultrastructural analysis. *Hum Pathol* 1988; **19**, 155–165.

66. Heitmiller RF, Marasco WJ, Hruban RH, Marsh BR. Endobronchial metastasis. *J Thorac Cardiovasc Surg* 1993; **106**, 537–542.

67. Gamble AR, Bell JA, Ronan JE, Pearson D, Ellis IO. Use of tumour marker immunoreactivity to identify primary site of metastatic cancer. *Br Med J* 1993; **306**, 295–298.

3

Pulmonary cytology

Stanley J Radio, James Linder

Introduction

The study of cytological specimens from the respiratory tract is integral to the evaluation of any patient with suspected primary or metastatic lung neoplasm or opportunistic infection. Examination of exfoliated cells of the respiratory tract was reported as early as 1845 by Donne.[1] Several other authors at the turn of the century described the presence of malignant tumour in sputum specimens. However, respiratory cytology did not develop into an established method of diagnosis until the late 1940s and early 1950s, following the publication of George Papanicolaou's monograph entitled *The Diagnosis of Uterine Cancer by the Vaginal Smear*. His method of fixation, originally for cervical-vaginal cells, allowed for discrimination of fine cytological detail. This staining method was quickly adapted to cellular material from the respiratory tract. During the 1950s and 1960s, numerous publications described new techniques for the detection of neoplastic cells and cytohistologic correlations involving respiratory cytology.[2-14] The development of fine needle aspiration biopsy of the lung paralleled respiratory cytology in general. Pioneering efforts included the diagnosis of pneumonia in the 1880s by Gunther[15] and the diagnosis of lung cancer by Kronig[16] and Menetrier.[17] Rejuvenation of the interest in transthoracic needle biopsy occurred in the 1960s when improvements in fluoroscopy, together with refined biopsy needles, helped achieve a diagnostic accuracy of 87%. The further emergence of fine needle aspiration in the past 25 years has enhanced the diagnostic usefulness of clinical cytology in lung cancer diagnosis. In this chapter, we will discuss respiratory cytology through the morphological study of sputum, bronchial washings, bronchial brushings, bronchoalveolar lavages, and transthoracic as well as transbronchial fine needle aspiration biopsies from the lung.

Cytopreparatory techniques

High diagnostic accuracy in cytology requires an unrelenting commitment to excellence in cytopreparatory techniques. The final goal is the clear demonstration of abundant, well-preserved and meticulously stained cells prepared in a clinically relevant time frame, and suitable for long-term storage. Several major techniques meet these criteria. While they are among the most traditional techniques, they also remain widely utilized today. They are the wet film preparation and fixation from fresh or prefixed cytological material, the Saccomanno blender technique,[18,19] membrane filtration[20-22] and cytocentrifugation.[23] The Papanicolaou method is the most popular staining technique in the USA. In many laboratories, especially those in Europe, air-dried smears stained by the Romanowsky method are equally popular, particularly in the evaluation of fine needle aspirates. In addition, the rapid acceptance and expansion of the use of immunohistochemistry in characterizing neoplasms has made paraffin-embedding and sectioning of tissue fragments obtained by needle biopsies a valuable adjunct. The following types of cellular specimens and related techniques for cellular preparation are those currently most commonly utilized in a contemporary clinical cytology laboratory. The ultimate pattern of cellular presentation recognized on the slide is closely determined by the type of preparation utilized.

Sputum

Sputum, because of its ease of sampling, is a useful means to examine the respiratory tract. Following collection of early morning sputum, the specimen may be prepared for examination by several techniques. If the specimen is brought directly to the laboratory without fixation, representative areas are made into smears and fixed in 95% ethyl alcohol.

When conditions dictate that the sputum specimen cannot be rapidly delivered to the cytology laboratory, the specimen may be collected in a wide-mouthed small jar which has been previously filled with 70% ethyl alcohol. The main disadvantage of this technique is the less efficient penetration of the specimen by the alcohol with less efficient fixation. Many laboratories utilize the Saccomanno method of sputum preparation in which the specimen is collected in a mixture of 50% ethyl alcohol and 2% polyethylene glycol (Carbowax). A blender is then utilized to homogenize the specimen and smears are prepared from a centrifuged cell button. While this technique provides concentration of cells, and the possibility of multiple slides from the same case,[19] there are also inherent artefacts present in tumour cells prepared from this technique, as will be discussed later.

Bronchial aspirates and washings

Fibreoptic bronchoscopy, in addition to allowing visualization of the lower respiratory tract, also provides the opportunity to sample suspect regions directly. A suction apparatus allows the operator to aspirate secretions from the area of interest. In addition, washings from this visualized area may be obtained and collected following instillation and reaspiration of 3–5 ml of a balanced salt solution through the bronchoscope. Cellular specimens derived from the aspiration or washing techniques may then be processed by cytocentrifugation; membrane filter preparation or smears may be prepared following centrifugation. In addition, cell buttons resulting from centrifugation may be embedded in paraffin for histological sectioning.

Bronchial brushings

In addition to aspirates and washings, fibreoptic bronchoscopy allows the operator to brush a suspected lesion and submit the collected cytological material for laboratory examination, utilizing techniques similar to those for bronchial aspirates and washings.

Bronchoalveolar lavage

Bronchoalveolar lavage (BAL) involves wedging of the fibreoptic bronchoscope in a distal segmental bronchus followed by the infusion and reaspiration of sterile saline solution. Much of the early developmental work in BAL centred around therapeutic uses in such diseases as pulmonary alveolar proteinosis, cystic fibrosis, asthma or pulmonary alveolar microlithiasis.[24] BAL has subsequently become an important tool of the pulmonologist in the detection of opportunistic infection in hosts immunocompromised in the setting of cancer therapy, solid organ or bone marrow transplantation, or acquired immune deficiency syndrome (AIDS).[24-27] Membrane filter preparation or cytocentrifugation are both utilized as cytopreparatory techniques of BAL specimens.

Fine needle aspiration biopsy

Although fine needle aspiration biopsy of the respiratory tract has been utilized since the turn of the century, improvements in radiological localization of lung masses as well as the low morbidity associated with small bore needles have greatly boosted the popularity of this procedure.[28-31] The technique is similar to fine needle aspiration of superficial masses, except that localization is provided by fluoroscopy or computerized axial tomography rather than by palpation. The resultant aspirated cellular material is examined after preparation by conventional cellular techniques, as described in the sections on bronchial specimens.

In general, fine needle aspiration (FNA) biopsy of the lung is indicated when a patient with suspected inoperable primary or metastatic lung cancer undergoes multiple consecutive deep cough specimens of sputum and a bronchial brushing or washing fails to provide the diagnosis of malignancy.[32,33] Similarly, FNA is indicated when an infectious process is suspected in an immunocompromised host.

The contraintrications for FNA biopsy are: (1)

debilitated, uncooperative patients or those with uncontrollable cough; (2) possible ecchinococcus cyst; (3) a haemorrhagic diaphysis, anticoagulant therapy or suspected vascular lesion. While the incidence is low, known complications include pneumothorax, haemoptysis, and haemothorax.[32,33]

Pertinent anatomy and cytology

Knowledge of basic pulmonary histology and anatomy is critical in interpreting normal and abnormal cells derived from the lung. The upper respiratory tract consists of at least five separate cell types, the three most common being ciliated cells, goblet cells and basal cells. The basal cells are capable of maturing into either ciliated or goblet cells. In addition, argyrophil or Kulchitsky-like cells and brush border cells of unknown function are also present in small numbers. The distribution of the cells varies throughout the respiratory tree. For example, ciliated cells become less prominent in the distal or terminal bronchioles while non-ciliated clear cells become numerous. In the alveoli, approximately 95% of the surface epithelium is composed of type 1 pneumocytes. The remainder are termed type 2 or granular pneumocytes, which when examined ultrastructurally are distinguished by prominent microvilli and osmophilic lamellated inclusion bodies within the cytoplasm. Type 2 pneumocytes are thought to be the chief cell type involved in repair of the alveolar epithelium as well as the likely source of surfactant. Alveolar macrophages, which originate from stem cell precursors within the bone marrow, are also found within alveolar air spaces as well as in the extracellular lining of the alveolar surface.[34,35]

The relative ratio of different cell types in the respiratory tract specimen depends on the method of sampling. For instance, the normal epithelial components of sputum are squamous epithelial cells from the pharynx and oral cavity, respiratory columnar epithelium from the tracheobronchial tree, terminal bronchiolar epithelium and alveolar epithelium.

In contrast, wash, brush or lavage specimens have more ciliated columnar cells. The ciliated columnar cell is recognized by its columnar or prismatic shape tapering distally into a tail. The nucleus is basally oriented with a finely granular chromatin pattern and one or more small nucleoli are present. Cilia and a terminal plate are characteristic findings present on the end of the cell

opposite the nucleus. The goblet, or mucus-producing bronchial cell, is a less commonly seen type of epithelial cell. It is recognized by either a single or multiple vacuole(s) which distend the cytoplasm and distort the nucleus shape. Typically, goblet cells are abundant in specimens from patients with chronic respiratory diseases such as asthmatic bronchitis, chronic bronchitis and bronchiectasis. In response to a wide variety of irritants including infection or environmental toxins, goblet cells may be present in such great numbers that their differentiation from bronchioloalveolar carcinoma may present a challenge to the cytopathologist. In this setting, the reactive cells are characterized by enlargement of the nucleus with a diameter 10–20 times that of a normal bronchial cell (Figs 3.1 and 3.2). In addition, these cells exhibit coarsening of the chromatin pattern and the presence of one or more enlarged nucleoli.

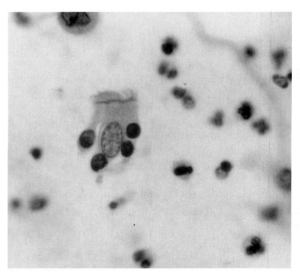

Fig. 3.1 Reactive bronchial epithelium. A reactive bronchial epithelial cell is present with enlarged nucleus and abnormal but uniformly distributed chromatin. These changes in the middle cell are in contrast to the normal-sized nuclei in the adjacent epithelial cells. The entire group of cells maintain their columnar shape as well as their ciliated border and terminal bar. Bronchoalveolar lavage specimen. Papanicolaou stain; × 360.

Hyperplasia of respiratory epithelial cells occurs in many pulmonary diseases such as tuberculosis,[36,37] bronchiectasis, chronic bronchitis and asthma.[38,39] Papillary tissue fragments of reactive epithelial cells may mimic carcinoma. These clusters, which are partially covered by well-

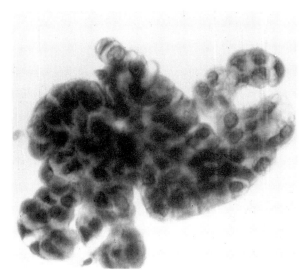

Fig. 3.2 Bronchial cell hyperplasia. A cluster of hyperplastic bronchial epithelial lining cells with prominent goblet cells, uniform chromatin and tight cell cohesion. Bronchial brushing specimen. Papanicolaou stain; × 360.

differentiated, ciliated respiratory epithelium, have been named 'Creola bodies' after the patient in whom they were first seen.[38] Even though chromatin and nuclear structures in Creola bodies may appear near normal, the presence of nuclear moulding between individual cells can be a problematic feature. Identifying the finely granular chromatin pattern, uniform nucleoli and the presence of cilia will assure the correct diagnosis and distinguish these cells from tumour.

Although much less frequently encountered, epithelial cells from the terminal bronchioles and alveoli none the less may pose a potential diagnostic pitfall, especially in fine needle aspirations. Terminal bronchiolar and alveolar cells are relatively small; when present they appear as rounded single cells with finely vacuolated cytoplasm and centrally placed nuclei with one or two small nucleoli. These cells possess a large potential to enlarge, proliferate, and may present with prominent distended cytoplasmic vacuoles. They also may be organized as small papillary tissue fragments. Such changes, which are usually secondary to repair or stimulation associated with infection, pulmonary fibrosis, thromboembolism with pulmonary infarction or organizing pneumonia, may resemble bronchioloalveolar carcinoma. Hyperplastic bronchial or alveolar cells in fine needle aspirates thus constitute one of the most dangerous

of diagnostic pitfalls if the cytopathologist does not require high cellularity, well-preserved three-dimensional clusters and appropriate cellular atypia as criteria for the diagnosis of carcinoma.

Severe morphological alteration of respiratory cells can occur in response to radiation therapy and anticancer chemotherapy. A knowledge of treatment history is essential in order to categorize these cells correctly as they may possess sufficient morphological atypia to be mistaken for carcinoma. Ionizing radiation may result in stimulating changes of both squamous and columnar cells characterized by cytomegaly with both cytoplasmic and nuclear enlargement, multinucleation, macronucleoli and cytoplasmic vacuolization (Fig. 3.3). Cancer chemotherapy tends to produce single cells with cytomegaly, hyperchromasia and macronucleoli. These changes can be seen in cells derived from the tracheobronchial epithelium, terminal bronchiolar epithelium and alveolar epithelium. Features that allow the recognition of these cells include sparse singular cells which are rectangular in shape and exhibit nuclear degeneration.

In addition to respiratory epithelial cell hyperplasia and changes secondary to irradiation and chemotherapy, another cytological pattern that is essential to recognize is the attempt of the host to repair damaged epithelium surface. The process of squamous metaplasia begins with a proliferation of reserve cells to a multi-layered epithelium that gradually matures, resulting in an epithelium that closely resembles a stratified squamous epithelium. Reserve cell hyperplasia appears as a group of small, uniform tightly coherent cells with hyperchromatic nuclei and a thin rim of faintly cyanophilic cytoplasm (Fig. 3.4). Because of the cell size, nuclear hyperchromasia and moulding, reserve cell hyperplasia is in the differential diagnosis of small cell undifferentiated carcinoma. The coherent nature of the cell clusters distinguishes reserve cell hyperplasia from such single cell malignancies as non-Hodgkin's lymphoma.

Squamous metaplasia appears cytologically as small tissue fragments or as single cells. Metaplastic cells are smaller and possess a higher nuclear to cytoplasmic ratio than mature squamous cells. Often groups of metaplastic cells are present in a monolayered cobble-stoned arrangement with a characteristic lack of cell-to-cell variance in either cytoplasmic or nuclear appearance. The cytoplasmic features in squamous metaplasia also mimic maturing squamous epithelium with deep cyanophilia or orangeophilia along with keratinization.

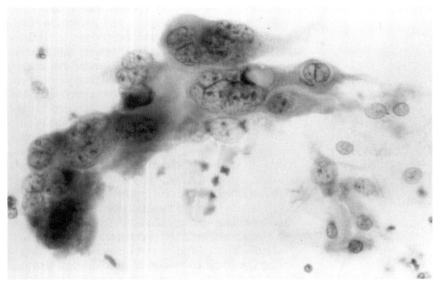

Fig. 3.3 Radiation change. Group of bronchial epithelial cells with enlarged nuclei with uniformly dispersed chromatin, prominent nucleoli and cytoplasmic vacuolization. The nuclear changes are similar from cell to cell. The epithelial cells maintain tight cohesiveness. Bronchial brushing. Papanicolaou stain; × 400.

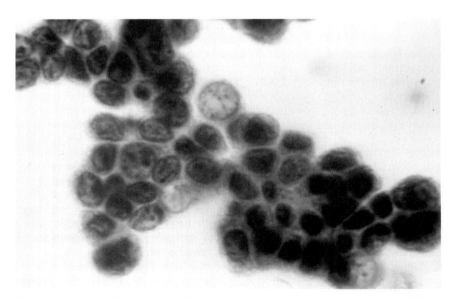

Fig. 3.4 Reserve cell hyperplasia. Group of reserve cells with hyperchromatic nuclei and small nucleoli in a tightly cohesive cluster with no single cells present. Bronchoalveolar lavage specimen. Papanicolaou stain; × 400.

Significant evidence suggests that squamous metaplasia is an early alteration in respiratory epithelium in the pathogenesis of bronchogenic carcinoma. In this respect, squamous metaplasia may exhibit differing degrees of nuclear abnormality, altered nuclear to cytoplasmic ratio, thickening of the nuclear membrane, increasing granularity and hyperchromasia of the chromatin, and the appearance of nucleoli. It should be noted, however, that in the majority of patients, these atypical metaplastic cells are actually associated with non-neoplastic conditions of the lung, especially pneumonia.[40]

Haematopoietic cells

Cells of haematopoietic origin can be seen in bronchopulmonary material. The most prevalent of these is the pulmonary alveolar macrophage which originates from peripheral stem cells within the bone marrow. Pulmonary macrophages can be recognized by the eccentric position of the nucleus, ample foamy cytoplasm, and an often phago-cytozed carbon material (Fig. 3.5). The nuclei may assume a bean-shape and show one or more nucleoli and cytoplasmic processes. Binucleated and multinucleated giant cell macrophages are common findings, especially in association with chronic lung disease due to sarcoidosis, tuberculosis, non-tuberculosis mycobacterial disease and other inflammatory diseases. Patients with aspiration pneumonia may have large fat-containing vacuoles within their pulmonary macrophages.

Lymphocytes, eosinophils, neutrophils and plasma cells also occur in bronchopulmonary material. The type and pattern of inflammatory cells present may provide a clue to the underlying disease. For example, lymphocytes predominate in sarcoidosis and hypersensitivity pneumonitis, neutrophils are numerous in idiopathic pulmonary fibrosis, and bacterial bronchopneumonia and eosinophils can be seen in association with asthmatic bronchitis or any of a number of allergy-related diseases.

Non-cellular inanimate components

Non-cellular structures in respiratory specimens may be endogenous to the lung or inspired. For instance, Curschmann spirals result from inspissated mucus from small bronchioles. Ferruginous bodies have been noted in tissues and cellular specimens from the lungs for many years and can result from a number of different inhaled mineral fibres. Various substances, including irons help form these bodies, which are incrusted upon a thin needle-like fibre (Fig. 3.6).

Psammoma bodies (calcospherites) are dark staining, rounded bodies with concentric rings and radial striations which are calcified bodies containing phosphates, iron, magnesium and sudanophilic material. Psammoma bodies can be seen in malig-

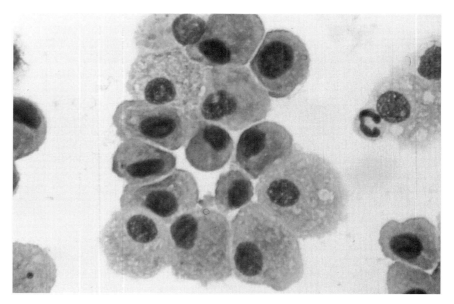

Fig. 3.5 Alveolar macrophages. A group of alveolar macrophages with eccentric nuclei and abundant foamy cytoplasm. Bronchoalveolar lavage specimen. Wright-Giemsa stain; × 1000.

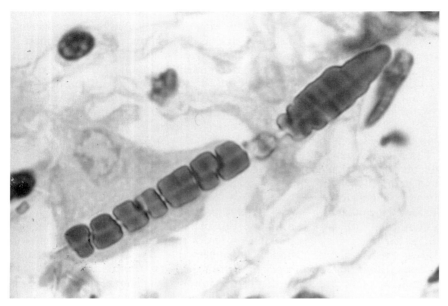

Fig. 3.6   Ferruginous body. Central fibre with surrounding incrustation in a patient with asbestosis. Wedge biopsy of lung. Haematoxylin and eosin stain; × 400.

nant neoplasms including bronchioloalveolar carcinoma, papillary adenocarcinoma, and small cell carcinoma, along with a rare benign disease known as pulmonary microlithiasis. Corpora amylacea, on the other hand, are composed of glycoproteins, do not calcify and can be seen in patients with heart failure, pulmonary infarction and chronic bronchitis.

Cellular manifestations of infectious diseases

Depending on the infectious agent, a relatively specific pattern of inflammatory cells and/or characteristic changes in bronchial, bronchiolar alveolar epithelial cells may be seen. In most situations the inflammatory component more commonly consists of a mixed infiltrate of neutrophils, lymphocytes, plasma cells and macrophages. When numerous neutrophils are the predominant inflammatory cell present, an acute suppurative bronchitis or bronchopneumonia should be suspected. When granulomatous inflammation including epithelioid and Langhans' giant cells are present, the differential diagnosis would include sarcoidosis, tuberculosis or fungal infections.

In the vast majority of bacterial pneumonias, the primary diagnosis does not rely on conventional cytological methods. Although it is common for bacteria to be present within sputum specimens, they are not a reliable marker of pulmonary infection as they commonly contaminate oral-derived cytological specimens. Since fine needle aspirates are not exposed to contamination of the oral pharynx, bacteria in these specimens represent a potentially significant finding. The likelihood of detecting acid-fast bacteria in a cytological specimen increases when there is other evidence of granulomata or the patient is at particularly high risk for such infections. In this setting, acid-fast stains may be performed on cell blocks or fluorescent microscopy utilizing Auramine-O may reveal the organisms.

Fungal infections

Cytological features of the more common respiratory fungal infections are presented below.

Species of *Candida* are the most frequently encountered fungi in cytological specimens. Morphologically, they are small, oval, budding yeasts measuring 2–4 μm in diameter, and may also include elongated non-septate pseudohyphae. Assessing the significance of *Candida* spp. in respiratory material is difficult, since the organism when present in pulmonary specimens often represents colonization of the oral pharynx or proximal respiratory tract. Actual infection is usually restricted to immunocompromised patients. The distinction between colonization and infection is difficult and may be resolved through open biopsy, fine needle aspiration, or the presence of large numbers of organisms in the alveolar component of BAL specimens.

Blastomyces dermatitidis is capable of systemic and/or localized infection. Importantly, the clinical and radiographical presentation of this organism frequently mimics those of lung carcinoma. *B. dermatitidis* has the cytological appearance of single or broad-based budding spherical cells 8–15 µm in diameter with thick refractile walls when stained by the Papanicolaou technique. No hyphae are seen. The accompanying inflammatory reaction with *B. dermatitidis* ranges from tuberculoid granulomata to numerous neutrophils and microabscess formation.

Cryptococcus neoformans is most frequently encountered as a secondary invader in the immunocompromised host. The cytological appearance is that of single budding yeast which pinches off or leaves a thin isthmus of attachment to the mother cell, thus projecting a tear-drop shape (Fig. 3.7). The cell is ovoid-to-spherical, thick-walled and 5–20 µm in diameter. A gelatinous capsule is

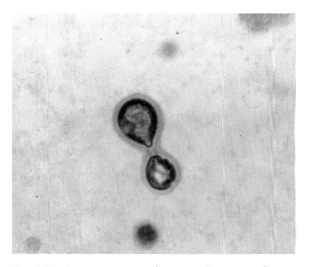

Fig. 3.7 *Cryptococcus neoformans.* Cryptococcal yeast with characteristic narrow-based budding and prominent staining of the capsule. Bronchial brushing specimen. Papanicolaou stain; × 1000.

readily highlighted by periodic acid Schiff, mucicarmine or Alcian blue stains. The capsule is often apparent with routine Papanicolaou technique, particularly in organisms that have been phagocytized by macrophages, where it appears as a clear zone in the cytoplasm. An important characteristic of *Cryptococcus* is its considerable variability in size and internal morphology. *Cryptococcus* may induce slight, granulomatous, or no inflammatory reaction at all.

Spherules and endospheres of *Coccidioides immitis* have been reported in sputum, bronchial washings and fine needle aspiration biopsies. The spherule is a round thick-walled structure of 20–60 µm diameter. The spherules may be empty or contain endospheres. The endospheres are round, non-budding structures ranging from 1 to 5 µm in diameter. The empty spherules must be distinguished from the non-budding forms of *B. dermatitidis*, while endospores resemble *Candida* spp.

Histoplasma capsulatum is not commonly seen in sputum, bronchial washings, or fine needle aspiration biopsies. Due to its small size, it is best visualized with methenamine silver where 1–5 µm round oval budding yeast can be seen within macrophages or neutrophils.

Aspergillus spp. may be seen in a number of presentations including allergic reactions, fungal growth in a pre-existing cavity, tracheobronchitis, and rapidly progressive fungal invasion of pulmonary parenchyma and vasculature. *Aspergillus* spp. is most often identified in sputum or BAL where its thick, septated, uniform hyphae with 45° angle branching provides strong morphological evidence of infection (Fig. 3.8). The presence of septa within the hyphae helps distinguish *Aspergillus* spp. from phycomycosis. *Aspergillus* derives its name from the resemblance of its conidiophores or fruiting heads to an aspergillum (a brush or perforated globe for sprinkling holy water) (Fig. 3.9). Unfortunately conidiophores are a rare finding in pulmonary specimens, except in needle biopsies of an aspergilloma.

Phycomycosis, also termed mucormycosis or zygomycosis, is an acute fungal infection usually occurring in immunocompromised patients with diabetes mellitus, leukaemia, lymphoma or those receiving steroids, chemotherapy, or antibiotics. The organisms of this order are ubiquitous and commonly seen in food moulds. Vascular thrombosis, secondary to blood vessel invasion, is a common feature of this disease and may result in

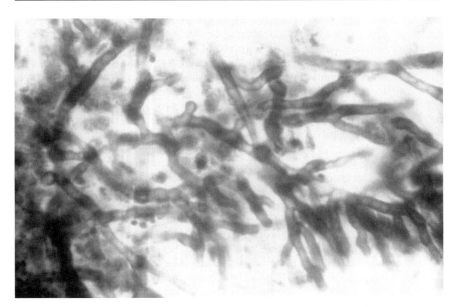

Fig. 3.8 *Aspergillus*. Numerous fungal septated hyphae with diagonal branching and parallel cell walls. Bronchoalveolar lavage specimen. Papanicolaou stain; × 400.

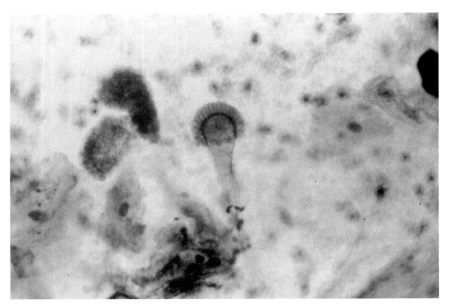

Fig. 3.9 *Aspergillus* with fruiting head displaying the resemblance to an aspergillum. Bronchoalveolar lavage specimen. Papanicolaou stain; × 400.

pulmonary infarcts. These fungi have ribbon-like, non-septate, irregular branching hyphae with considerable variability. The width of the septa may vary from 6 up to 50 μm. Cultures are necessary to identify which specific fungus is present.

Additional fungi recognized in pulmonary specimens include *Actinomyces* spp., paracoccidioidomycosis, *Alternaria* spp. (Fig. 3.10), *Sporothrix schenckii*, *Penicillium* spp., *Phaeohyphomycosis* spp. and *Trichosporon* spp.[41]

Fig. 3.10 Alternaria. Elongated brown-pigmented conidia with tapered ovate beak as well as longitudinal and transverse septations. Bronchoalveolar lavage specimen. Papanicolaou stain; × 400.

Viral infections

Cytological manifestations of pulmonary viral infections by adenovirus, herpes simplex, measles, cytomegalovirus, parainfluenza and respiratory syncytial virus include both non-specific reactive changes and more definitive features associated with certain individual viruses. Non-specific changes include ciliocytophthoria, which refers to the degenerative separation of the cilia-bearing cytoplasm and the nucleated cytoplasm, resulting in an anucleated mass of cytoplasm-bearing cilia and a degenerating nucleus with cytoplasm. In addition, atypical regenerative changes may be seen in the respiratory epithelium as tissue fragments with enlarged hyperchromatic nuclei and prominent nucleoli. Distinguishing these cell groups from carcinoma is based on their tight cohesion and the absence of single atypical cells. In a few instances specific cellular changes permit a diagnosis on cytological grounds. Such is the case with herpes simplex, where multiple moulded nuclei are present with either eosinophilic inclusion bodies or a homogenized, slate-grey appearance to the nuclear contents, sometimes referred to as 'ground glass nuclei' (Fig. 3.11). Infection by cytomegalovirus results in cytomegaly as well as large

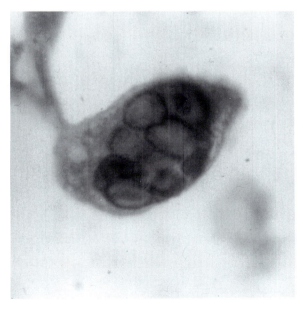

Fig. 3.11 Herpes simplex. Typical features of herpes simplex infection including multi-nucleation, nuclear moulding, and 'ground glass' appearance of the chromatin. In addition, eosinophilic nuclear inclusions are present in two of the nuclei. Sputum specimen. Papanicolaou stain; × 1000.

amphophilic, smooth, intranuclear inclusions which are typically surrounded by a prominent peri-inclusion clearing or halo (Fig. 3.12). Margination of the nuclear chromatin on the inner surface of the nuclear membrane is also present. In contrast to herpes simplex, infection by cytomegalovirus does not produce nuclear moulding but may also result in multiple small cytoplasmic inclusions. Infection with adenovirus may produce either a small red body surrounded by a well-circumscribed clear halo or a homogeneous basophilic mass which nearly replaces the nucleus.[42] The most characteristic cytological finding in measles pneumonia is the presence of multinucleated cells containing eosinophilic inclusions that are present within the nucleus and cytoplasm.[43] A proliferation of multinucleated cells with cytoplasmic basophilic inclusions surrounded by haloes can also be seen in patients with pneumonia due to respiratory synctial virus.[44]

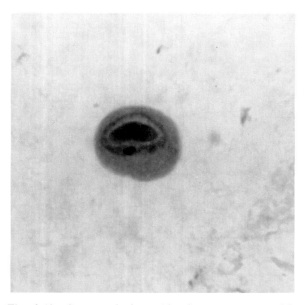

Fig. 3.12 Cytomegalovirus. Alveolar pneumocyte with cytomegaly, large eosinophilic nuclear inclusion and a margination of nuclear chromatin producing a halo surrounding the inclusion. Bronchoalveolar lavage specimen. Papanicolaou stain; × 400.

Parasitic infection

Due to increasing numbers of patients who are immunocompromised secondary to solid organ or bone marrow transplantation therapy or cancer chemotherapy, the detection of pulmonary parasitic infection has become an increasingly important task for the cytology laboratory. A prime example is *Pneumocystis carinii* pneumonia, which prior to the AIDS epidemic was largely limited to an interstitial pneumonia seen in premature and debilitated infants. Utilizing Papanicolaou stained material, the organisms may be difficult to identify, although typically a mass of partially eosinophilic amorphous material is seen. The presence of such material should alert the cytopathologists to evaluate the specimen closely with other stains (Fig. 3.13). Methenamine stains clearly outline the organism as a spherical cyst measuring 6–8 μm, which may be cup-shaped, crescent-shaped, or variably wrinkled. Small interior structures may also be seen in the form of rings, dots, or commas. Toluidine blue O and Gram–Weigert also stain the cyst wall, although the latter also tends to be taken up by cellular debris. Giemsa, or other Romanowsky's-type stains are alternatives which allow identification of the 0.15–1.0 μm trophozoites within the cyst. Although early efforts with polyclonal antibodies did not enhance morphological identification of *P. carinii*, more recently a number of monoclonal antibodies have been shown to increase sensitivity of diagnosing *P. carinii* in induced sputum specimens or in extrapulmonary tissues.[45-47] While such antibodies may aid the detection of *P. carinii* when the number of microorganisms is scant, in the setting of AIDS, where microorganisms are usually abundant, they may provide little advantage over standard histochemical stains. The latter are less expensive, less time-consuming and provide the opportunity to identify fungal organisms at the same time.

Pulmonary infection with *Strongyloides stercoralis* in industrialized countries is almost entirely limited to patients receiving prolonged steroid therapy.[48,49] The haemorrhagic pneumonia results from migration of the filiform larvae through the intestinal wall into the bloodstream and eventual penetration of the alveolar spaces. Numerous filiform larvae measuring 400–500 μm with a closed gullet and slightly notched tail can be readily identified in blood or sputum from these patients. Differential diagnosis would include *Ascaris lumbricoides, Necator americanus* and *Ancylostoma duodenale*. Less common but well recognized parasites include *Echinococcosis, Paragonimus kellicotti*, trichomoniasis, *Entamoeba gingivalis, Paragonimus westermanii, Dirofilaria immitis* and microfilariae.[41,49]

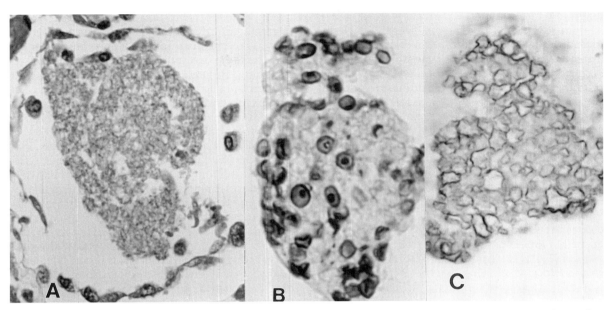

Fig. 3.13 *Pneumocystis carinii.* (A) Alveolar eosinophilic foamy exudate characteristic of *Pneumocystis carinii* in a patient with AIDS. Haematoxylin and eosin stain. (B) Gomori methenamine silver staining walls of spherical-to-cup-shaped cysts with typical central dark-staining region. (C) Monoclonal antibody to *Pneumocytis carinii* (DAKO) staining outline of cyst walls. Wedge biopsy of lung, × 352. (Reproduced with permission from ref. 47.)

Cytological diagnosis of lung cancer

The major types of lung carcinoma arise from the respiratory lining epthelium. A modification of the World Health Organization (WHO) histological classification of malignant epithelial tumours of the lung is presented in Table 3.1. *In situ* carcinomas of both squamous cell and adenocarcinoma have also been described. Many believe that lung carcinomas arise from precursor cells of endodermal origin, capable of expressing one or more patterns of cellular differentiation. This theory would help explain the heterogeneity demonstrated by combined small cell/large cell carcinomas and adenosquamous carcinomas.[50–52]

The morphology of lung tumours obtained by different cytological sampling methods is similar, with a few important differences. Fine needle aspiration specimens, because of direct sampling of the tumour, typically contain large numbers of well-preserved cancer cells as well as tissue fragments. Such abundant cellular specimens permit processing of cell blocks to perform histochemical and immunohistochemical stains. Likewise, since FNAs generally provide highly cellular specimens, extreme caution should be advised before rendering a conclusive diagnosis of cancer when only

Table 3.1 Histological classification of major neoplasms of the lung

Squamous cell carcinoma
  Well-differentiated
  Moderately differentiated
  Poorly differentiated
Adenocarcinoma
  Acinar
  Papillary
  Bronchioloalveolar
Large cell undifferentiated carcinoma
  Giant cell
Small cell undifferentiated carcinoma
  Oat cell
  Intermediate
  Combined
Adenosquamous carcinoma
Bronchial gland carcinoma
  Adenoid cystic
  Mucoepidermoid
Carcinoid
Metastatic tumours

limited numbers of suspicious tumour cells are present in a fine needle aspiration biopsy.

Squamous cell carcinoma classically arises in the large airways of the central lung. Large numbers of diagnostic malignant cells are usually exfoliated in sputum, bronchial washings or brushings. Histologically, they recapitulate normal squamous cell differentiation with keratin pearls or individually keratinized cells, and so-called intercellular bridges, which are spicules of cytoplasm present at the sites of desmosomal junctions. Areas of squamous metaplasia and/or carcinoma *in situ* may occur in adjacent bronchial epithelium. Cells shed from these areas may appear in the cytological sample. In sputum, the neoplastic cells occur singly and in loose clusters. Squamous cell carcinoma is characterized by marked cellular pleomorphism in both size and shape including classic caudate or 'tadpole' cells as well as spindle cells. Marked hyperchromasia of the nuclei is present with frequent karyopyknosis. Alternatively, chromatin is arranged into irregular, sharp-bordered clumps with abnormal clearing of the parachromatin. Nucleoli, while not common, may be conspicuously large in poorly differentiated forms. Often nucleoli are obliterated by dense karyopyknotic chromatin. Keratinization of the cytoplasm imparts an intense hyaline appearance with a bright orangeophilic or intense cyanophilic stain (Fig. 3.14). Ecto-endoplasmic ringing within the cytoplasm is another feature of dense keratinization. In

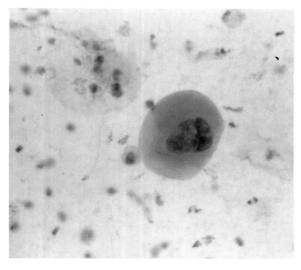

Fig. 3.14 Squamous cell carcinoma. Single large tumour cell with hyperchromatic, and angular nucleus with irregular chromatin distribution along with keratinization of cytoplasm imparting a hyalinized appearance with bright organeophilic staining. Bronchoalveolar lavage specimen. Papanicolaou stain; × 352.

poorly differentiated squamous cell carcinomas, intercellular bridges, pearl body formation and individual cell keratinization may be absent, so that the diagnosis must rest on the monolayer of tumour cells and sharp cytoplasmic borders. Squamous cell carcinomas often reach a large size and tend to undergo central necrotizing cavitation. Thus, in either sputum or fine needle aspirations, the cytological picture may be dominated by features of necrotic tumour with large masses of keratinized squamous ghosts and no well-preserved tumour cells.

Adenocarcinomas have in recent years significantly increased in frequency so that, in some series, they are the most commonly diagnosed type of lung carcinoma.[53-55] Acinar adenocarcinomas arise from the more peripherally situated respiratory bronchi and bronchioles and may frequently involve the overlying pleura. Histologically they may demonstrate glandular structures, papillary structures or a bronchioloalveolar pattern. In less well-differentiated forms, mucin production may be the lone diagnostic feature.[56]

Most adenocarcinomas exfoliate large numbers of diagnostic cells into the respiratory specimen as both single cells and tissue fragments. Typically, the nuclei of adenocarcinomas are bland with extremely fine granular chromatin occasionally imparting a ground glass appearance. Individual nuclei may be round and regular. Centrally placed macronucleoli are a key diagnostic feature of acinar adenocarcinoma, but not necessarily of the other types. The cytoplasm may have a foamy quality due to multiple small vacuoles or may contain larger vacuoles that distort the nucleus. Intranuclear cytoplasmic invaginations or inclusions may be present, particularly in papillary and bronchioloalveolar types. In well-differentiated adenocarcinoma, the cells may assume a columnar shape while poorly differentiated tumours lack any resemblance to their glandular origin. Tissue fragments often demonstrate ball-like clusters of cells or true acinar structures with internal lumens (Fig. 3.15). Ultrastructurally, all types of pulmonary adenocarcinomas may exhibit features such as goblet cells, bronchiolar (Clara) cells, or type 2 pneumocytes.[57]

Bronchioloalveolar carcinomas are a subtype of adenocarcinoma that arise from the terminal bronchoepithelium or from type 2 pneumocytes of the alveolar epithelium and produce a growth pattern which follows the anatomical outline of the alveolar spaces. Bronchioloalveolar carcinomas may

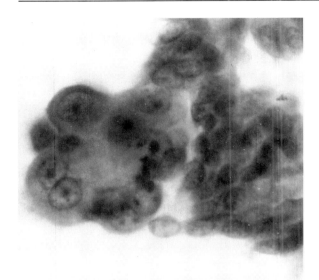

Fig. 3.15 Adenocarcinoma. Large clusters of tumour cells with three-dimensional cluster at one end displaying acinus with a central lumen. Prominent central nucleoli are present along with variation in nuclear size and chromatin distribution. Bronchial brushing specimen. Papanicolaou stain; × 380.

form single peripheral masses, multiple nodules, or they may massively involve the entire lung.

Bronchioloalveolar carcinomas usually exfoliate both single cells and cell groups with round-to-oval, uniform nuclei. Their chromatin is finely granular or powdery with small inconspicuous nucleoli. Nuclear folds are a common feature and occasionally nuclear cytoplasmic inclusions or prominent central nucleoli are seen. The cytoplasm can vary from granular to finely vacuolated to multiple small or a single large vacuole. Although bronchioloalveolar carcinoma often consist of rather complex three-dimensional clusters and sometimes a radiating 'cartwheel' pattern, the tumour may also present as numerous single cells with abundant cytoplasm. It is this pattern that may be difficult to distinguish from alveolar macrophages.[58,59] In addition, bronchial goblet cell hyperplasia or reactive alveolar epithelium can be confused with this tumour (Fig. 3.16).

Large cell undifferentiated carcinoma refers to primary non-small cell lung carcinomas which lack histological evidence of either glandular or squamous differentiation by light microscopy. Ultrastructural examination, however, commonly demonstrates cytoplasmic mucin or microvilli or less commonly squamous differentiation.[60-62] These tumours arise in large bronchi and exfoliate large numbers of diagnostic cells into the specimen, these cells occurring both as single cells and

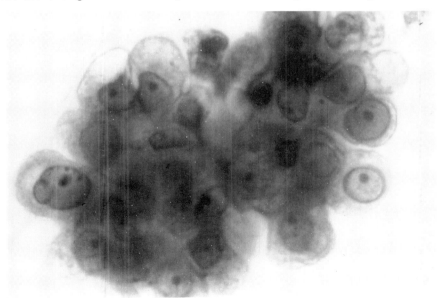

Fig. 3.16 Bronchioloalveolar carcinoma. Cell cluster with variation of nuclear chromatin from fine powdery to slightly darker and occasional nuclei with irregular nuclear borders and folds. Cytoplasmic vacuoles and prominent nucleoli are also present. Sputum specimen. Papanicolaou stain; × 1000.

tissue fragments. Large cell undifferentiated carcinomas, regardless of their ultrastructural characteristics, have a poorer prognosis than either adenocarcinoma or squamous cell carcinoma.[60]

As their name implies, the tumour cells are large and possess a high nuclear to cytoplasmic ratio, marked aberrations in chromatin patterns, and multiple and large irregular nucleoli (Fig. 3.17). Cytoplasm may be homogenous or wispy with a tendency towards cyanophilia. Tissue fragments may be present without any recognizable architectural pattern being demonstrated by the cells. Giant cell carcinoma is a variant of large cell carcinoma and contains multiple multinucleated tumour giant cells.

Small cell undifferentiated carcinoma of the lung occurs primarily in major central bronchi and has been subdivided into the oat cell type, intermediate cell type and combined small cell/large cell type. The neoplasm is characterized by a round cell with centrally placed nucleus and uniform, hyperchromatic chromatin. Nucleoli are only occasionally visible. In sputum, large numbers of tumour cells can be found entrapped within strands of mucus. Clusters of small tumour cells often exhibit prominent moulding, superimposed upon irregular nuclear outlines. Tumour necrosis is common; thus

the cellular specimen will often contain areas of karyopyknosis, cytoplasm disruption and dense cyanophilic masses of necrotic debris. Preparation of sputum by the Saccomanno method may result in greater dispersion of cells and thus less nuclear moulding. Cells of the oat cell type are approximately twice the diameter of small lymphocytes, possess scanty cytoplasm, a high nuclear to cytoplasmic ratio, nuclear hyperchromasia with finely dispersed granular chromatin and absent or inconspicuous nucleoli (Fig. 3.18). Mitoses and necrosis are common.

The cells of intermediate cell variant have a larger nucleus but chromatin pattern similar to the oat cell with more abundant cytoplasm and may have distinct nucleoli. Spindle-shaped cells may occasionally predominate.[63]

Ultrastructural examination reveals membrane-bound cytoplasmic granules known as neurosecretory granules. Extensive sampling and ultrastructural examination may reveal evidence of squamous or glandular differentiation in some small cell carcinomas which has imparted a favourable prognosis by some authors. Thus, small cell carcinoma alone may be classified into oat cell carcinoma, intermediate cell carcinoma, or combined cell type including either squamous cell

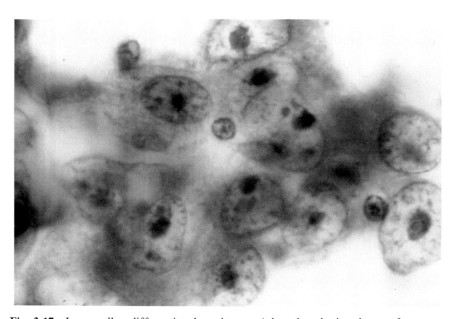

Fig. 3.17 Large cell undifferentiated carcinoma. A loosely cohesive cluster of tumour cells with a high nuclear to cytoplasmic ratio, irregularly dispersed chromatin, and large prominent nucleoli. Tumour cells lack features of glandular or squamous differentiation. Fine needle aspiration of lung. Papanicolaou stain; × 1000.

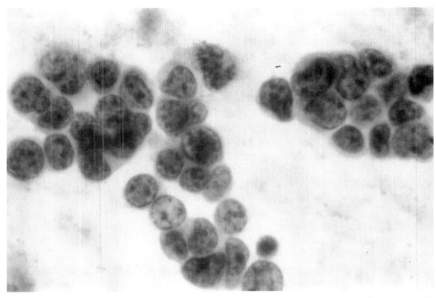

Fig. 3.18 Small cell undifferentiated carcinoma, oat cell type. A group of tumour cells with hyperchromatic nuclei with granular chromatin and occasional small nucleoli, along with focal nuclear moulding. Sputum specimen. Papanicolaou stain; × 1000.

carcinoma or adenocarcinoma. In the combined type of small cell carcinoma, a combination of malignant cells diagnostic for small cell carcinoma and either squamous cell carcinoma or adeno-carcinoma may be present; alternatively, the specimen may contain only cells representing small cell carcinoma, squamous cell carcinoma, or adeno-carcinoma.

Related to small cell carcinoma is primary carcinoid tumour of the lung. This tumour arises in the submucosa, with resultant exophytic endo-bronchial lesions with intact overlying mucosa. Because cells are uncommonly shed into respiratory secretions, fine needle aspiration is the major means for cytological diagnosis. Cytologically, carcinoid tumours consist of single cells, sheets and three-dimensional ball-like clusters of cells with small, round-to-oval uniform nuclei. The cells have stippled, granular chromatin pattern and small nucleoli with scant cytoplasm.[64]

In the spectrum of neuroendocrine bronchial tumours that ranges from carcinoid tumour to small cell undifferentiated carcinoma, the middle ground is occupied by the atypical carcinoids which are usually found in the lung periphery. Cytologically they demonstrate greater pleomorphism, nuclear hyperchromasia, mitotis, necrosis and nuclear moulding than typical carcinoids, but more cytoplasm and less nuclear atypia than small cell carcinoma.[65,66] In general, it is not possible to distinguish reliably atypical carcinoid from small cell undifferentiated carcinoma with regularity.

Other primary neoplasms of the lung

Fine needle aspiration or exfoliative cytology of the lung can also provide the diagnosis of less common neoplasms including mucoepidermoid tumour, leiomyosarcoma, pulmonary blastoma, Hodgkin's and non-Hodgkin's lymphoma, carcinosarcoma, malignant fibrous histiocytoma and bronchial gland carcinoma. Benign neoplasms of the lung are rare, including the clear cell tumour and pulmonary hamartomas.[40,41,67,68]

Tumours metastatic to the lung

The lung is a common site of metastasis for many tumours and in fact metastatic tumours outnumber primary lung cancer.[69] Since they spread via lymphatics, vascular spaces and secondarily to alveolar spaces, they are not as frequently present in sputum or bronchial specimens. They are more readily accessible by fine needle aspiration. While on occasion metastatic tumours may demonstrate a pattern characteristic enough as to suggest a primary site, just as commonly they may be indistinguishable from primary lung squamous cell or adenocarcinoma. The most common primary sites of metastases to the lung are the breast, kidney and colon. A colonic origin may be diagnosed in some metastases to the lung due to tumour fragments with columnar differentiation and a palisading of tumour cells with hyperchromatic nuclei and macronucleoli. Occasionally ductal carcinoma from the breast may be recognized by single cells and small clusters of large polygonal cells or lobular carcinoma will demonstrate linear arrays (Fig. 3.19). Fine needle aspiration of peripheral nodules has permitted sampling of many types of malignant tumours that may demonstrate features that enable the cytopathologist to establish the origin. One such tumour is malignant melanoma, which may present with large nuclei, often with large nucleoli and sometimes with intranuclear cytoplasmic invaginations. The presence of mela-

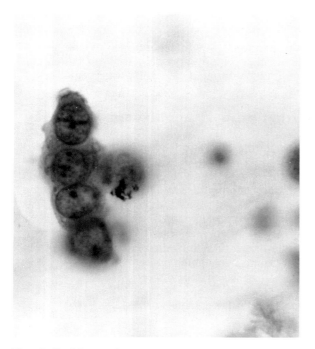

Fig. 3.19 Metastatic breast carcinoma. Linear array (Indian filing) of tumour cells metastatic to lung in a patient with lobular breast carcinoma. Bronchoalveolar lavage specimen. Monoclonal antibody B72.3 stain, immunoperoxidase; × 400. (Reproduced with permission from ref. 80.)

nin pigment within tumour cells helps to establish the diagnosis.

Fine needle aspiration usually provides a larger amount of well-preserved cellular material than is generally available in sputum or bronchial specimens. This increased amount of cellular material also allows for paraffin-embedding of cell blocks and subsequent histochemical or immunohistochemical stains that may be helpful in determining the cell origin.

Diagnostic accuracy

The accuracy of cytological specimens in diagnosing lung cancer has been documented since 1964 when Koss[70] reported a sensitivity of 89% in a series of histologically proven carcinomas in which three or more satisfactory sputum specimens had been examined. The value of multiple sputum specimens has been reiterated in subsequent studies by Erozan and Frost[71] and Ng and Horak.[72,73] In a series of 9892 consecutive sputum, bronchial washings and brushings collected over a ten-year period, Johnston reported that while the first cytological specimen detected cancer in 41.7% of 424 patients with carcinoma of the lung, the submission of three specimens increased the detection rate to 69.5%. If five satisfactory specimens were obtained, a cytological diagnosis of cancer was provided in 86.6% of patients.[68,70,74] While examination of sputum has shown a high level of sensitivity in patients with centrally located tumours, this is not the case in peripheral tumours where bronchial brushing bolsters the diagnostic accuracy to 70–88%.[75] Popp and colleagues reviewed bronchial brush, forceps biopsy and touch imprints and found that utilizing all three methods provided a diagnostic sensitivity of 97.3% and a specificity of 100%.[76]

In addition to providing a reliable diagnosis of lung cancer, sputum and bronchial specimens accurately predict the histological type. The accuracy is highest with squamous cell carcinoma, reported as 96% in bronchial washing by Ng and Horak[72,73] compared with 86% for adenocarcinoma, 77% for large cell carcinoma, and less than 50% for bronchioloalveolar carcinoma with comparable rates with sputum specimens. A later study by Johnston reported similar results including a 93% predictive accuracy for small cell carcinoma.[77]

The diagnostic accuracy of FNA of lung is reported as a sensitivity of 75–95% and a specificity of 99%. In a series of 288 consecutive patients with both histological and fine needle aspiration specimens available reported by Johnston,[77] complete agreement was present in 85.4%. In two patients, the cytological diagnosis of cancer was not confirmed by histology for a false-positive rate of 0.7%. In both of these cases the authors report that the diagnosis was erroneously based upon overinterpretation of a few atypical cells. This underscores the principle that the presence of only small numbers of putative tumour cells in an FNA should caution the cytopathologist in rendering a conclusive diagnosis. Correlations between fine needle aspiration and histological classification have been reported of up to 81% in 1967[67] and 86% in 1987.[78] In one large series the highest level of correlation was 96% for adenocarcinoma, 95% for small cell carcinoma and 80% for squamous cell carcinoma.[77] In 1990 a College American Pathologists survey of 436 institutions, based on 11 922 satisfactory fine needle aspiration specimens, reported a 99% sensitivity of the pathologist's diagnosis and a 96% specificity.[79]

Summary

In many laboratories the volume of cytology specimens will be second only to that of gynaecological tract specimens. This, coupled with the broad spectrum of inflammatory, infectious and neoplastic conditions that occur in the lung, provide ample motivation for cytologists to develop skills in respiratory cytology. The growing popularity and efficacy of techniques such as bronchoalveolar lavage and fine needle aspiration will increasingly provide greater opportunities to evaluate pulmonary lesions that would otherwise require more invasive and costly procedures. While a wonderfully wide array of special techniques are now applicable to cytological specimens such as immunocytochemistry, image analysis of ploidy status as well as a growing number of molecular tools, it is likely that standard cytological features will continue to be the mainstay of the diagnosis provided by the cytopathologist.

References

1. Donne A. *Cours de microscopie*. Baillière 1845.
2. Archer PG, Koprowska I, McDonald JR, Naylor B, Papanicolaou GN, Umiker WO. A study of variability in the interpretation of sputum cytology slides. *Cancer Res* 1966; **26**, 2122–2144.
3. Bamforth J. The examination of the sputum and pleural fluid in the diagnosis of malignant diseases of the lung. *Thorax* 1946; **1**, 118–127.
4. Grunze H. A critical review and evaluation of cytodiagnosis in chest diseases. *Acta Cytol* 1960; **4**, 175–198.
5. Koss L. Cellular changes simulating bronchogenic carcinoma. *Acta Unio Int Cancer* 1958; **14**, 501–503.
6. Koss L, Richardson H. Some pitfalls of cytological diagnosis of lung cancer. *Cancer* 1955; **8**, 937–947.
7. Koss LG. Progress in cytologic diagnosis of lung cancer. *Proc Natl Cancer Conf* 1970; **6**, 817–819.
8. Koss LG, Wolley RC, Schreiber K, Mendecki J. Flow-microfluorometric analysis of nuclei isolated from various normal and malignant human epithelial tissues. A preliminary report. *J Histochem Cytochem* 1977; **26**, 565–572.
9. Koss MN, Hochholzer L, Nichols PW, Wehunt WD, Lazarus AA. Primary non-Hodgkin's lymphoma and pseudolymphoma of lung: a study of 161 patients. *Hum Pathol* 1983; **14**, 1024–1038.
10. Papanicolaou G. Degenerative changes in ciliated cells exfoliating from the bronchial epithelium as a cytologic criterion in the diagnosis of diseases of the lung. *NY State J Med* 1956; **56**, 2647–2650.
11. Papanicolaou G, Cromwell H. Diagnosis of cancer of the lung by the cytologic method. *Dis Chest* 1949; **15**, 412–418.
12. Russell W, Neidhardt H, Mountain C, Griffith K, Chang J. Cytodiagnosis of lung cancer. A report of a four-year laboratory, clinical, and statistical study with a review of the literature on lung cancer and pulmonary cytology. *Acta Cytol (Baltimore)* 1963; **7**, 1–44.
13. Wandall H. A study on neoplastic cells in sputum as a contribution to the diagnosis of primary lung cancer. *Acta Chir Scand* 1944; **91**, 1–143.
14. Woolner L, McDonald J. Diagnosis of carcinoma of the lung; the value of cytologic study of sputum and bronchial secretions. *J Am Med Assoc* 1949; **139**, 497–502.
15. Feldman P, Covell J. Breast and lung. In: *Fine Needle Aspiration Cytology and its Clinical Applications*. Chicago: American Society of Clinical Pathologists, 1985.
16. Kronig G. Diagnosticher beitrag zur herz und lungen pathologie. *Berl Klin Wochenschr* 1887; **24**, 961–967.
17. Menetrier P. Cancer primitif du poumon. *Bull Soc Anat Paris* 1886; **11**, 643.
18. Hajdu S. A note on the history of carbowax in cytology. *Acta Cytol (Baltimore)* 1983; **27**, 204–206.
19. Saccomanno G, Saunders R, Ellis H, Archer V, Wood B, Beckler P. Concentration of carcinoma or atypical cells in sputum. *Acta Cytol (Baltimore)* 1963; **7**, 305–310.
20. Chang J, Aken M, Russell W. Sputum cell concen-

tration by membrane filtration for cancer diagnosis: a preliminary report. *Acta Cytol (Baltimore)* 1961; **5**, 168–172.

21. Fields MJ, Martin WF, Young BL, Tweeddale DN. Application of the Nedelkoff–Christopherson millipore method to sputum cytology. *Acta Cytol* 1966; **10**, 220–222.
22. Suprum H. A comparative filter technique study and the relative efficiency of these sieves as applied in sputum cytology for pulmonary cancer cyto-diagnosis. *Acta Cytol* 1974; **18**, 248–251.
23. Fleury FJ, Escudier E, Pocholle MJ, Carre C, Bernaudin JF. The effects of cytocentrifugation on differential cell counts in samples obtained by bronchoalveolar lavage. *Acta Cytol* 1987; **31**, 606–610.
24. Linder J, Rennard S. *Bronchoalveolar Lavage*. Chicago: American Society of Clinical Pathologists, 1988.
25. Linder J, Vaughan WP, Armitage JO *et al.* Cytopathology of opportunistic infection in bronchoalveolar lavage. *Am J Clin Pathol* 1987; **88**, 421–428.
26. Stover DE, White DA, Romano PA, Gellene RA. Diagnosis of pulmonary disease in acquired immune deficiency syndrome (AIDS). Role of bronchoscopy and bronchoalveolar lavage. *Am Rev Respir Dis* 1984; **130**, 659–662.
27. Stover DE, Zaman MB, Hajdu SI, Lange M, Gold J, Armstrong D. Bronchoalveolar lavage in the diagnosis of diffuse pulmonary infiltrates in the immunosuppressed host. *Ann Intern Med* 1984; **101**, 1–7.
28. Johnston W. Percutaneous FNAB of the lung. *Acta Cytol (Baltimore)* 1984; **28**, 218–224.
29. Malberger E, Lemberg S. Transthoracic fine needle aspiration cytology: a study of 301 aspirations from 221 cases. *Acta Cytol* 1982; **26**, 172–178.
30. Mitchell M, King D, Bonfiglio T, Patten S. Pulmonary fine needle aspiration cytopathology: a five-year correlation study. *Acta Cytol (Baltimore)* 1984; **28**, 72–76.
31. Nordenstrom BE. Technical aspects of obtaining cellular material from lesions deep in the lung. A radiologist's view and description of the screw-needle sampling technique. *Acta Cytol* 1984; **28**, 233–42.
32. Heaston DK MS, Moore AV, Johnston WW. Percutaneous thoracic needle biopsy. In: *Pulmonary Disease*, Putman C (ed.). New York: Appleton-Century-Crofts, 1981.
33. Stitik F. Percutaneous lung biopsy. In: *Multiple Imaging Procedures: Pulmonary System, Practical Approaches to Pulmonary Diagnosis*, vol 1, Seigelman FS, Summer WR (eds). New York: Grune & Stratton, 1979.
34. Murray J. Postnatal growth and development of the lung. In: *The Normal Lung: The Basis for Diagnosis and Treatment of Pulmonary Disease*. Philadelphia: WB Saunders, 1976.
35. Wang N-S. Anatomy. In: *Pulmonary Pathology*, Dail D, Hammar S (eds). New York: Springer-Verlag, 1988.
36. Garret M. Cellular atypias in sputum and bronchial secretions associated with tuberculosis and bronchiectasis. *Am J Clin Pathol* 1960; **34**, 237–246.
37. Marchevsky A, Nieburgs HE, Olenko E, Kirschner P, Teirstein A, Kleinerman J. Pulmonary tumorlets in cases of 'tuberculoma' of the lung with malignant cells in brush biopsy. *Acta Cytol* 1982; **26**, 491–494.
38. Naylor B, Railey C. A pitfall in the cytodiagnosis of sputum of asthmatics. *J Clin Pathol* 1964; **7**, 84–89.
39. Sanerkin N, Evans D. The sputum in bronchial asthma: pathognomonic patterns. *J Pathol Bacteriol* 1965; **89**, 535–541.
40. Johnston WW. Ten years of respiratory cytopathology at Duke University Medical Center. III. The significance of inconclusive cytopathologic diagnoses during the years 1970 to 1974. *Acta Cytol* 1982; **26**, 759–766.
41. Johnston W, Elson C. Respiratory Tract. In: *Comprehensive Cytopathology*, Bibbo M (ed.). Philadelphia: WB Saunders Co, 1991, pp. 320–369.
42. Koss L. *Diagnostic Cytology and Its Histopathologic Basis*, 3rd edn. Philadelphia: JB Lippincott, 1979.
43. Beale A, Campbell W. A rapid cytological method for the diagnosis of measles. *J Clin Pathol* 1959; **12**, 335–337.
44. Naib Z, Stewart J, Dowdle W, Casey H, Marine W, Nahmias A. Cytological features of viral respiratory tract infections. *Acta Cytol (Baltimore)* 1968; **12**, 162–171.
45. Blumenfeld W, Kovacs JA. Use of a monoclonal antibody to detect *Pneumocystis carinii* in induced sputum and bronchoalveolar lavage fluid by immunoperoxidase staining. *Arch Pathol Lab Med* 1988; **112**, 1233–1236.
46. Kovacs J, Ng V, Masar H *et al.* Diagnosis of *Pneumocystis carinii* pneumonia: improved detection in sputum with use of monoclonal antibodies. *New Eng J Med* 1988; **318**, 589.
47. Radio S, Hansen S, Goldsmith J, Linder J. Immunohistochemical characterization of *Pneumocystis carinii* infection in patients with AIDS. *Mod Pathol* 1990; **3**, 462–469.
48. Chaudhuri B, Nanos S, Soco J, McGrew E. Disseminated *Strongyloides stercoralis* infestation detected by sputum cytology. *Acta Cytol (Baltimore)* 1980; **24**, 360–362.
49. Warren K. Diseases due to helminths. In: *Principles and Practice of Infectious Disease*, Mandel G, Douglas Jr R, Bennett J (eds). New York: John Wiley & Sons, 1979.
50. Yesner R. Spectrum of lung cancer and ectopic hormones. *Am J Surg Pathol* 1978; **13**, 217–240.

51. Reid J, Carr A. The validity and value of histological and cytological classification of lung cancer. *Cancer* 1961; **14**, 673–698.

52. Roggli V, Vollmer R, Greenberg S, McGavran M, Spjut H, Yesner R. Lung cancer heterogeneity: a blinded and randomized study of 100 consecutive cases. *Hum Pathol* 1985; **16**, 569–579.

53. Cox J, Yesner R. Adenocarcinoma of the lung; recent results from the Veteran's Administration Lung Group. *Am Rev Respir Dis* 1979; **120**, 1025–1029.

54. Valaitis J, Warren S, Gamble D. The increasing incidence of adenocarcinoma of the lung. *Cancer* 1981; **47**, 1042–1046.

55. Vincent R, Pickren J, Lane W *et al.* The changing histopathology of lung cancer: a review of 1682 cases. *Cancer* 1977; **39**, 1647–1655.

56. Kreyberg L. *Histological Typing of Lung Tumours* 2nd edn. Geneva: World Health Organization, 1981.

57. Kimula Y. A histochemical and ultrastructural study of adenocarcinoma of the lung. *Am J Surg Pathol* 1978; **2**, 253–264.

58. Roger V, Nasiell M, Linden M, Enstad I. Cytologic differential diagnosis of bronchiolo-alveolar carcinoma and bronchogenic adenocarcinoma. *Acta Cytol (Baltimore)* 1976; **20**, 303–307.

59. Elson CE, Moore SP, Johnston WW. Morphologic and immunocytochemical studies of bronchioloalveolar carcinoma at Duke University Medical Center, 1968–1986. *Anal Quant Cytol Histol* 1989; **11**, 261–274.

60. Albain K, True L, Golomb H, Hoffman P, Little A. Large cell carcinoma of the lung: ultrastructural differentiation and clinicopathologic correlations. *Cancer* 1985; **56**, 1618–1623.

61. Churg A. The fine structure of large cell undifferentiated carcinoma of the lung: evidence for its relation to squamous cell carcinomas and adenocarcinomas. *Hum Pathol* 1978; **9**, 143–156.

62. Delmonte V, Alberti O, Saldiva P. Large cell carcinoma of the lung: ultrastructural and immunohistochemical features. *Chest* 1986; **90**, 524–526.

63. Carter D. Small cell carcinoma of the lung. *Am J Surg Pathol* 1983; **7**, 787–795.

64. Gephardt GN, Belovich DM. Cytology of pulmonary carcinoid tumours. *Acta Cytol* 1982; **26**, 434–438.

65. Paladugu R, Benfield J, Pak H, Ross R, Teplitz R. Bronchopulmonary Kulschitzky cell carcinomas: a new classification scheme for typical and atypical carcinoids. *Cancer* 1985; **55**, 1303–1311.

66. Szyfelbein WM, Ross JS. Carcinoids, atypical carcinoids, and small-cell carcinomas of the lung: differential diagnosis of fine-needle aspiration biopsy specimens. *Diagn Cytopathol* 1988; **4**, 1–8.

67. Dahlgren S. Aspiration biopsy of intrathoracic tumours. *Acta Pathol Microbiol Scand (B)* 1967; **70**, 566–576.

68. Johnston WW, Bossen EH. Ten years of respiratory cytopathology at Duke University Medical Center. I. The cytopathologic diagnosis of lung cancer during the years 1970 to 1974, noting the significance of specimen number and type. *Acta Cytol* 1981; **25**, 103–107.

69. Kern W, Schweizer C. Sputum cytology of metastatic carcinoma of the lung. *Acta Cytol (Baltimore)* 1976; **20**, 514–520.

70. Koss L, Melamed M, Goodner J. Pulmonary cytology – a brief survey of diagnostic results from July 1st, 1952 until December 31st, 1960. *Acta Cytol* 1964; **8**, 104–113.

71. Erozan Y, Frost J. Cytopathologic diagnosis of cancer in pulmonary material: a critical histopathologic correlation. *Acta Cytol (Baltimore)* 1970; **14**, 560–565.

72. Ng AB, Horak GC. Factors significant in the diagnostic accuracy of lung cytology in bronchial washing and sputum samples. I. Bronchial washings. *Acta Cytol* 1983; **27**, 391–396.

73. Ng AB, Horak GC. Factors significant in the diagnostic accuracy of lung cytology in bronchial washing and sputum samples. II. Sputum samples. *Acta Cytol* 1983; **27**, 397–402.

74. Johnston WW, Bossen EH. Ten years of respiratory cytopathology at Duke University Medical Center. II. The cytopathologic diagnosis of lung cancer during the years 1970 to 1974, with a comparison between cytopathology and histopathology in the typing of lung cancer. *Acta Cytol* 1981; **25**, 499–505.

75. Bibbo M, Fennessy JJ, Lu C-T, Strauss F, Variakojis D, Weid G. Bronchial brushing technique for the cytologic diagnosis of perpheral lung lesions. *Acta Cytol* 1973; **17**, 245–251.

76. Popp W, Rauscher H, Ritschka L, Redtenbacher S, Zwick H, Dutz W. Diagnostic sensitivity of different techniques in the diagnosis of lung tumours with the flexible fiberoptic bronchoscope: comparison of brush biopsy, imprint cytology of forceps biopsy and histology of forceps biopsy. *Cancer* 1991; **67**, 72–75.

77. Johnston W. Cytologic correlations. In: *Pulmonary Pathology*, Dail D, Hammar S (eds). New York: Springer-Verlag, 1988.

78. Bonfiglio T. Transthoracic thin needle aspiration biopsy. In: *Masson Series in Diagnostic Cytopathology*, vol. 4, Johnston W (ed). Paris: Masson, 1983.

79. Zarbo R, Fenoglio-Preiser C. Interinstitutional database for comparison of performance in lung fine needle aspiration cytology: a College of American Pathologists Q-probe study of 5264 cases with histologic correlation. *Arch Pathol Lab Med* 1992; **116**, 463–470.

80. Reproduced with permission from Radio SJ. Breast carcinoma in bronchoalveolar lavage: a cytologic and immunocytochemical study. *Arch Pathol Lab Med* 1989; **113**, 333–336.

4

Pulmonary complications of acquired immunodeficiency syndrome (AIDS)

Alberto M Marchevsky

Introduction

The acquired immunodeficiency syndrome (AIDS) has emerged as the most serious epidemic in developed countries of the late twentieth century.[1-3] It was originally described for epidemiological surveillance in 1982 by the Centers for Disease Control (CDC) as a syndrome characterized by the presence of defective cell-mediated immunity of unknown cause, associated with opportunistic infections and/or Kaposi's sarcoma.[4] Since 1984 the list of illnesses included in the case definition for AIDS has been greatly expanded to include *Pneumocystis carinii* pneumonia, mycobacterial infections, cytomegalovirus (CMV) pneumonia, herpetic tracheobronchitis and many other conditions listed in Table 4.1.[3,5,6]

AIDS is now interpreted as the most severe manifestation of a spectrum of conditions caused by the human immunodeficiency viruses (HIV) 1 and 2.[3, 7-9] These lymphotropic retroviruses were recognized as the aetiological agent for AIDS in 1983 by Barre-Sinoussi and associates and Gallo and associates and are members of the subfamily of lentiviruses.[3]

The spectrum of HIV infections includes:

1. the acute non-specific viral-like illness associated with the initial infection;
2. asymptomatic carrier state with serology that usually becomes positive within 6 months of the contact;
3. AIDS-related complex (ARC); and
4. AIDS affecting approximately 3% of infected patients.[3,6,10,11]

Patients with ARC present with persistent general-

Table 4.1 Selected indicator diseases of AIDS[a]

No serological evidence of HIV infection
Candidiasis (oesophagus, trachea, lungs)
Extrapulmonary cryptococcosis
Cryptosporidiosis with persistent diarrhoea (over 1 month)
Cytomegalovirus disease (other than in the liver, spleen or lymph nodes)
Herpes simplex infections: mucocutaneous over 1 month, bronchitis/pneumonitis, oesophagitis over 1 month
Kaposi's sarcoma affecting patients younger than 60 years
Brain lymphoma
LIP/PLH complex[b] in children younger than 13 years
Disseminated *Mycobacterium avium* or *M. kansasii* infection
Pneumocystis carinii pneumonia
Progressive multifocal leukoencephalopathy
Brain toxoplasmosis in patients older than 1 month

Serological evidence of HIV infection
Multiple/recurrent bacterial infections in children younger than 13 years (except otitis media, superficial skin, mucosal abscess)
Disseminated coccidioidomycosis
HIV encephalopathy
Histoplasmosis, disseminated
Isosporiasis with diarrhoea over 1 month
Kaposi's sarcoma
Brain lymphoma
Certain non-Hodgkin's malignant lymphomas
Disseminated mycobacterial disease, other than *M. tuberculosis*
Extrapulmonary *M. tuberculosis* infection
Septicaemia by *Salmonella* sp.
HIV wasting syndrome

[a] The table does not include all conditions.
[b] Lymphoid interstitial pneumonia/pulmonary lymphoid hyperplasia (LIP/PLH complex).

ized lymphadenopathy, minor opportunistic infections such as oral candidiasis and herpes zoster, persistent diarrhoea, haematological problems and other conditions.[3] A staging system for HIV infections was reported by the Walter Reed Army Institute.[12] HIV 1 has been isolated in most patients with AIDS and ARC in the USA and Western countries.[7-9] HIV 2 is the aetiological agent of a serious epidemic in West Africa.[7]

The lung is a frequent site of involvement of the many infectious, neoplastic and other conditions reported in patients with AIDS. Indeed, over 30% of patients with AIDS experience pulmonary complications during their disease.[13,14] *Pneumocystis carinii* is the most frequent clinical problem, as the organism is diagnosed in approximately 60-80% of patients with AIDS.[15-20] The organism frequently coexists with CMV, *Mycobacterium avium* complex (MAI), *Cryptococcus neoformans* and other infections (Table 4.2).[3,6,10,11] The non-infectious pulmonary manifestations of AIDS are listed in Table 4.3.

Table 4.2 Pulmonary infections in patients with AIDS

Parasitic
Pneumocystis carinii[a]
Toxoplasma gondii
Cryptosporidium spp.
Strongyloides stercoralis

Viral
Cytomegalovirus
Herpes zoster
Herpes simplex
Epstein–Barr
Influenza
HIV?

Mycobacterial
M. avium complex
M. tuberculosis
Other non-tuberculous mycobacteria

Fungal
Cryptococcus neoformans
Coccidioides immitis
Histoplasma capsulatum
Blastomyces dermatitides
Candida spp.
Aspergillus spp.

Pyogenic bacteria
Streptococcus pneumonia
Hemophilus influenzae

[a] The taxonomic classification of *P. carinii* is controversial, with strong evidence indicating fungal origin.

Table 4.3 Non-infectious complications of AIDS

Neoplastic
Kaposi's sarcoma
Malignant lymphoma
  Non-Hodgkin's lymphomas
  Hodgkin's disease
  T cell lymphoma/leukaemia
  ? Carcinomas

Non-neoplastic
Diffuse alveolar damage
Non-specific interstitial pneumonitis
Lymphocytic interstitial pneumonitis/pulmonary
  lymphoid hyperplasia (LIP/PLH) complex
Secondary alveolar proteinosis

Clinical findings of the various pulmonary complications of AIDS

A detailed discussion of the clinical manifestations of the various pulmonary complications of AIDS is beyond the scope of this chapter. Patients can be asymptomatic or present with a variety of symptoms including cough, dyspnoea, pleuritic chest pain, haemoptysis, fever, weight loss, fatigue and others.[6,21] Chest roentgenographic findings vary from a normal chest X-ray to diffuse interstitial infiltrates, focal alveolar densities, interstitial densities involving the upper lobes, cavities, single or multiple nodules, pleural effusion, pneumothorax, mediastinal lymphadenopathy and others.[22,23]

Laboratory methods for the diagnosis of the pulmonary complications of AIDS

Many laboratory diagnostic modalities can be used in clinical practice for the diagnosis of the various pulmonary manifestations of AIDS, as listed in Table 4.4.[16,24-33]

Table 4.4 Diagnostic modalities for the detection of the pulmonary complication of AIDS

Induced sputum
Bronchoalveolar lavage
Endobronchial brush biopsy
Percutaneous needle aspiration biopsy
Fibreoptic transbronchial biopsy
Percutaneous needle aspiration
Thoracoscopic lung biopsy
Open-lung biopsy

Induced sputum analysis

Induced sputum analysis is a rapid, inexpensive, non-invasive technique that has gained an increasing role for the diagnosis of the infectious complications of AIDS.[25,34-36] The sensitivity of the procedure is approximately 56% for the detection of *P. carinii* in patients at risk for HIV infection that present with either normal or abnormal chest roentgenograms, respiratory symptoms, low single breath diffusing capacity and/or abnormal gallium-67 scans.[13] Direct smears are prepared from the mucoid materials obtained by induced sputum, stained with Giemsa and/or Gomori's methenamine silver and screened for the presence of *P. carinii*.[37] Studies of induced sputum using a rapid immunofluorescent technique for the detection of the organism report an increased detection rate for *P. carinii* of up to 92%.[34,37-39]

Induced sputum preparations can also be stained and cultured for the detection of mycobacteria and other microorganisms. Various diagnostic systems have been utilized for the detection of mycobacteria in sputum, with the Bactec system providing the most effective method for recovery of the organisms within 11 days, as compared with 26 days with conventional culture techniques.[37]

Induced sputum analysis has a low sensitivity for the detection of pulmonary fungal infections in patients with AIDS.[37]

Fibreoptic bronchoscopy with bronchoalveolar lavage and bronchial brushing

This is the preferred diagnostic procedure in many medical centres for the diagnosis of pulmonary infections in patients with AIDS.[13] Patients undergo flexible fibreoptic bronchoscopy with bronchoalveolar lavage (BAL) and bronchial brushing. Examination of cell blocks, smears and brushings from these materials can yield a diagnostic sensitivity for *P. carinii* of up to 97%.[37] Also, BAL fluid samples are studied with microbiological methods for the detection of viral, fungal, bacterial and mycobacterial infections.

Transbronchial biopsy

Most AIDS patients undergo transbronchial biopsy (TBB) at the time of fibreoptic bronchoscopy, with a diagnostic yield for the detection of *P. carinii* of 95%. This diagnostic yield can be improved to 98% when TBB is used in combination with BAL. TBB can also be very valuable for the diagnosis of fungal infections, CMV pneumonia, tumours and other pulmonary manifestations of AIDS.[10,28,33,37,40] The sensitivity of TBB for the diagnosis of CMV pneumonia ranges from 55 to 75%.[5,41-43] TBB combined with BAL offers a yield of approximately 85% for the overall diagnosis of pulmonary infections in AIDS patients, comparable with that of open lung biopsies.[37]

Percutaneous fine needle aspiration biopsy

Percutaneous fine needle aspiration biopsy is not the procedure of choice for the routine management of the pulmonary complications of most patients with AIDS but can be useful for the diagnosis of individuals with pulmonary nodules. However, this procedure is particularly useful for the evaluation of focal pulmonary lesions and is associated with a high rate of pneumothorax (up to 50%).[37]

Open-lung biopsy and video-assisted thoracoscopic biopsy

Open-lung biopsy has a selected role in the management of patients with AIDS, as the diagnostic yield of TBB is low for the identification of Kaposi's sarcoma, lymphoid interstitial pneumonia/pulmonary lymphoid hyperplasia and/or malignant lymphoma.[37] The indications for this invasive diagnostic procedure in these patients is still controversial, as individuals with malignant lymphoma or Kaposi's sarcoma do not survive for long time periods, in spite of current therapeutic modalities.

The value of video-assisted thoracoscopc biopsy in the management of the pulmonary complications of patients with AIDS is still unclear, to my knowledge. This procedure has a lower morbidity than open-lung biopsy.

Pulmonary infections in patients with AIDS

Pneumocystis carinii pneumonia

Pneumocystis carinii pneumonia (PCP) is the most frequent opportunistic infection in patients with AIDS and affects up to 80% of individuals with

HIV infections.[2,37,40]. Patients with AIDS and PCP can present with a variety of symptoms ranging from only mild cough and dyspnoea to respiratory failure and death in up to 20% of patients.[18,19] The infection can be controlled in most patients with appropriate therapy. However, it is difficult to eradicate *P. carinii* completely from pulmonary tissues and the pneumonia due to this organism recurs in up to 65% of AIDS patients.[20,44-46]

Pathogenesis

Pneumocystis carinii is an organism of uncertain taxonomy. It has been classified for many years as a parasite. However, recent studies of ribosomal RNA sequences have shown that the organism has a greater degree of homology to fungi than to protozoa.[47] In spite of this evidence, there is reluctance in the literature to classify *P. carinii* definitively as a fungus, as its structural features and antimicrobial sensitivity differ from those of most pathogenic fungi.

Pneumocystis carinii is an extracellular organism that has its natural habitat in the alveoli of humans and other mammals.[20,48,49] It can present in the human lung as: (a) 2–8 µm in diameter, pleomorphic trophozoites (Fig. 4.1) that tend to occur in intra-alveolar clumps; (b) 5–7 µm in diameter,

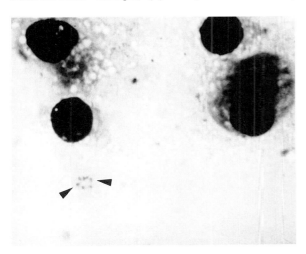

Fig. 4.1 Trophozoites of *P. carinii* in a BAL preparation (arrowheads). Papanicolaou stain; × 980.

thick-walled, cup-shaped, ovoid or crescentic cysts (Fig. 4.2); and (c) intermediate forms such as diploid trophozoites and precysts with eight intracystic sporozoites.[48,49] The trophozoites become

attached to the pneumocytes type 1, before developing into precysts.

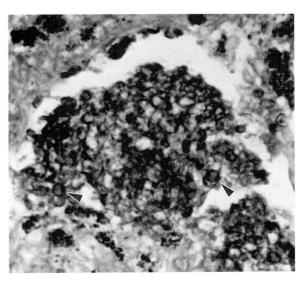

Fig. 4.2 Cysts of *P. carinii* with round and crescentic forms with intracystic bodies (arrowheads). GMS stain; × 400.

Pneumocystis carinii is transmitted aerogenously and colonizes up to 80% of children by the age of 4 years, as demonstrated by the presence of positive serological tests in normal populations.[13,17] The organism remains latent through adulthood and becomes reactivated in patients with AIDS and other immunodeficiency syndromes.[13,37] PCP epidemics have been described in children and hospital patients.[50]

In patients with HIV infection, PCP is unlikely to develop until the CD4 counts fall to below 200/mm³.[13,14] The attack rate for *P. carinii* in HIV-infected individuals is up to 30% per year, compared with 1% per year for patients with renal transplants or lymphoproliferative disorders.[18,20,46,51]

Clinical manifestations of PCP in patients with AIDS and other conditions

Several studies have demonstrated significant differences in the clinico-pathological features of PCP in patients with AIDS and those with other causes of immunosuppression such as lymphoproliferative disorders, malignancies, chemotherapy, and/or transplantation.[51,52] In general, the course of PCP is longer, more insidious, more difficult to resolve and more likely to recur in patients with HIV infections.[13] The most important differences in the clinico-pathological manifestations of PCP in immunocompromised patients with AIDS and other conditions are summarized in Table 4.5.

Poor prognostic findings in AIDS patients with PCP

AIDS patients with severe clinical abnormalities on initial chest roentgenograms, concurrent CMV infection, V-Q mismatches greater than 30 mm Hg requiring mechanical ventilation, neutrophilia in BAL and/or high serum levels of lactate dehydrogenase have a poor prognosis.[18,20]

Pathology

Table 4.6 summarizes the spectrum of histopathological features of PCP in patients with HIV infections and other causes of immunosuppression.[51]

The most typical histopathological change in patients with *P. carinii* pulmonary infections is the presence of an interstitial pneumonia with foamy intra-alveolar eosinophilic infiltrates (Fig. 4.3).[10,28,53,54] Similar collections of 'foamy' material can be observed in BAL preparations stained with Papanicolaou stain (Fig. 4.4) The interstitium shows mild to moderate infiltration by lymphoid cells and plasma cells. The alveolar walls show pneumocyte type 2 hyperplasia. The alveolar

Table 4.6 Spectrum of histopathological findings in AIDS patients with *P. carinii* pneumonia

Minimal interstitial inflammation
Interstitial pneumonia with intra-alveolar foamy
  exudates
Diffuse alveolar damage
  Exudative phase
  Organizing phase
Granulomatous inflammation
Calcification
Necrotizing pneumonia with cavitation
Lymphocytic interstitial pneumonitis (LIP)-like reaction

spaces contain a foamy, eosinophilic material. Within the spaces of this material, basophilic 'dots' corresponding to the sporozoites of *P. carinii* can be seen in routine haematoxylin and eosin stained preparations (Fig. 4.5). In some patients, the degree of inflammation is minimal and the intra-alveolar exudations can be seen only focally (Fig. 4.6). This may present a diagnostic problem, particularly in transbronchial biopsies where sampling problems are often critical.

Pathological diagnosis of *P. carinii*

The diagnosis of *P. carinii* pneumonia can usually be suspected in haematoxylin stained trans-

Table 4.5 *P. carinii* pneumonia in immunocompromised patients with AIDS and other conditions

	AIDS	Other causes of immunosuppression
Attack rate	30% per year	1–2% per year
Clinical course	Subacute	Acute
	Longer time for resolution	
	Higher rate of progression to respiratory failure	
Therapeutic course	Longer	Shorter
Adverse reactions to therapy	Over 50%	Approximately 10%
Recurrence rate	Up to 65%	2–3 times less frequent
Atypical pulmonary manifestations	Localized upper lobes disease	
	Pulmonary nodules	
	Cavitation	
	Spontaneous pneumothorax	
	Rapid progression too respiratory failure	
Extrapulmonary dissemination	More frequent	Very unusual

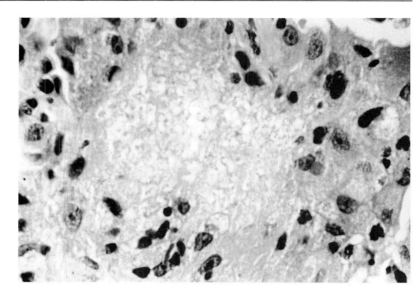

Fig. 4.3 *P. carinii* pneumonia with characteristic intra-alveolar 'foamy' exudate. The interstitium shows a mild lymphoplasmacytic infiltrate. Haematoxylin and eosin stain; × 368.

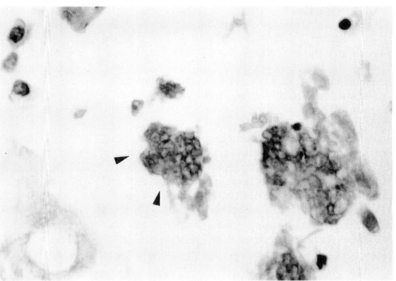

Fig. 4.4 Foamy exudate with *P. carinii* in BAL preparation (arrowheads). Papanicolaou stain; × 1000.

bronchial biopsies or TBB but needs to be confirmed by special stains. Giemsa stain can be used to visualize trophozoites.[53,55] However, the diagnosis of PCP is difficult to establish with certainty in TBB stained with Giemsa stain, as nuclear debris and other background structures can simulate the organisms. The stain is most helpful, in our experience, for the diagnosis of PCP in touch imprints or other cytological preparations (Fig. 4.1).

More often, the diagnosis of *P. carinii* pneumonia is established by visualizing cysts with silver stains such as Gomori's methenamine silver, Grocott's stain or others in TBB biopsies or cytological preparations. The walls of the cysts stain dark black with these stains and the cysts appear as round, oval or crescentic shaped with a 1-2 µm focal area of capsular thickening ('intracystic bodies') that is characteristic for the organism (Fig. 4.2).[53,55] In patients who have received previous treatment, the cysts become degenerated with blurred walls and are aggregated in a dark black, granular material.[56,57]

The cysts of *P. carinii* need to be differentiated on silver stained preparations from fungal spores,

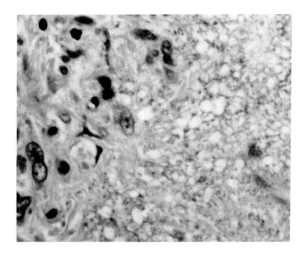

Fig. 4.5 *P. carinii* pneumonia with 'foamy' intra-alveolar exudate. The foamy material has basophilic 'dots' which correspond to trophozoites of the microorganism. Haematoxylin and eosin stain; × 320.

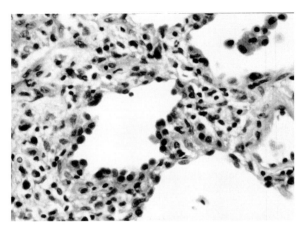

Fig. 4.6 *P. carinii* pneumonia with mild interstitial lymphocytic infiltrates. The intra-alveolar foamy material is absent in this biopsy. Special stains need to be performed in all lung biopsies from patients with AIDS, as atypical histopathological features are not infrequent. Haematoxylin and eosin stain; × 160.

particularly those of *Histoplasma capsulatum* and *Candida* species. The spores usually do not exhibit crescentic forms, lack areas of capsular thickening and show focal budding forms (Fig. 4.7).

Immunostains with monoclonal antibodies that react specifically with the trophozoites and cyst walls have been developed for the specific diagnosis of *P. carinii* pneumonia.[34,39,58] Several studies have demonstrated that the use of these immunopathological methods increase the sensitivity for the diagnosis of *P. carinii* in bronchial biopsies. The use of monoclonal antibody MAB-3F6 also increases the sensitivity of detection of *P. carinii* in bronchial lavage specimens from patients with PCP and without clinically apparent pneumonia.[39]

However, in our experience the number of additional patients diagnosed with these expensive and time-consuming immunopathological methods is too small to warrant the routine utilization of these stains for the diagnosis of all AIDS patients. However, immunopathological staining methods with monoclonal antibodies can be very helpful for the diagnosis of *P. carinii* extrapulmonary infections or in instances when the diagnosis is unclear.[18]

Recent studies have described the use of DNA hybridization and polymerase chain reaction techniques for the specific diagnosis of *P. carinii* infection.[20,59]

Fig. 4.7 The histopathological distinction between *P. carinii* and *Histoplasma capsulatum* can be difficult in silver-stained preparations. The left photograph demonstrates cysts of *P. carinii*, with intracystic bodies (arrowheads). Note that the cysts are slightly larger than the yeasts seen in the right photograph. The latter photograph shows budding yeasts of *H. capsulatum*. Both microorganisms were present in different areas from the lung of a patient with AIDS at autopsy. GMS stain; × 400.

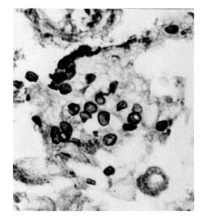
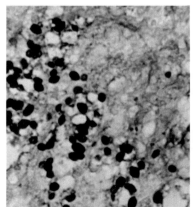

'Atypical' histopathological features of *Pneumocystosis* in patients with AIDS

Atypical pathological manifestations can be observed in over 70% of cases of *P. carinii* pneumonia in patients with AIDS.[60]

Pneumocystis carinii pulmonary infections can present with histopathological changes of diffuse alveolar damage, with hyaline membranes and pneumocyte type 2 hyperplasia (exudative phase) and intra-alveolar organization by spindle cells with early fibrosis (organizing phase).[56,51] Patients with extensive alveolar damage in the exudative phase usually present with rapidly progressive respiratory failure.[13,20]

Other histopathological changes listed in Table 4.6 are encountered less frequently in AIDS patients with PCP and are very unusual in patients with immunosuppression unrelated to HIV infections.

Granulomatous inflammation has been reported in 5% of *P. carinii* pneumonias, with the formation of epithelioid granulomas composed of focal collections of epithelioid histiocytes with multinucleated giant cell formation and/or foreign-body-type granulomas with giant cell reaction to the alveolar foam.[60] Epithelioid or foreign body granulomas due to *P. carinii* seldom exhibit caseation but can become confluent and undergo necrosis, resulting in cavitary lesions. The diagnosis of granulomatous *P. carinii* pneumonia is established by exclusion of other aetiological agents that can cause granulomatous inflammation, such as mycobacterial and fungal infections. The histological sections also need to be studied with polarized light to rule out the presence of foreign material, commonly present in intravenous drug abusers with AIDS.[60]

Foci of interstitial calcification, often associated with granulomatous formation, and calcification of subpleural cavities and intrapulmonaries cavities has been described in PCP.[60] Several histopathological patterns have been described, including 'bubbly' conchoidal, plate-like and elongate calcifications (Fig. 4.8).[62]

Pulmonary nodules, with or without intrapulmonary parenchymal cavities, have been described in 2% of patients with AIDS and *P. carinii* pneumonia.[60] The nodular lesions are difficult to distinguish on chest roentgenograms from neoplasms and fungal infections. Cystic *P. carinii* pneumonia is more frequent in the upper lobes and results from the destruction of areas of the pulmonary parenchyma in patients with severe infections (Fig. 4.9). Patients with cavitary pulmonary lesions may present with spontaneous pneumothorax. Histopathologically, biopsies or autopsy specimens taken from patients with cavitary PCP frequently exhibit necrosis, focal granulomas, calcification and numerous trophozoites of *P. carinii* within intra-alveolar exudates and invading the interstitium (Fig. 4.8). Necrotizing *P. carinii* vasculitis has also been described in association with lung

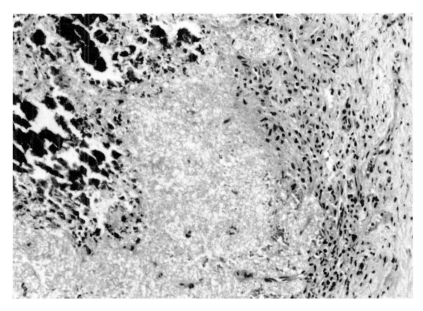

Fig. 4.8 *P. carinii* pneumonia showing foamy intra-alveolar exudates and plaque-like areas of calcification. Haematoxylin and eosin stain; × 100.

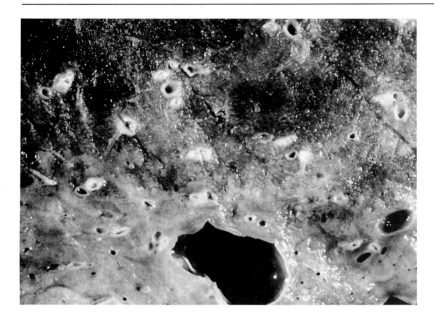

Fig. 4.9 Cystic *P. carinii* pneumonia.

necrosis and cavitation and it has been suggested that ischaemia due to vascular involvement may play a role in the pathogenesis of cavitary PCP.[63] The diagnosis of cavitary *P. carinii* can only be established by exclusion of other aetiological agents.

Pneumocystis carinii pneumonia can also present, in patients treated with pentamidine for active infections or receiving the drug for prophylaxis, as localized areas of pneumonia involving only the upper lung fields.[60]

A lymphocytic/plasmacytic interstitial pneumonitis with clinico-pathological features similar to those of lymphoid interstitial pneumonia (LIP) has been described in a small number of patients with *P. carinii* pneumonia. Associated infections such as cytomegalovirus pneumonia and others are also frequent in patients with PCP.[3,20,64]

Extrapulmonary and disseminated *P. carinii* infections
In a small number of patients with AIDS, *P. carinii* can spread from the lung into other organs through the bloodstream or lymphatics.[18,65] Extrapulmonary *P. carinii* infections can be (a) *disseminated*, involving multiple organs such as the bone marrow, spleen, thyroid, liver, kidney, adrenal gland and others, or (b) *localized* to the hilar lymph nodes or other tissues. In extrapulmonary locations the diagnosis of *P. carinii* infection can be suspected in the presence of a foamy tissue reaction similar to that described in the lung. Silver stains demonstrate the presence of the cysts characteristic for the organism. Immunopathological stains can be particularly useful to confirm the diagnosis of *P. carinii* infection in these unusual locations.[39]

Other parasitic infections in AIDS patients

Patients with AIDS can develop several parasitic infections listed in Table 4.2.

Toxoplasmosis is a frequent infection in AIDS patients that usually affects the central nervous system but seldom results in a pneumonia.[66,67] Pulmonary toxoplasmosis is usually associated with disseminated infection and is seldom diagnosed during life, as the cysts and trophozoites of *Toxoplasma gondii* can be sparse in pulmonary tissue and difficult to distinguish from nuclear debris on biopsy materials stained with haematoxylin and eosin or Giemsa stains. Immunostains with a monoclonal antibody specific for the organism can be helpful in patients where the infection is suspected.[66,67]

Pulmonary cryptosporidiosis is unusual, but the organism, which usually infects the gastrointestinal tract, can colonize the bronchial mucosa and the alveolar epithelium.[68] The pathogenic role of cryptosporidiosis in the lung is controversial, as most patients with intrapulmonary organisms have had other concomitant infections.[13]

Viral pneumonias in AIDS patients

Patients with AIDS frequently develop viral pneumonias caused by the various organisms listed in Table 4.2.

Cytomegalovirus pulmonary infection

Cytomegalovirus (CMV) is the most frequent viral infection in patients with AIDS.[5,42,69,70] Approximately 50–95% of adults and over 95% of homosexual men exhibit positive serology to the virus, either because of childhood primary airborne infections or as a result of sexually transmitted infections acquired during adulthood.[5,42,69,70] In immunocompetent hosts the virus becomes latent.

Cytomegalovirus infections have a high prevalence in patients with AIDS, particularly at autopsy where the organism in found in 70–90% of patients[3,13] The incidence during life is lower.

Histopathologically, CMV pneumonia presents as an interstitial pneumonia with infiltration of the interstitium by lymphoplasmacytic cells, pneumocyte type 2 hyperplasia, hyaline membranes and intra-alveolar oedema. The diagnosis is established in the presence of enlarged pneumocyte type 2 cells, bronchial cells or endothelial cells that measure up to 40–50 µm in diameter and exhibit the typical intranuclear and intracytoplasmic in-

clusions of the virus (Fig. 4.10). The intranuclear inclusions of CMV measure up to 17–18 µm in diameter, are basophilic or eosinophilic and are surrounded by a clear halo. The intracytoplasmic inclusions are smaller, multiple and usually basophilic. These inclusions are composed of packages of viral particles with characteristic ultrastructural features. In instances when the inclusions are sparse, immunopathological stains with monoclonal antibodies to CMV or *in situ* hybridization studies with specific DNA probes are diagnostic of infection.[5,43,71,72]

It is difficult to assess with certainty the role of CMV as the aetiological agent of a pulmonary infection in a particular patient with AIDS. Most of the patients with the syndrome have positive serology for CMV and viral inclusions characteristic of the virus are frequently seen in BAL and other cytological preparations.[5] Occasionally, scanty CMV inclusions are seen in transbronchial biopsies from patients with co-existent infections by *P. carinii* or other organisms, sepsis, diffuse alveolar damage or other conditions or in BAL fluids, bronchial wash preparations or transbronchial biopsies that exhibit pulmonary tissue with only minimal inflammation. It is difficult to assess with certainty the significance of CMV cellular inclusions in these patients: is the virus a pathogen

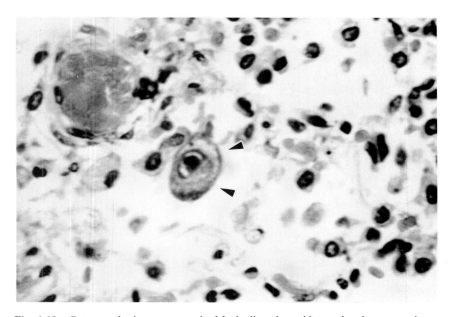

Fig. 4.10 Cytomegalovirus pneumonia. Markedly enlarged intra-alveolar macrophage with amphophilic intranuclear inclusion (arrowheads). Haematoxylin and eosin stain; × 400.

or a commensal organism in an immunosuppressed patient? In general, it is accepted that the presence of characteristic viral inclusions in a biopsy specimen is diagnostic of CMV pneumonia.[5,42,70] The significance of the presence of CMV inclusions in cytological preparations is less certain.[69] These conclusions are controversial. Murray and Mills have suggested as criteria for the clinical diagnosis of CMV pneumonia (a) the presence of a positive culture of CMV from the respiratory tract; (b) the identification of inclusions in a BAL or lung biopsy specimen; (c) the absence of another pathogen; and (d) progressive clinical signs and symptoms of pneumonia.[73] The diagnosis is confirmed in the presence of a specific clinical response to antiviral therapy.[69]

Varicella-zoster virus pneumonia in AIDS patients
The lung can become involved in patients with AIDS that develop disseminated varicella-zoster infections but isolated pneumonia by this virus is unusual.[74] Pathologically, the lungs show an interstitial pneumonia with multiple foci of necrosis. The characteristic, basophilic, intranuclear inclusions (Cowdry type A inclusions) are present.[13,74]

Herpes simplex pneumonia
Although herpes simplex infections (HSV) are frequent in the general population and in patients with AIDS, pneumonia caused by this virus has only been reported in a few patients with AIDS.[74,75] These patients presented with localized laryngotracheobronchitis, bronchopneumonia and/or disseminated infection involving multiple organs in addition to the lungs. Histopathologically, the virus causes ulceration of the airways epithelium, followed by focal squamous metaplasia, multinucleated giant cell formation and development of cells with characteristic intranuclear, basophilic inclusions (Cowdry type A) surrounded by a clear halo. The cells with intranuclear inclusions are not enlarged, as seen in CMV infections.[74,75] The pulmonary parenchyma shows areas of bronchopneumonia with foci of necrosis. The diagnosis is suspected in the presence of cells with the characteristic inclusions and can be confirmed by using immunostains with monoclonal antibodies to HSV or *in situ* hybridization with specific probes.[13]

Epstein-Barr virus pulmonary infections
The role of Epstein–Barr (EBV) virus in the de-velopment of AIDS has been controversial.[3,13,14] It is now recognized that EBV is probably not involved with the development of the syndrome. EBV has been detected with polymerase chain reaction (PCR) in lung tissues from AIDS patients exhibiting clinico-pathological features of lymphoid interstitial pneumonia (LIP) and non-specific interstitial pneumonitis (NIP).[64]

Other viral infections in AIDS patients
Influenza pneumonia has been described in patients with AIDS.[13] The role of HIV in the development of pulmonary disease remains controversial. Recently, Travis and associates described the presence of HIV RNA within macrophages in lung biopsies from AIDS patients with LIP and NIP.[64]

Mycobacterial infections in AIDS patients

Infections by non-tuberculous mycobacteria and *Mycobacterium tuberculosis* are frequent in AIDS patients.[13,76] *Mycobacterium avium-intracellulare* (MAI) complex is the most frequent mycobacterium isolated from HIV-infected patients. It can be cultured from bronchoscopic specimens in approximately 17% of AIDS patients.[77–79] Disseminated infection by this organism, a very unusual complication in the general population, is present in approximately 5% of AIDS patients.[13] This complication usually occurs late in the course of HIV infection and is associated with high morbidity and mortality. MAI infection is identified in up to 50% of AIDS patients at autopsy.[13]

In immunocompromised patients without AIDS, MAI induces granulomatous pneumonia with or without cavitation. It can also present as solitary pulmonary nodules that resemble neoplasms on chest roentgenograms.[40] In AIDS patients, MAI infections present histopathologically as focal interstitial collections of histiocytes with striated pale or light blue cytoplasm that resemble Gaucher cells (pseudo-Gaucher cells) or with ill-defined nodular infiltrates with necrosis.[13] The diagnosis is readily made with AFB stains that demonstrate a large number of intracytoplasmic bacteria. These mycobacteria also frequently stain positively with GMS and PAS stains.

Tuberculosis in AIDS patients
The HIV epidemic has increased the incidence of tuberculosis, including resistant strains of *M. tuberculosis*.[80–82] The prevalence of tuberculosis in

the AIDS population is significantly higher in intravenous drug users and in Haitians. In contrast to MAI infections, tuberculosis is an earlier complication HIV-related disease.[80-82].

Pathologically, tuberculosis presents in AIDS patients with nodular or cavitary lesions with caseating epithelioid granulomas or ill-formed granulomatous reaction. Atypical forms such as intrathoracic adenopathy, lower lung fields infiltration, diffuse pulmonary infiltrates without cavitation and miliary forms with dissemination have been described.[13] Histologically, *M. tuberculosis* can be suspected in the presence of positive AFB stains and granulomatous reaction, unusual in AIDS patients with non-tuberculous mycobacterial infections.

Pulmonary fungal infections in AIDS patients

Cryptococcosis, histoplasmosis, coccidioidomycosis, blastomycosis and other recurrent fungal infections caused by organisms that thrive in immunocompromised patients are frequent in individuals with AIDS.[3,13] These patients have impaired T cell, macrophage and granulomatous responses and develop reactivation of latent fungal infections.

Cryptococcal pneumonia in AIDS patients
Cryptococcosis is the most frequent cause of fungal pneumonia in AIDS patients.[13,83] *Cryptococcus neoformans* is a saprophytic yeast that spreads aerogenously through soil contaminated with bird droppings.[84] In immunocompetent individuals the infection results in a self-contained granulomatous reactions. In AIDS and other immunosuppressed patients *C. neoformans* results in meningoencephalitis, disseminated cryptococcosis, single or multiple granulomatous nodules, bronchopneumonia with ill-formed granulomas, an interstitial pattern with mild non-specific inflammation and organisms in the interstitium and capillaries (Fig. 4.11) and mixed forms.[13] The fungus can be identified with silver stains as 4–7 μm in diameter round, budding yeasts (Fig. 4.11). The buds are usually single and have a narrow base. *C. neoformans* has a characteristic mucopolysaccharide capsule that stains positively with Mayer's mucicarmine and diastase-PAS stains (Fig. 4.12). However, this capsule can be absent or deficient in patients with AIDS.[13] In the absence of a capsule, *C. neoformans* can be difficult to distinguish on light microscopy from *Histoplasma capsulatum*. Electron microscopy can be helpful to demonstrate an attenuated capsule in *C. neoformans*. *Blastomyces dermatitides* usually has larger yeasts with a thick wall.

Pulmonary coccidioidomycosis in AIDS patients
Coccidioidomycosis is a frequent pulmonary in-

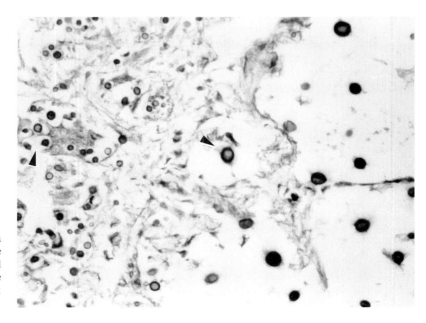

Fig. 4.11 Cryptococcal pneumonia in an AIDS patient, with multiple round, intra-alveolar and interstitial yeasts (arrowheads). Note the sparse inflammatory reaction. PAS stain; × 400.

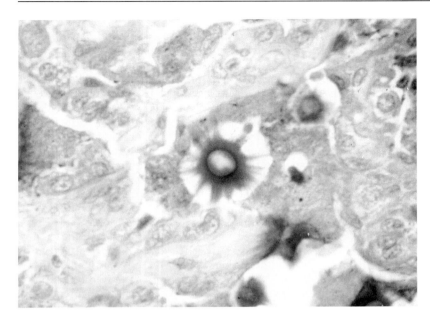

Fig. 4.12 Encapsulated yeast of *Cryptococcus neoformans*. Muci-carmine stain; × 100.

fection in AIDS patients who live in areas where this fungus is endemic, such as the southwestern USA.[85] *Coccidioides immitis* is a dimorphic fungus that thrives in the soil contaminated with bird or bat droppings.[13] In the soil the fungus grows as a mycelial form. In infected individuals, *C. immitis* occurs as characteristically large, 25–30 µm in diameter, round, endosporulating, thickwalled spherules. These spherules contain numerous round endospores that measure 2–5 µm in diameter.

AIDS patients reactivate latent *C. immitis* infections and present with diffuse interstitial infiltrates, ill-formed nodules or disseminated disease.[13] Granuloma formation is poor in HIV-infected patients with coccidioidomycosis. The diagnosis of *C. immitis* infection can be established by BAL, brush cytology or biopsies by visualizing the characteristic spherules of the organism or by culture.

Histoplasmosis in AIDS patients

Disseminated histoplasmosis is a severe complication of AIDS.[86,87] *Histoplasma capsulatum* is a dimorphic fungus that thrives as a mycelial form in soil contaminated with bird or bat droppings.[86,87] It is endemic in the Mississippi and Ohio river valleys, and in other areas with temperate climate.[13] In infected individuals, the fungus grows as 1–5 µm in diameter ovoid yeasts with infrequent budding (Fig. 4.13). Clinicopathological features

are similar to those described in coccidioidomycosis.

H. capsulatum can be confused in silver-stained preparations with *P. carinii* (Fig. 4.7). The absence of crescentic forms, lack of intracystic bodies, and the presence of occasional budding are characteristic of *H. capsulatum* and rule out *P. carinii*.

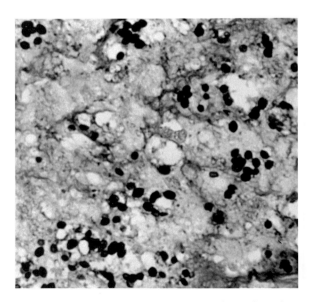

Fig. 4.13 Yeasts of *Histoplasma capsulatum* in a lung biopsy from a patient with AIDS and pneumonia. GMS stain; × 400.

Other fungal infections in AIDS patients
Infections by opportunistic fungi such as *Candida* spp. and *Aspergillus* spp. which are common in immunocompromised patients with neoplasms and other conditions, are less frequent in HIV-infected individuals.[13,20,40]

Pulmonary bacterial infections in AIDS patients

Although the various opportunistic infections described above are frequent in HIV-infected individuals, bacterial pneumonias caused by *Streptococcus pneumoniae*, *Hemophilus influenzae*, *Pseudomonas aeruginosa*, *Klebsiella pneumoniae* and other Gram-positive and Gram-negative organisms can occur at any stage of AIDS.[12,20,40] Patients with AIDS have an increased susceptibility to infections by pyogenic bacteria as a result of impaired humoral immunity, deficient macrophage functions, and abnormal phagocytic and bactericidal activity of neutrophils.[88,89] Bacterial pneumonias in AIDS patients tend to recur, can be difficult to diagnose and are associated with high mortality rates.[88,89]

Neoplastic complications of AIDS

Kaposi's sarcoma, Hodgkin's and non-Hodgkin's malignant lymphomas and carcinomas such as neuroendocrine pulmonary tumours have been described in patients with AIDS.[90-93] However, no aetiological or epidemiological relationships between lung carcinomas and HIV infections have been identified.[13]

Pulmonary involvement by Kaposi's sarcoma in AIDS patients

Kaposi's sarcoma is a frequent complication of the late stages of AIDS and has been described in about half of these patients at autopsy.[13] Pulmonary involvement by the sarcoma is approximately 13% at autopsy.[13] Patients with AIDS and pulmonary involvement by Kaposi's sarcoma present clinically with symptoms similar to those of opportunistic infections. They develop cough, fever, dyspnoea, hoarseness, stridor, hypoxaemia and/or respiratory failure. Pulmonary Kaposi's sarcoma is an important cause of death in AIDS patients.[90-93] On chest roentgenograms, patients with Kaposi's sarcoma develop diffuse reticulonodular infiltrates, nodules or ill-defined densities.

Pathologically, the tumour can present as endobronchial, red, submucosal lesions, intrapulmonary infiltrates in an interstitial–lymphatic distribution, and/or multiple ill-defined nodules involving the pleura, interlobular septa, bronchovascular sheaths and alveolar septa.[90-93] Pleural involvement by the tumour can result in bloody pleural effusions.[13] Microscopically, the nodular lesions of Kaposi's sarcoma are composed of hyperchromatic spindle cell infiltrates with cleft-like spaces containing red blood cells and hyaline globules (Fig. 4.14). Intrapulmonary Kaposi's sarcoma can be very difficult to diagnose by transbronchial biopsies and even open-lung biopsies, owing to sampling problems, and particularly in its early stages. The early forms of the disease present as polymorphous lesions with chronic inflammation, scanty spindle cells and proliferating capillaries. Bronchial lesions can be diagnosed endoscopically and confirmed histologically.[13]

Pulmonary involvement by malignant lymphoma in AIDS patients

Involvement of the lung parenchyma by non-Hodgkin's B cell high-grade malignant lymphomas usually occurs late in the course of AIDS and frequently is established only at autopsy (Fig. 4.15).[94] It is unusual for an HIV-infected patient to present with an isolated lymphoma of the lung. Most patients develop pulmonary involvement at the stage of disseminated malignant lymphoma.[13] Hodgkin's disease of the lung is unusual in AIDS, but has been described in a few patients. Peripheral T cell lymphoma/leukaemia has also been described in a few patients with HIV infection.[13]

Non-neoplastic complications of AIDS

Diffuse alveolar damage

Diffuse alveolar damage (DAD) is present in 65% of AIDS patients at autopsy.[13] The incidence in biopsies is probably lower. It can be associated with cytomegalovirus or other infections or can be the result of other aetiological factors that are more difficult to diagnose morphologically, such as sepsis, oxygen toxicity, shock or others. Patients

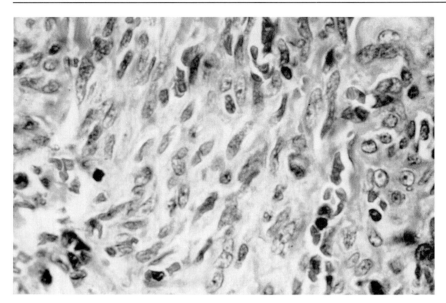

Fig. 4.14   Intrapulmonary Kaposi's sarcoma in an AIDS patient. Haematoxylin and eosin stain; × 200.

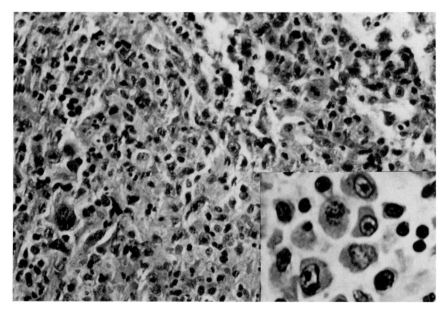

Fig. 4.15   Pulmonary malignant lymphoma (large cell type, immunoblastic) in an AIDS patient. Haematoxylin and eosin stain; × 100. The inset shows large atypical lymphoid cells with immunoblastic features. Haematoxylin and eosin stain; × 400.

present with respiratory failure and diffuse roentgenographic infiltrates that start in the hilar areas and progress to complete lung field 'white-out'.

Pathologically there is an exudative phase of DAD with hyaline membranes, intra-alveolar oedema and alveolar cell hyperplasia, and an organizing phase of DAD with more prominent hyperplasia of pneumocytes type 2 and intra-alveolar fibroblastic proliferation.

Non-specific interstitial pneumonitis

In approximately 30–40% of transbronchial biopsies from patients with AIDS no aetiological diagnosis can be established. The pulmonary parenchyma shows mild to moderate interstitial infiltrates by lymphocytes and occasional plasma cells and focal lymphoid aggregates (non-specific pneumonitis or NIP).[13,20,64,95] The aetiology of this non-specific pneumonitis is unclear. Possible aetiological agents that have been suggested include reactions to illicit drugs or medications and viral infections by HIV, EBV and other organisms.[13,64,95]

Lymphocytic interstitial pneumonitis/pulmonary lymphoid hyperplasia (LIP/PLH) complex

Children with AIDS frequently develop a variety of pulmonary infiltrates related to lymphoid hyperplasia.[64,95,96] They include pulmonary lymphoid hyperplasia (PLH), lymphoid interstitial pneumonitis (LIP), and polyclonal polymorphic B cell lymphoproliferative disorder (PBLD). These changes have also been described, albeit less frequently, in adult patients with HIV infections.[33,64,97]

Pulmonary lymphoid hyperplasia is characterized by the presence of multiple lymphoid nodules in a peribronchial and peribronchiolar location (Fig. 4.16). Lymphoid interstitial pneumonitis is characterized by multinodular aggregates of lymphocytes, mononuclear cells and plasma cells in a lymphatic distribution along the bronchovascular bundles, interlobular septa and the interstitium (Fig. 4.17). No bronchial destruction is seen. Occasional granulomas can be present, and the possibility of a concomitant mycobacterial infection needs to be ruled out in these instances. Considerable overlap exists between LIP and PLH. Children with the LIP/PLH complex usually develop progressive respiratory insufficiency, although the syndrome can regress with steroid therapy. An aetiological role for EBV has been postulated in patients with LIP/PLH and AIDS, as DNA from the virus has been recovered from children with the syndrome.

Other non-neoplastic pulmonary reactions

Alveolar proteinosis-like reactions associated with cytomegalovirus, *P. carinii* and other pulmonary

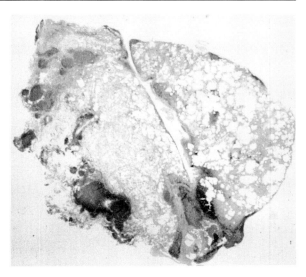

Fig. 4.16 Pulmonary lymphoid hyperplasia (PLH) in a child with AIDS. Haematoxylin and eosin stain; × 1.8.

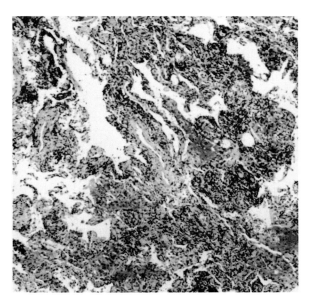

Fig. 4.17 Lymphoid interstitial pneumonitis (LIP) in an adult AIDS patient. Haematoxylin and eosin stain; × 40.

infections have been reported.[98] Pneumonitis with features resembling desquamative interstitial pneumonia (DIP-like reactions), drug reactions, foreign body granulomas in drug addicts and other non-neoplastic pulmonary reactions have also been described in patients with AIDS.[13]

References

1. Haseltine WA, Wong-Staal F. The molecular biology of the AIDS virus. *Sci Am* 1988; **259**, 52.
2. Nash G, Fligiel S. Pathologic features of the lung in the acquired immune deficiency syndrome (AIDS): an autopsy study of seventeen homosexual males. *Am J Clin Pathol* 1984; **81**, 6.
3. Nash G. Spectrum of HIV infection. In: *Pathology of AIDS and HIV Infection*, Nash G, Said J (eds). Philadelphia: WB Saunders Co, 1992, p. 8.
4. Center for Disease Control. Update on acquired immune deficiency syndrome (AIDS) – United States. *Mortality Morbidity Weekly Reports* 1982; **31**, 507.
5. Aukrust P, Farstad IN, Froland SS, Holter E. Cytomegalovirus (CMV) pneumonitis in AIDS patients: the result of intensive CMV replication? *Eur Respir J 1992;* **5**, 362.
6. Mohar A, Romo J, Salido F et al. The spectrum of clinical and pathological manifestations of AIDS in a consecutive series of autopsied patients in Mexico. *Aids* 1992; **6**, 467.
7. Besnier J-M, Baillou BF, Liard F, Choutet P, Goudeau A. Symptomatic HIV-2 primary infection. *Lancet* 1990; **335**, 798.
8. Busch MP, Zahawa EA, Sheppard HW, Ascher MS, Lang W. Primary HIV-1 infection. *New Engl J Med* 1991; **325**, 733.
9. Haseltine WA. Silent HIV infections. *New Engl J Med* 1989; **320**, 1487–1489.
10. Luna MA, Cleary KR. Spectrum of pathologic manifestations of *Pneumocystis carinii* pneumonia in patients with neoplastic diseases. *Semin Diagn Pathol* 1989; **6**, 262.
11. Stover DE, White DA, Romano PA et al. Spectrum of pulmonary diseases associated with the acquired immune deficiency syndrome. *Am J Med* 1985; **78**, 429.
12. Redfield RR, Wright DC, Tramont EC. The Walter Reed staging classification for HTLV III/LAV infection. *New Engl J Med* 1986; **314**, 131.
13. Nash G. Respiratory system. In: *Pathology of AIDS and HIV Infection*, Nash G, Said J (eds). Philadelphia: WB Saunders Co, 1992, p. 60.
14. Said JW. Pathogenesis of HIV infection. In: *Pathology of AIDS and HIV Infections*, Nash G, Said J (eds). Philadelphia: WB Saunders Co, 1992, p. 15.
15. Abouya YL, Beaumel A, Lucas S et al. *Pneumocystis carinii* pneumonia. An uncommon cause of death in African patients with acquired immunodeficiency syndrome. *Am Rev Respir Dis* 1992; **145**, 617.
16. Bedrossian CWM, Mason MR, Gupta PK. Rapid cytologic diagnosis of *Pneumocystis*: a comparison of effective techniques. *Semin Diagn Pathol* 1989; **6**, 245.
17. Bier S, Halton K, Krivisky B et al. *Pneumocystis carinii* pneumonia presenting as a single pulmonary nodule. *Pediatr Radiol* 1986; **16**, 59.
18. Cote RJ, Rosenblum M, Telzak EE et al. Disseminated *Pneumocystis carinii* infection causing extrapulmonary organ failure: clinical, pathologic and immunohistochemical analysis. *Mod Pathol* 1990; **3**, 25.
19. Engelberg LA, Lerner CW, Tapper ML. Clinical features of *Pneumocystis* pneumonia in the acquired immune deficiency syndrome. *Am Rev Respir Dis* 1984; **130**, 689.
20. Gal AA, Koss MN, Strigle S et al. *Pneumocystis carinii* infection in the acquired immune deficiency syndrome. *Semin Diagn Pathol* 1989; **6**, 287.
21. Fox R, Eldred LJ, Fuchs EJ et al. Clinical manifestations of acute infection with human immunodeficiency virus in a cohort of gay men. *AIDS* 1987; **1**, 35.
22. Suster B, Akerman M, Orenstein M et al. Pulmonary manifestations of AIDS: review of 106 episodes. *Radiology* 1986; **161**, 87.
23. Cohen BA, Pomeranz S, Rabinowitz JG et al. Pulmonary complications of AIDS: radiologic features. *Am J Radial* 1984; **143**, 115.
24. Del Rio C, Guarner J, Honig EG et al. Sputum examination in the diagnosis of *Pneumocystis carinii* pneumonia in the acquired immunodeficiency syndrome. *Arch Pathol Lab Med* 1988; **112**, 1229.
25. Bigby TD, Margolskee D, Curtis JL et al. The usefulness of induced sputum in the diagnosis of *Pneumocystis carinii* pneumonia in patients with the acquired immunodeficiency syndrome. *Am Rev Respir Dis* 1986; **133**, 515.
26. Hackman RC, Myerson D, Meyers JD et al. Rapid diagnosis of cytomegaloviral pneumonia by tissue immunofluorescence with a murine monoclonal antibody. *J Infect Dis* 1985; **151**, 325.
27. Hanson PJV, Harcourt-Webster JN, Gazzard BG, Collins JV. Fiberoptic bronchoscopy in diagnosis of bronchopulmonary Kaposi's sarcoma. *Thorax* 1987; **42**, 269.
28. Kovacs JA, Ng VL, Masur H. Diagnosis of *Pneumocystis carinii* pneumonia: improved detection in sputum with use of monoclonal antibodies. *New Engl J Med* 1988; **318**, 589.
29. Mann JM, Altus CS, Webber CA et al. Non-bronchoscopic lung lavage for diagnosis of opportunistic infection in AIDS. *Chest* 1987; **9**, 319.
30. Myerson D, Hackman RC, Meyers JD. Diagnosis of cytomegaloviral pneumonia by *in situ* hybridization. *J Infect Dis* 1984; **150**, 272.
31. Pitchenik AE, Ganjei P, Torres A et al. Sputum examination for the diagnosis of *Pneumocystis carinii* pneumonia in the acquired immunodeficiency syndrome. *Am Rev Respir Dis* 1986; **133**, 226.
32. Rorat E, Garcia RJ, Skolom J. Diagnosis of

Pneumocystis carinii pneumonia by cytologic examination of bronchial washings. *J Am Med Assoc* 1985; **254**, 1950.

33. Rosen MJ, Tow TWY, Teirstein AS, Chuang MT, Marchevsky AM, Bottone EJ. Diagnosis of pulmonary complications of the acquired immune deficiency syndrome. *Thorax* 1985; **40**, 571.

34. Blumenfeld W, Kovacs JA. Use of a monoclonal antibody to detect *Pneumocystis carinii* in induced sputum and bronchoalveolar lavage fluid by immunoperoxidase staining. *Arch Pathol Lab Med* 1988; **112**, 1233.

35. Ng VL, Gartner I, Weymouth LA *et al*. The use of mycolysed induced sputum for the identification of pulmonary pathogens associated with human immunodeficiency virus infection. *Arch Pathol Lab Med* 1989; **113**, 488.

36. Wolfson JS, Waldron MA, Luz SS. Blinded comparison of a direct immunofluorescent monoclonal antibody staining method and a Giemsa staining method for identification of *Pneumocystis carinii* in induced sputum and bronchoalveolar lavage specimens of patients infected with human immunodeficiency virus. *J Clin Microbiol* 1990; **28**, 2136–2138.

37. Sittler SY, Ross DJ, Mohsenifar Z, Marchevsky AM. Pulmonary complications in patients with the acquired immunodeficiency syndrome: diagnostic methods. *Prog AIDS Pathol* 1990; **2**, 73.

38. Ghali VS, Garcia RI, Skolom J. Fluorescence of *Pneumocystis carinii* in Papanicolaou stained smears. *Hum Pathol* 1984; **15**, 907.

39. Homer KS, Wiley EL, Smith AL *et al*. Monoclonal antibody to *Pneumocystis carinii*. Comparison with silver stain in bronchial lavage specimens. *Am J Clin Pathol* 1992; **97**, 619.

40. Marchevsky AM, Rosen MJ, Chrystal G *et al*. Pulmonary complications of the acquired immunodeficiency syndrome: a clinicopathologic study of 70 cases. *Hum Pathol* 1985; **16**, 659.

41. Klatt EC, Shibata D. Cytomegalovirus infection in the acquired immunodeficiency syndrome. *Arch Pathol Lab Med* 1988; **112**, 540.

42. Smith CB. Cytomegalovirus pneumonia: state of the art. *Chest* 1989; **95**, 182.

43. Paradis IL, Grgurich WF, Dummer JS *et al* Rapid detection of cytomegalovirus pneumonia from lung lavage cells. *Am Rev Respir Dis* 1988; **138**, 697.

44. Brenner M, Ognibene FP, Lack EE *et al*. Prognostic factors and life expectancy of patients with acquired immunodeficiency syndrome and *Pneumocystis carinii* pneumonia. *Am Rev Respir Dis* 1987; **136**, 1199.

45. DeLorenzo LJ, Maguire GP, Wormser GP *et al*. Persistence of *Pneumocystis carinii* pneumonia in the acquired immunodeficiency syndrome: evaluation of therapy by follow-up transbronchial lung biopsy. *Chest* 1985; **88**, 79.

46. Phair J, Munoz A, Detels R *et al*. The risk of *Pneumocystis carinii* pneumonia among men infected with immunodeficiency virus type I. *Am J Med* 1990; **76**, 501.

47. Edman JC, Kovacs JA, Masur H *et al*. Ribosomal RNA sequence shows *Pneumocystis carinii* to be a member of the fungi. *Nature* 1988; **334**, 519.

48. Campbell WG. Ultrastructure of *Pneumocystis* in human lung: life cycle in human pneumocystosis. *Arch Pathol* 1972; **93**, 312.

49. Gutierrez Y. The biology of *Pneumocystis carinii*. *Semin Diagn Pathol* 1989; **6**, 203.

50. Ruebush TK, Weinstein RA, Baehner RL *et al*. An outbreak of *Pneumocystis* pneumonia in children with acute lymphocytic leukemia. *Am J Dis Child* 1978; **132**, 143.

51. Kovacs JA, Hiemenz JW, Macher AM *et al*. *Pneumocystis carinii* pneumonia: a comparison between patients with the acquired immunodeficiency syndrome and patients with other immunodeficiencies. *Ann Intern Med* 1984; **100**, 663.

52. Sterling RP, Bradley BB, Khalil KG *et al*. Comparison of biopsy-proven *Pneumocystis carinii* pneumonia in acquired immune deficiency sndrome patients and renal allograft recipients. *Ann Thorac Surg* 1984; **38**, 494.

53. Price RA, Hughes WT. Histopathology of *Pneumocystis carinii* infestation and infection in malignant disease in childhood. *Hum Pathol* 1974; **5**, 737.

54. Rosen PP, Martini N. *Pneumocystis carinii* pneumonia: diagnosis by lung biopsy. *Am J Med* 1975; **58**, 794.

55. Radio SJ, Hansen S, Goldsmith J *et al*. Immunohistochemistry of *Pneumocystis carinii* infection. *Mod Pathol* 1990; **3**, 462.

56. Saldana MJ, Mones JM, Martinez GR. The pathology of treated *Pneumocystis carinii* pneumonia. *Semin Diagn Pathol* 1989; **6**, 300.

57. Shelhamer JH, Ognibene FP, Macher AM. Persistence of *Pneumocystis carinii* in lung tissue of acquired immunodeficiency syndrome patients treated for *Pneumocystis* pneumonia. *Am Rev Respir Dis* 1984; **130**, 1161.

58. Ng VL, Yajko DM, PcPhaul LW *et al*. Evaluation of an indirect fluorescent-antibody stain for detection of *Pneumocystis carinii* in respiratory specimens. *J Clin Microbiol* 1990; **28**, 975.

59. Blumenfeld W, McCook O, Helodniy M, Katzenstein DA. Correlation of morphologic diagnosis of *Pneumocystis carinii* with the presence of *Pneumocystic* DNA amplified by the polymerase chain reaction. *Mod Pathol* 1992; **5**, 103.

60. Travis WD, Pittaluga S, Lipschik GY *et al*. Atypical pathologic manifestations of *Pneumocystis carinii* pneumonia in the acquired immune deficiency syndrome. *Am J Surg Pathol* 1990; **14**, 615.

61. Saldana MJ, Mones JM. Cavitation and other

atypical manifestations of *Pneumocystis carinii* pneumonia. *Semin Diagn Pathol* 1989; **6**, 273.

62. Lee MM, Schinella RA. Pulmonary calcification caused by *Pneumocystis carinii* pneumonia. A clinicopathological study of 13 cases in acquired immune deficiency syndrome patients. *Am J Surg Pathol* 1991; **15**, 376.

63. Liu YC, Tomashefski JF, Tomford JW *et al.* Necrotizing *Pneumocystis carinii* vasculitis associated with lung necrosis and cavitation in a patient with acquired immunodeficiency syndrome. *Arch Pathol Lab Med* 1989; **113**, 494.

64. Travis WD *et al.* Lymphoid pneumonitis in 50 adult patients infected with the human immunodeficiency virus: lymphocytic interstitial pneumonitis versus nonspecific interstital pneumonitis. *Hum Pathol* 1992; **23**, 529.

65. Unger PD, Rosenblum M, Krown SE. Disseminated *Pneumocystis carinii* infection in a patient with acquired immune deficiency syndrome. *Hum Pathol 1988;* **19**, 113.

66. Mendelson MH, Finkel LJ, Meyers BR *et al.* Pulmonary toxoplasmosis in AIDS. *Scand J Infect Dis* 1987; **19**, 703.

67. Tschirhart D, Klatt EC. Disseminated toxoplasmosis in the acquired immunodeficiency syndrome. *Arch Pathol Lab Med* 1988; **112**, 1237.

68. Brady EM, Margolis ML, Korzeniowski OM. Pulmonary cryptosporidiosis in acquired immunodeficiency syndrome. *J Am Med Assoc* 1984; **252**, 89–90.

69. Miles PR, Baughman RP, Linnemann CCJ. Cytomegalovirus in the bronchoalveolar lavage fluid of patients with AIDS. *Chest* 1990; **97**, 1072.

70. Millar AB, Patou G, Miller RF *et al.* Cytomegalovirus in the lungs of patients with AIDS: respiratory pathogen or passenger? *Am Rev Respir Dis* 1990; **141**, 1474.

71. Mintz L, Drew WL, Miner RC *et al.* Cytomegalovirus infections in homosexual men: an epidemiologic study. *Ann Intern Med* 1983; **99**, 326.

72. Hilborne LH, Nieberg RK, Cheng L *et al.* Direct *in situ* hybridization for rapid detection of cytomegalovirus in bronchoalveolar lavage. *Am J Clin Pathol* 1987; **87**, 766.

73. Murray JF, Mills J. Pulmonary infectious complications of human immunodeficiency virus infection. *Am Rev Respir Dis* 1990; **141**, 1356.

74. Cohen PR, Beltrani VP, Grossman ME. Disseminated herpes zoster in patients with human immunodeficiency virus infection. *Am J Med* 1988; **84**, 1076.

75. Corey L, Spear PG. Infections with herpes simplex viruses. *New Engl J Med* 1986; **314**, 686.

76. Horsburgh DRJ, Selik RM. The epidemiology of disseminated nontuberculous mycobacterial infection in the acquired immune deficiency syndrome (AIDS). *Am Rev Respir Dis* 1989; **139**, 4.

77. Hawkins CC, Gold JWM, Whimbey E *et al.* *Mycobacterium avium* complex infections in patients with the acquired immune deficiency syndrome. *Ann Intern Med* 1986; **105**, 184.

78. Klatt EC, Jensen DF, Meyer PR. Pathology of *Mycobacterium avium-intracellulare* infection in acquired immunodeficiency syndrome. *Hum Pathol* 1987; **18**, 709.

79. MacDonell KB, Glassroth J. *Mycobacterium avium* complex and other nontuberculous mycobacteria in patients with HIV infection. *Semin Respir Infect* 1989; **4**, 123.

80. Ankobiah WA, Finch P, Powell S, Heurich A, Shivaram I, Kamholz SL. Pleural tuberculosis in patients with and without AIDS. *J Assoc Acad Min Phys* 1990; **1**, 20.

81. Handwerger S, Mildvan D, Senie R *et al.* Tuberculosis and the acquired immunodeficiency syndrome at a New York City hospital: 1978–1985. *Chest* 1987; **91**, 176.

82. Hill AR, Premkumar S, Brustein S *et al.* Disseminated tuberculosis in the acquired immunodeficiency syndrome era. *Am Rev Resp Dis* 1991; **144**, 1164.

83. Gal AA, Koss MN, Hawkins J *et al.* The pathology of pulmonary cryptococcal infections in the acquired immunodeficiency syndrome. *Arch Pathol Lab Med* 1986; **110**, 502.

84. Bottone EJ, Toma M, Johansson BE *et al.* Poorly encapsulated *Cryptococcus neoformans* from patients with AIDS. I. Preliminary observations. *AIDS Res* 1986; **2**, 211.

85. Graham AR, Sobonya RD, Bronnimann DA *et al.* Quantitative pathology of coccidioidomycosis in acquired immunodeficiency syndrome. *Hum Pathol* 1988; **19**, 800.

86. Johnson PC, Hamill RC, Sarosi GA. Clinical review: progressive disseminated histoplasmosis in the AIDS patient. *Semin Respir Infect* 1989; **4**, 139.

87. Wheat LJ, Slama TG, Zeckel ML. Histoplasmosis in the acquired immune deficiency syndrome. *Am J Med* 1985; **78**, 203.

88. Nichols L, Balogh K, Silverman M. Bacterial infection in the acquired immune deficiency syndrome: clinicopathologic correlations in a series of autopsy cases. *Am J Clin Pathol* 1989; **92**, 787.

89. Polsky B, Gold JWM, Whimbey E *et al.* Bacterial pneumonia in patients with the acquired immunodeficiency syndrome. *Ann Intern Med* 1986; **104**, 38.

90. Ognibene FP, Steis RG, Macher AM *et al.* Kaposi's sarcoma causing pulmonary infiltrates and respiratory failure in the acquired immunodeficiency syndrome. *Ann Intern Med* 1985; **102**, 471.

91. Meduri GU, Stover DE, Lee M *et al.* Pulmonary Kaposi's sarcoma in the acquired immune deficiency syndrome: clinical, radiographic and pathologic manifestations. *Am J Med* 1986; **81**, 11.

92. Nash G, Fligiel S. Kaposi's sarcoma presenting as

pulmonary disease in the acquired immuno-deficiency syndrome: diagnosis by lung biopsy. *Hum Pathol* 1984; **15**, 999.

93. Weitberg AB, Mayer K, Miller ME *et al*. Dysplastic carcinoid tumour and AIDS-related complex [letter]. *New Engl J Med* 1986; **314**, 1455.

94. Loureiro C, Gill PS, Meyer PR *et al*. Autopsy findings in AIDS-related lymphoma. *Cancer* 1988; **62**, 735.

95. Malamou-Mitsi V, Tsai MM, Gal AA, Koss MN, O'Leary TJ. Lymphoid interstitial pneumonia not associated with HIV infection: role of Epstein Barr virus. *Mod Pathol* 1992; **5**, 487.

96. Joshi VV, Kauffman S, Oleske JM *et al*. Polyclonal polymorphic B-cell lymphoproliferative disorder with prominent pulmonary involvement in children with acquired immune deficiency syndrome. *Cancer* 1987; **59**, 1455.

97. Morris JD, Rosen MJ, Marchevsky AM, Teirstein AS. Lymphocytic interstitial pneumonia in patients at risk for the acquired immune deficiency syndrome. *Chest* 1987; **91**, 63.

98. Israel RH, Magnussen CR. Are AIDS patients at risk for pulmonary alveolar proteinosis? *Chest* 1989; **96**, 641.

5

Lung transplantation

S Stewart

Introduction

Lung transplantation is now well established as a therapeutic procedure for end-stage pulmonary vascular and parenchymal disease.[1] Combined heart–lung transplantation was initially developed for patients with pulmonary hypertension and right heart failure and was performed most commonly for Eisenmenger's syndrome and primary pulmonary hypertension. Heart–lung transplantation is also performed for advanced parenchymal pulmonary disease with right heart failure (Table 5.1). Some end-stage pulmonary conditions, including cystic fibrosis, have normal or adequate cardiac function. Under these circumstances it is possible to perform a combined heart–lung graft with donation of the recipient's own heart to a patient requiring cardiac transplantation. This is the so-called 'domino' procedure. Single lung transplantation is performed particularly for end-stage pulmonary fibrosis and emphysema. Clearly single lung transplantation is not suitable for patients with bilateral pulmonary sepsis or chronic obstructive lung disease as the remaining contralateral native lung may infect the transplanted lung, particularly in the face of immunosuppression. Where these patients have adequate or recoverable right ventricular function, double lung transplantation is possible. It avoids the complications of concomitant cardiac transplantation, namely rejection and coronary occlusive disease. Ischaemic complications of the airways, particularly at the anastomosis, are greater in double and single lung transplants than in combined heart-lung blocks. In the latter the tracheal anastomosis is more vascular and there are also bronchial/

Table 5.1 Diagnosis of patients transplanted for parenchymal disease

Cystic fibrosis
Emphysema including AIAT[a]
Sarcoidosis
Cryptogenic fibrosing alveolitis
Langerhan's cell histiocytosis
Bronchiectasis
Leiomyosarcoma of pulmonary artery
Bronchiolitis obliterans[b]
Churg–Strauss vasculitis
Systemic lupus erythematosus
Scleroderma
Haemosiderosis
Lymphangioleiomyomatosis

[a] AIAT, α-1-antitrypsin deficiency.
[b] Including rare retransplants for chronic lung rejection.

coronary collateral vessels which may reduce local airway ischaemia. Various operative manoeuvres including omental pedicles to the bronchial anastomoses have been developed to reduce the risk of ischaemia to the donor bronchi and, thus the rate of dehiscence. The problem of airway ischaemia is however still considerable in single and double lung grafts and many patients require stents. Rarely the lungs have been transplanted with the heart and liver (so-called triple transplant) for conditions such as cystic fibrosis complicated by cirrhosis or advanced cirrhosis associated with plexogenic pulmonary hypertension.

The availability of suitable donor organs continues to limit the development of lung transplantation. There are strict selection criteria. Heart-lung transplantation requires a donor with a combination of suitable cardiac and pulmonary function. Double and single lung transplantation permit the use of the donor heart for cardiac transplantation, maximizing the use of scarce donor organs. The organs are harvested from patients who have suffered brainstem death; this means that by definition, the heart and lungs are not normal. The donor lungs can be preserved by a variety of techniques allowing distant procurement. A commonly used perfusate is cold colloid flushed into the pulmonary artery, preceded by prostacyclin. Prior to harvesting the lungs have been mechanically ventilated and a study of unused donor lungs has shown almost universal acute bronchitis and bronchiolitis.[2] Unused partner lungs of single lung transplants and occasional unplaced heart–lung blocks have shown other pathological abnormalities including broncho-pneumonia, adult respiratory distress syndrome, thromboemboli, bone marrow emboli and evidence of aspiration. Ventilation may also cause interstitial emphysema. Skilful immediate postoperative management can overcome some of the effects of preservation injury and inevitable donor lung abnormalities. However some patients are still lost through primary donor organ failure.

The pathology of lung transplantation centres mainly on rejection and infection.[5] These conditions pose the major threat both in the short- and long-term to the survival of both the graft and patient.[4,5] Post-transplant lymphoproliferative disease occurs in lung transplant recipients with a predeliction for the graft. The diagnosis of these postoperative complications is usually confined to specialist transplant centres but with the expansion of lung transplant programmes it becomes increasing likely that biopsies may have to be performed at other institutions. The practical difficulties of diagnosing each of the major complications will be discussed in this chapter. Many of these are of relevance to the management of immunosuppressed patients in general, excepting, of course, rejection.

Acute pulmonary rejection

Rejection of pulmonary allografts shows identical features in single, double and combined grafts.[3] It shares many features in common with rejection of other solid organ grafts and has been extensively studied in both animal and clinical settings.[3,4,7] Using inbred rats, Prop *et al.* demonstrated an initial latent phase of rejection followed by the vascular phase with prominent perivascular and peribronchial mononuclear cell infiltrates.[7] These extended into the interstitium in the alveolar phase, advancing to necrosis and acute inflammation in the final destructive phase. Similar histopathological features are seen in transbronchial, open biopsy and autopsy material in human lung transplants, with some modification by immunosuppression.[3,4,8] The commonly used combination of cyclosporin, azathioprine and steroids may be enhanced by antithymocyte globulin and OKT3.

The earliest pathological change in acute pulmonary rejection is the presence of perivascular and peribronchiolar mononuclear infiltrates, which consist largely of small lymphocytes with macrophages, plasma cells and, less commonly, neutrophils.[3,4] These cuffing infiltrates may be very sparse and require adequate sampling for their diagnosis (Fig. 5.1). As rejection increases in severity, the infiltrates become more obvious at scanning magnification. As well as becoming more frequent, they are more cellular with larger lymphoid cells admixed with increased numbers of neutrophils and eosinophils. Lymphocytes can be seen within the intima infiltrating the endothelium with the appearance of endotheliitis (Fig. 5.2). This is associated with endothelial cell hyperplasia. Advancing rejection involves extension of perivascular and peribronchiolar infiltrates into the alveolar walls and spaces (Fig. 5.3). Finally, the infiltrates become confluent and there may be accompanying haemorrhage, necrosis and hyaline membrane formation (Fig. 5.4). The bronchi and bronchioles also progress from non-infiltrative cuffing to lymphocytic infiltration of the epithelium, polymorph infiltration, ulceration and epithelial necrosis. Open biopsy and autopsy material may reveal fibrinoid vasculitis involving vessels of larger calibre than those included in transbronchial lung biopsies.

The histological features of pulmonary rejection have been classified and graded by the lung rejection study group of the International Society for Heart and Lung Transplantation.[9] This Working Formulation aims to be a reasonably simple, uniform and reproducible grading system allowing comparison of data from various institutions and incorporating the essential points of other grading systems in use. The study group recommended a minimum of five transbronchial specimens of lung parenchyma examined at a minimum of three levels, with mandatory haematoxylin and eosin,

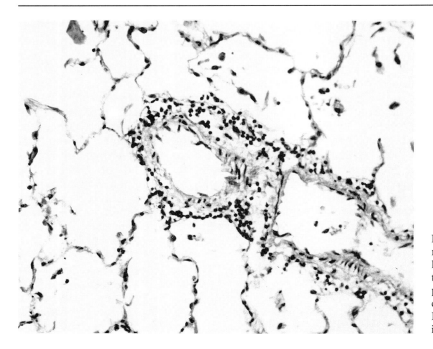

Fig. 5.1 Minimal acute pulmonary rejection showing scanty mature small lymphocytes around a venule. Endothelialitis is not present. Adjacent parenchyma and accompanying bronchioles were normal. Grade A1b. Haematoxylin and eosin stain; medium power. Autopsy lung.

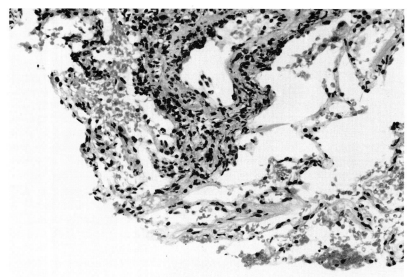

Fig. 5.2 Mild acute pulmonary rejection showing larger perivascular infiltrates with small and large lymphocytes and conspicuous endothelialitis. Perivascular adventitia is expanded but the infiltrate does not extend into adjacent parenchyma. Alveoli do not contain inflammatory cells. Bronchioles elsewhere in the biopsy showed mononuclear cell infiltration. Grade A2a. Haematoxylin and eosin stain; medium power. Transbronchial biopsy.

connective tissue and silver stains. In practice many institutions exceed these minimum standards. Serial sections are much more time-consuming to examine but do have a higher diagnostic yield. An average of 10 transbronchial biopsies from the transplanted lung are submitted for pathological examination at Papworth either when there is a clinical indication or for routine surveillance.[3,4] The usual indications are fever, cough, sputum, radiological abnormality and decline in pulmonary function tests. Radiological changes are less frequent and less specific in the months following transplantation, making clinical and functional changes more important.[10] Routine protocol biopsies are also performed, although not at all institutions; as yet there is no clear guide as to

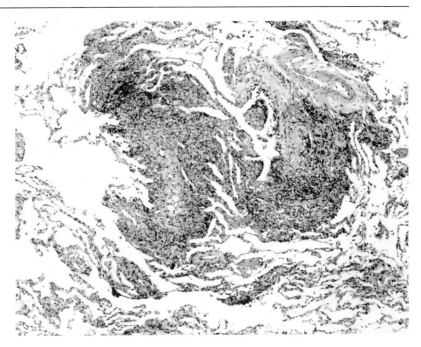

Fig. 5.3 Moderate acute pulmonary rejection with more extensive mononuclear cell infiltration of perivascular and interstitial distribution. Areas of normal parenchyma are seen indicating that this is still a discrete process. Accompanying bronchioles also showed acute rejection. Grade A3a. Haematoxylin and eosin stain; low power. Transbronchial biopsy.

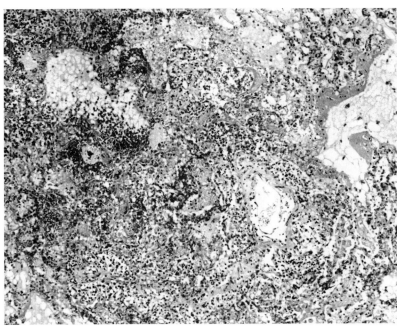

Fig. 5.4 Severe acute pulmonary rejection with confluent acute and chronic inflammatory cell infiltrates, endothelialitis, haemorrhage and hyaline membranes. Airways were also involved. Grade A4a. Haematoxylin and eosin stain; medium power. Autopsy section.

the optimum frequency of these. They do however yield an alarming number of positive findings, including significant rejection and infection.[4,11]

Acute rejection (grade A) is divided into four grades in the Working Formulation rejection cor-responding to minimal, mild, moderate and severe (Table 5.2).[9] The presence or absence of bronchiolar inflammation and large airway inflammation is indicated by the suffix a, b or c, respectively. Suffix d indicates that no bronchioles or bronchial

Table 5.2 Working formulation for the classification and grading of pulmonary rejection (Lung Rejection Study Group)[9]

Grade A: Acute rejection
1. Minimal acute rejection
2. Mild acute rejection
3. Moderate acute rejection
4. Severe acute rejection
   a. With evidence of bronchiolar inflammation
   b. Without evidence of bronchiolar inflammation
   c. With large airway inflammation
   d. No bronchioles are present

Grade B: Active airway damage without scarring
1. Lymphocytic bronchitis
2. Lymphocytic bronchiolitis

Grade C: Chronic airway rejection
1. Bronchiolitis obliterans – subtotal
2. Bronchiolitis obliterans – total
   a. Active
   b. Inactive

Grade D: Chronic vascular rejection

Grade E: Vasculitis

tissue were included. Grade A1 rejection is not obvious at scanning magnification with very infrequent small perivascular infiltrates composed mainly of mature lymphocytes (Fig. 5.1). Grade A2 shows large infiltrates easily recognizable at low magnification and accompanied by endotheliitis in many biopsies (Fig. 5.2). Although the perivascular and peribronchiolar adventitia may be expanded by the infiltrate, the differentiation between this and grade A3 is the lack of extension into the alveolar septae and spaces (Fig. 5.3). Grade A3 shows more frequent polymorphs and eosinophils but its lack of confluence distinguishes it from grade A4 rejection (Fig. 5.4). Frequent and high-grade acute rejection with inflammatory involvement of the airways is associated with development of obliterative bronchiolitis, a fibrosing occlusive disorder of small airways which limits the long-term survival of lung grafts.

Follow-up transbronchial biopsies are designated as (i) ongoing rejection if no change has occurred since the previous biopsy; (ii) resolving rejection if the infiltrates are reduced but still present; and (iii) resolved rejection if the infiltrates are eliminated.

Lymphocytic bronchitis (grade B1) and lymphocytic bronchiolitis (grade B2) are conditions where lymphocytic infiltration of airways is not accompanied by perivascular infiltration and therefore an A grade cannot be assigned (Fig. 5.5) It is not associated with fibrosis but may well represent the progenitor lesion of obliterative bronchiolitis.[9] Connective tissue stains are often essential for distinguishing grade B from grade C, obliterative bronchiolitis in which fibrosis must be present.

The grading of acute pulmonary rejection is based on perivascular mononuclear cell infiltrates and their semiqualitative and semiquantitative assessment. These infiltrates are not specific for rejection and have a wide differential diagnosis, the most important of which are infections, particularly cytomegalovirus and pneumocystis.[9] The grading of acute rejection therefore requires the rigorous exclusion of concomitant infection. This can be achieved by both histological and microbiological means.[12] Where rejection and infection appear to coexist and cannot be reliably distinguished, both increased immunosuppression and antimicrobial treatments may be required with re-biopsy to classify the episode correctly.

Lung allograft recipients have a very high infection rate both by common bacterial pathogens and by opportunistic organisms.[3,4,13,14] These infections could potentially limit the usefulness of a histological classification of acute rejection. One hundred consecutive transbronchial biopsies obtained from 43 heart–lung and single lung patients for clinical reasons and routine surveillance were analysed in conjunction with microbiological and serological data in order to assess the scale of the problem.[12] Serological studies and cultures for toxoplasma, adenovirus, influenza virus, mycoplasma, legionella and Epstein–Barr virus (EBV) were performed according to clinical suspicion and lung tissue and lavage fluid were also sent for viral culture. Microbiological data for all specimens taken 2 days before and 2 days following the transbronchial biopsy were also reviewed including sputa, blood cultures and throat swabs. Seventy-six of the 100 biopsies were assigned a rejection grade at the time of histological reporting, the commonest grade being A2a. The histological appearances of purulent inflammation in the airways and/or excessive numbers of polymorphs in the lung parenchyma were suspicious of infection in 22 of these biopsies (Fig. 5.6). Eight of these showed significant positive cultures. Of the 54 biopsies confidently graded as showing rejection with no histological suspicion of infection, 14 had evidence of positive cultures or serology. These did not show an excess of airways inflammation (suffix

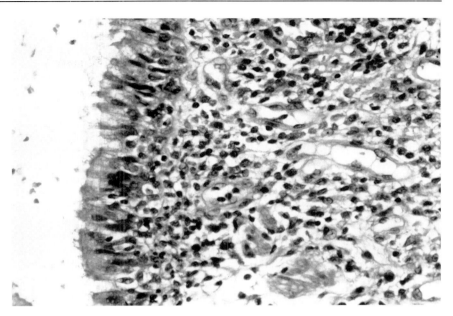

Fig. 5.5 Lymphocytic bronchiolitis with mononuclear cell infiltration of epithelium and subepithelial tissue. Bronchial epithelial cells remain intact. No perivascular infiltrates were present in this biopsy nor airway fibrosis. Grade B2. Haematoxylin and eosin stain; high power. Transbronchial biopsy.

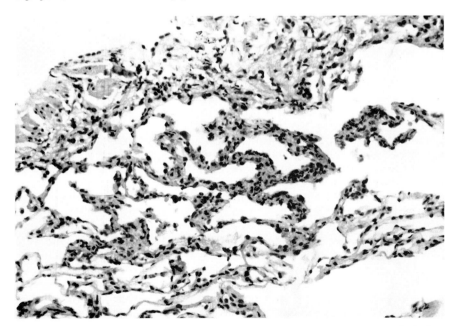

Fig. 5.6 Cellular infiltration of interstitium by mixed inflammatory cells, including frequent polymorphs. Alveolar epithelial cell hyperplasia is also present. The infiltration is not perivascular in distribution. These appearances are strongly suggestive of infection rather than rejection. Viral inclusions typical of cytomegalovirus were present in other lung fragments and in deeper serial sections of this piece. Haematoxylin and eosin stain; medium power. Transbronchial biopsy.

a or c). Where the respiratory infection was thought to involve mainly the upper airways and not involve bronchioles and lung parenchyma, a final rejection grade was not assigned in strict accordance with the Working Formulation. However the *initial* labelling of the episode as histologically proven rejection for the purposes of treatment was not inappropriate. In 24 transbronchial biopsies not graded for rejection two were inadequate (no lung parenchyma or previous biopsy site) and 22 had histological evidence of infection. This infection was confirmed by the appropriate culture or serology in 20 cases. The two cases appearing infective histologically but not confirmed by culture or serology showed non-specific pneumonitis and features suggestive of viral pneumonitis, respectively. Fifty-four biopsies in this small study were therefore able to be assigned a final rejection grade retrospectively, having excluded concomitant infection. The study also highlights the usefulness of transbronchial biopsy in the diagnosis of pulmonary infections in lung transplant recipients and the high specificity of the histological appearances of infection in ungradeable biopsies.[12]

Chronic pulmonary rejection

Again in common with other solid organ grafts, chronic rejection is a major factor in limiting graft survival.[4,5,8,15] Chronic rejection appears to be due to persistent, progressive and irreversible host immunological attack on the graft, even in the presence of maintenance immunosuppression. Improvements in surgical and perioperative techniques as well as early postoperative diagnosis of acute rejection and infections by transbronchial lung biopsy may have some beneficial effect in the long-term survival of the graft[4,5] As yet, however, there are few cohorts of long-term survivors and factors influencing prognosis are still being evaluated. Obliterative bronchiolitis is the manifestation of chronic airways rejection in the grafted lung.[4,5,8,9,15] There is growing evidence to support this aetiology, including its association with severe and frequent episodes of acute rejection.[5,16] There is also an association with acute rejection involving the small and large airways.[16] Persistent acute rejection which requires prolonged periods of augmented immunosuppression is a risk factor for the development for obliterative bronchiolitis.[5]

The distal respiratory epithelium is an immune target which has been shown to express class 2 antigens.[17] Increased expression of these antigens has been linked to the development of obliterative bronchiolitis, but it must be noted that the induction of class 2 antigen expression is not specific to rejection and may be caused by infectious agents including viruses and pneumocystis. There is a relationship between the extent of bronchiolar damage and the degree of histocompatibility mismatch in animal studies and also a clinical report that the longest surviving lung transplant recipient who did not develop obliterative bronchiolitis had the closest HLA match when compared with others from that institution.[18] In keeping with the theory that obliterative bronchiolitis is the manifestation of chronic rejection, augmented immunosuppression has been shown to slow the progression of this fibrosing condition.[5] Accurate diagnosis of acute rejection and other complications by transbronchial biopsy and prompt instigation of appropriate treatment have been shown to improve graft survival.[4,11] There are many causes of obliterative bronchiolitis in non-transplant patients, some of which may be relevant to lung allografts.[19]

Cytomegalovirus (CMV) can directly cause obliterative bronchiolitis. CMV infection has been reported as a risk factor in some lung transplant centres for development of obliterative bronchiolitis, particularly in seropositive recipients and mismatched recipients.[5,20,21] The association is strongest where there has been well-documented CMV pneumonitis.[21] The virus may release cytokines, increasing class 2 antigen expression on epithelial and endothelial cells and thus predisposing to further immune damage. Direct viral cytopathic effect on the graft and the formation of cross-reactive antibodies also have to be considered. Other viruses associated with obliterative bronchiolitis include respiratory syncytial virus, adenovirus and also mycoplasma and chlamydia.[19] Obliterative bronchiolitis is well described in the collagen vascular diseases and has also been reported in bone marrow transplant recipients experiencing graft versus host disease, both suggesting an immunological mechanism.[19] Other relevant non-immunological causes may include the ligation of bronchial circulation at operation. This may predispose to inadequate airway healing after either acute rejection or infection. Denervation and lymphatic interruption may also be contributory, particularly through the loss of the cough reflex increasing mucus retention and pulmonary infections.

Morphologically, obliterative bronchiolitis consist of fibrotic narrowing of the bronchiolar lumen which may be eccentric, concentric or totally obliterating.[8,9] The smooth muscle is often destroyed with extension of the fibrosis into the peribronchiolar interstitium (Fig. 5.7). In the early active phases there is mononuclear cell infiltration with epithelial damage including ulceration. As the process becomes more chronic the fibrosis is more or less acellular. The classification of chronic airway rejection (Table 5.1) takes account of the subtotal (C1) or total (C2) nature of the obliteration and the presence (a) or absence (b) of an inflammatory infiltrate (Table 5.2). Obliterative bronchiolitis can be diagnosed on transbronchial biopsy[22] but may be missed because of its patchy nature, particularly in the early stages of the disease. It may be missed altogether if trichrome or other connective tissue stains are not performed to demonstrate the airway fibrosis.[9] In open biopsies, re-transplants and autopsy material the small fibrosed bronchioles may only be evident on elastic stains which demonstrate their proximity to the accompanying pulmonary arterioles. Obliterative bronchiolitis in its pure form as a manifestation of chronic rejection does not extend into the distal airspaces.[23] The latter appearance favours a diagnosis of chronic organizing pneumonia or bronchiolitis obliterans organizing pneumonia, which is more often related to infection or aspiration with intra-alveolar granulation tissue. Obliterative bronchiolitis is, however, associated with an increased incidence of pulmonary infections. Finally, the two conditions may co-exist.

Chronic rejection affects the large airways distal to the tracheal or bronchial anastomoses. Bronchitis and bronchiectasis are the commonest findings with cylindrical dilatation of cartilage-containing airways and viscid retained secretions.[8] Squamous metaplasia usually accompanies the acute-on-chronic inflammation with lymphocytic infiltration of mucosa, submucosa and adventitia. Scarring of the mucosa and submucosa gradually replace submucosal glands. Fibrous replacement of smooth muscle is a likely factor in the development of bronchiectasis. Leu-7-positive lympho-

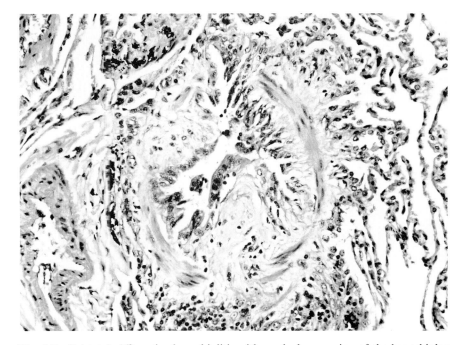

Fig. 5.7 Subtotal obliterative bronchiolitis with marked narrowing of the bronchiolar lumen by fibrous tissue internal to the bronchiolar smooth muscle, which has also been destroyed in places. The bronchiolar epithelium is cuboidal and non-ciliated. Few chronic inflammatory cells remain in the peribronchiolar adventitia indicating an inactive phase. Grade C1b. Haematoxylin and eosin stain; medium power. Open-lung biopsy.

cytes have been demonstrated in the donor tracheal and proximal bronchial epithelium in association with individual cell necrosis. In severe cases of chronic rejection the bronchi may be obliterated distally giving the appearance of obliterative bronchitis.

Chronic lung rejection is manifest by increased shortness of breath, cough, excessive sputum production, decreased exercise tolerance and irreversibly deteriorated pulmonary function. These symptoms and signs can be explained on the basis of progressive fibrosis of small airways. Chronic vascular rejection also occurs in pulmonary grafts but does not assume the clinical importance of chronic airways rejection.[9,15,24] It consists of fibro-intimal cellular proliferation of vessels of all sizes ranging from the large elastic to the smaller muscular pulmonary arteriolar vessels (Fig. 5.8). Initially it is a patchy process. Pulmonary veins are also involved and, like the chronic airways rejection, the process may be active or burnt out.[9] Active chronic vascular rejection shows marked cellularity with subendothelial and transmural infiltrate (Fig. 5.9). Veins tend to show more hyaline sclerosis with fewer cells. The changes can be detected in small vessels, particularly venules in transbronchial biopsy material, but the extent of involvement is best appreciated on open biopsy, re-transplant or autopsy material. Pulmonary vascular disease of this nature is associated with frequent and high-grade acute rejection and also with CMV infection.[24] The arteriosclerosis does not always correlate with obliterative bronchiolitis which is more common, but may correlate with coronary occlusive disease in combined heart–lung grafts. For the purpose of grading and classification a single category (D) is assigned to all pulmonary vascular occlusive disease. Although not of major clinical significance, grade D chronic vascular rejection is a common finding in transbronchial biopsies, even a few months after transplantation. We have an incidence of over 10% in transbronchial biopsies which includes grade D co-existent with grade C.

The main differential diagnosis of obliterative bronchiolitis in its active cellular phase is acute on chronic inflammation due to infection by a variety of organisms. Similar appearances may also be seen due to aspiration where there may be foreign material associated with a giant cell reaction and a distal organizing pneumonia.[23] Ischaemic injury to the small airways can also produce inflammatory fibrosis. The differential diagnosis of the chronic vascular changes include thrombosis and thromboembolism. Emboli are well documented in donor lungs.[2] The role of pre-existing donor

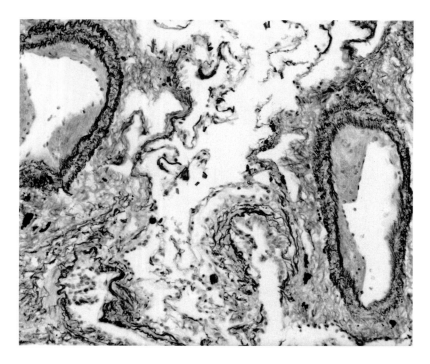

Fig. 5.8 Pulmonary artery branches showing fibrointimal proliferation without active inflammatory component. Chronic vascular rejection grade D. Elastic van Gieson (EVG) stain; low power. Transbronchial biopsy.

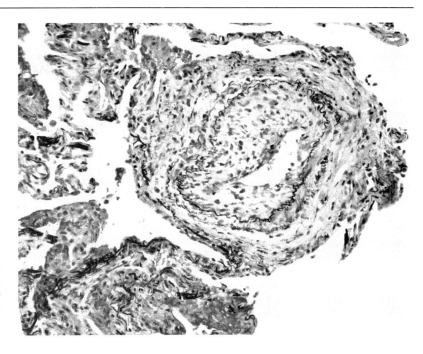

Fig. 5.9 Chronic vascular rejection showing transmural cellular infiltration and mild endothelialitis. Focal destruction and duplication of the internal elastic lamina is seen. EVG stain; medium power. Transbronchial biopsy.

pathology and preservation injury in the development of chronic rejection in the lung has not yet been fully evaluated. Endothelium insulted by cold, hypoxia, embolism or infection may subsequently be further damaged by acute and/or chronic rejection.

Vasculitis

Vasculitis reflects vessel injury by a transmural infiltrate of inflammatory cells – mainly mononuclear cells but including occasional neutrophils. This inflammatory infiltrate may be associated with necrosis of portions of the vessel wall. Under the classification of pulmonary rejection this separate category of grade E vasculitis is only used when this form of injury is disproportionate to other inflammatory changes in proximity to the vessels involved.[9] A further distinction from the vascular changes of acute pulmonary rejection is that the mononuclear and fibrinoid vasculitis usually involves larger vessels than venules and arterioles. Consequently this pathological change is rarely seen on transbronchial biopsy material, being more likely to be picked up on open biopsy or autopsy material. It should be noted, however, that the use of open biopsy for diagnosing acute complications of lung transplants has diminished.

Infection in the grafted lung

Lung transplant recipients are particularly prone to infectious complications by both common and opportunistic pathogens.[13,14] The common bacterial infections are increased after augmented immunosuppression, viral infections and establishment of obliterative bronchiolitis. Their diagnosis does not usually require transbronchial biopsy material. Bacterial infections, usually Gram-negative, constitute the largest group of infections but are less often fatal than viral and fungal infections.[13]

Cytomegalovirus

Cytomegalovirus is one of the most important opportunistic pathogens of lung transplants, causing morbidity and mortality as well as modulating rejection through immune mechanisms.[4,5,20,21] The prevalence of CMV pulmonary infection has been reported to exceed 75% amongst lung transplant recipients who survive at least 2 weeks.[21] Such infection can be primary (donor-transmitted) or due to reactivation.[20] Donor and recipient matching for CMV serological status has reduced the incidence of primary CMV considerably. Occasional transmission from unscreened blood products may occur. Prior to CMV matching policies, primary CMV infection caused a fatal pneumonitis in most cases with evidence of systemic involvement.[5,20] A much more common problem is CMV reactivation in a seropositive recipient. It is important to distinguish CMV viral inclusions indicating seropositivity or infection from active CMV pneumonitis where the viral inclusions are associated with acute inflammation.[3,25] This distinction is very important. It is not uncommon to see an isolated cell in a transbronchial biopsy containing intranuclear and intracytoplasmic CMV inclusions not associated with any inflammatory reaction. In the absence of any associated evidence for active CMV disease, e.g. fever, leucopenia, GI symptoms, this probably does not require specific antiviral treatment.[20] In contrast, CMV pneumonitis shows the features of a diffuse viral alveolitis with perivascular oedema, neutrophil microabcessess, alveolar cell hyperplasia and sometimes hyaline membranes (Fig. 5.10). The characteristic intranuclear owl's eye and

cytoplasmic granular inclusions enable a firm diagnosis to be made. A pre-inclusion stage can be recognized by the above features of viral alveolitis and distinguished from rejection by the presence of perivascular oedema and neutrophil microabcesses (Table 5.3). Tight perivascular mononuclear cell cuffing is not a feature of CMV pneumonitis in the lung transplant recipient. The presence of such infiltrates indicates coexistent rejection which, by definition, is ungradeable in the presence of infection. Sampling error and small transbronchial biopsies showing intermediate features can be confusing and CMV therefore remains a challenging diagnostic problem. Patients receiving antiviral therapy with ganciclovir or acyclovir do not show well-developed viral inclusions. The infected cell nuclei appear rather eosinophilic and degenerate without the owl's eye and halo. Cytoplasmic inclusions may also be inconspicuous. The degenerate inclusions may be difficult to recognize as viral in origin and may also be confused with those

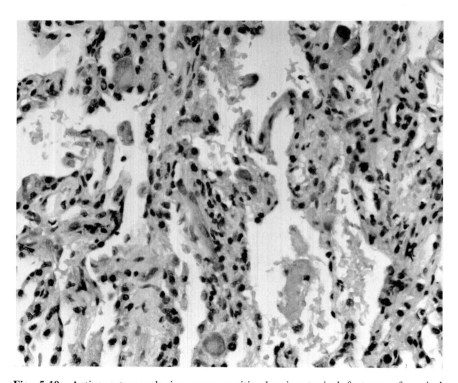

Fig. 5.10 Active cytomegalovirus pneumonitis showing typical features of a viral alveolitis. Frequent polymorphs, lack of perivascular distribution and the presence of characteristic viral inclusions distinguish from acute rejection. Some intranuclear inclusions appear degenerate – a common finding in patients who have had ganciclovir prophylaxis. Haematoxylin and eosin stain; high power. Transbronchial biopsy.

Table 5.3 Histological differentiation of acute pulmonary rejection and CMV pneumonitis

Rejection	CMV Pneumonitis
Tight perivascular infiltrates mainly mononuclear cells	Infiltrates not tightly perivascular; usually n external adventitia only
Polymorphs and eosinophils in moderate – severe grade	Polymorphs common and may predominate; eosinophils rare; polymorphs in alveolar walls
Perivascular oedema absent or mild	Perivascular oedema prominent
No microabscesses	Microabscesses common
Alveolar cell hyperplasia in high-grade rejection	Alveolar cell hyperplasia common
No viral inclusions	Intranuclear and intracytoplasmic viral inclusions in macrophages, alveolar and endothelial cells
Lymphocytic endotheliitis	Lymphocytes and polymorphs in endothelium, often related to inclusions

of herpes simplex virus (HSV). Special techniques including immunohistochemistry and *in situ* hybridization may be useful in these circumstances.[26,27] However, techniques which increase sensitivity of detection do not solve the dilemma of differentiating simple infection with the virus from active viral pneumonitis. An additional confusion is that some patients suffer CMV and HSV infections concurrently.[28] Follow-up biopsies in patients with CMV pneumonitis may show persistent degenerate inclusions with decreasing severity of pneumonitis. When the CMV pneumonitis has been successfully treated there is an increased risk of respiratory super-infections and mortality.[22] Patients who have biopsy-proven CMV pneumonitis are at significant increased risk of bacterial and fungal pneumonias and also of developing chronic rejection. Recent evidence shows that ganciclovir treatment of CMV infections decreased respiratory super-infections and improved patient survival but did not prevent the development of chronic rejection. Ganciclovir prophylaxis is now being used in several institutions with a reduction in the incidence of CMV infections and this may have a beneficial effect on long-term complications.[21] It may also decrease the unexpectedly high incidence of CMV pneumonia in surveillance biopsies.[11]

Pneumocystis carinii

Pneumocystis carinii is an important opportunistic pathogen in lung transplants with an unexpectedly high incidence[29] until prophylaxis was introduced to cover periods of augmented immunosuppression. Transbronchial biopsies have not shown the typical histological picture of *Pneumocystis carinii* pneumonia (PCP) with abundant intra-alveolar foamy exudate and variable interstitial inflammation as seen in AIDS and other immunosuppressed patients.[30] In contrast, lung recipients show a granulomatous response to pneumocystis as the most common histological pattern in which relatively small numbers of cysts can be identified only on silver staining and meticulous screening. Well-defined epithelioid granulomas may be present but often the granulomatous inflammation is less distinct. A useful pointer is the presence of numerous plasma cells and focal organizing intra-alveolar granulation tissue. Occasionally small amounts of foamy exudate can be seen in association with the giant cells and granulomas. PCP can show perivascular infiltrates exactly mimicking acute pulmonary rejection.[31] This is the reason for the mandatory use of a silver stain in the recommendations of the lung rejection study group.[9]

Aspergillus

Aspergillus infection has emerged as a significant problem in lung transplant recipients. It produces a range of pathological manifestations with some similarity to those seen in non-immunosuppressed patients.[32] It can be divided into saprophytic and invasive disease (Table 5.4).[33] Large airways are frequent sites of colonization, particularly when these become bronchiectatic as a result of repeated infections and the development of obliterative bronchiolitis. In this circumstance fungus may be cultured from sputum or identified in aspirates. Another manifestation of non-invasive *Aspergillus* disease of the large airways is tracheobronchial obstruction due to a mass of fungal hyphae. This obstructive tracheobronchial aspergillosis (OTBA) can be recognized at bronchoscopy, lavaged and eradicated by appropriate antifungal therapy.[33] *Aspergillus* can also cause bronchocentric granulomatous mycosis in lung transplant patients.[34] The

Table 5.4   Histopathological classification of *Aspergillus* disease in transplanted lungs

Classification	Large airways	Small airways	Parenchyma
Non-invasive	Saprophytic colonization	Saprophytic colonization	Colonization of cavity
	Obstructive tracheobronchial aspergillosis (OTBA)		
Minimally invasive	Bronchocentric granulomatous mycosis (BCG)	BCG	Colonization of cavity; focal invasion of wall
Invasive	Pseudomembranous tracheobronchitis		Suppurative pneumonia with or without cavitation and abscess formation
	Invasion of ischaemic tracheobronchial wall		
	Aspergillus bronchitis	*Aspergillus* bronchiolitis	

fungal fragments are associated with a giant cell granulomatous response which may be associated with a distal eosinophilic pneumonia (Fig. 5.11). The small airways involved may show replacement of the mucosa by granulomatous inflammatory tissue and it may be associated with bronchiectasis. It has been successfully treated in one of our patients by removal of the involved upper lobe. In

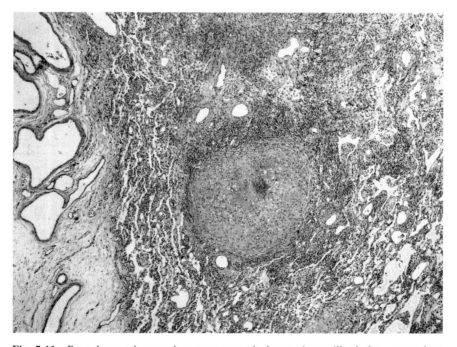

Fig. 5.11   Bronchocentric granulomatous mycosis due to *Aspergillus* in lung transplant recipient. The bronchiolar wall is replaced by granulomatous tissue with necrotic material in the lumen which contained sparse fungal hyphae on a Grocott methanamine silver stain. Adjacent parenchyma is infiltrated by numerous eosinophils. This lesion is subpleural adjacent to pleural adhesions. Haematoxylin and eosin stain; medium power. Resected transplanted right upper lobe.

the setting of immunosuppression, bronchocentric granulomatous mycosis has the potential for becoming invasive and such patients should be monitored carefully. The most lethal manifestations of *Aspergillus* infections are those that involve tissue invasion. In the airways *Aspergillus* may become locally invasive, particularly at the site of the anastomosis, with ulceration and pseudomembrane formation.[35] Ischaemic cartilage is prone to invasion and this is a dreaded complication of airways ischaemia, particularly in single and double lung transplant recipients with less well vascularized anastomoses than combined heart–lung patients. Progression of the fungal invasion and necrosis may lead to anastomotic dehiscence, erosion of the pulmonary artery

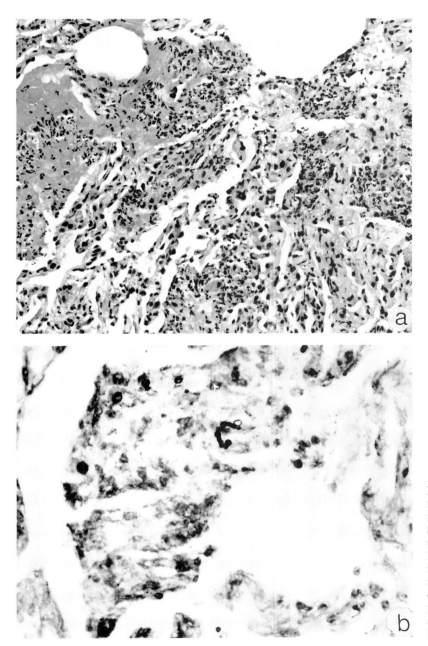

Fig. 5.12a,b Suppurative pneumonia with necrosis, fibrin and polymorphs in interstitium and air spaces. Special stains revealed hyphae characteristic of *Aspergillus*. Grocott stain; high power. (a) The patient had ischaemia of the bronchial anastomosis with invasive *Aspergillus* bronchitis, bronchiolitis and bronchopneumonia. Haematoxylin and eosin stains; medium power. (b) Transbronchial biopsy.

and/or systemic dissemination. In single lung transplants this locally invasive disease is confined to the grafted side, emphasizing the importance of local defence factors. The dissemination from ischaemic airways can lead to suppurative *Aspergillus* pneumonia in the ipsilateral lung with parenchymal necrosis, a tendency to vascular invasion and extrathoracic dissemination (Fig. 5.12). Invasive *Aspergillus* disease can arise on the basis of colonization of a parenchymal cavity either due to infarction or possibly a previous biopsy site. The very different prognosis of the various manifestations of *Aspergillus* infection in lung transplants requires accurate diagnosis when fungal hyphae are found in sputum, lavage and transbronchial biopsy specimens. This is particularly important in view of the toxicity of antifungal therapy. Some cases of *Aspergillus* infection are only detected at autopsy and in others the extent of dissemination is underestimated during life.[36]

Herpes simplex virus

In the lung transplant population HSV infection can be largely avoided by the use of prophylactic acyclovir in seropositive patients.[28] It is much less common than CMV, usually follows a period augmented immunosuppression, and is associated with characteristic intra-oral lesions. The mucosal herpetic lesions are not present in all patients however and the diagnosis of unsuspected HSV pneumonia can be difficult. Herpes simplex causes ulcerating tracheobronchitis and necrotizing bronchopneumonia (Fig. 5.13) with dissemination to other organs including the brain.[28] The presence of necrosis with abundant polymorph debris centred around airways with a bronchopneumonic distribution is very helpful in making the diagnosis. Viral inclusions are less easy to find than in CMV pneumonia but when present the lack of both cytomegaly and cytoplasmic inclusions is useful in making the distinction between these two viral infections (Table 5.5). Immunohistochemical and DNA probe methods can further differentiate the two. Intranuclear ground-glass inclusions are not usually as well developed as those in oral or genital herpes and multinucleation is uncommon. Virally infected cells are more readily found in the accompanying bronchoalveolar lavage where the intensely purulent and necrotic nature of the material should promote a search for HSV. The pattern of parenchymal necrosis in biopsies should be distinguished from the coagulative necrosis preserving cell outlines in lymphoproliferative disease.[3] Both of these conditions may produce pulmonary nodules, the former centred on airways, and both may show concomitant CNS symptoms and signs.

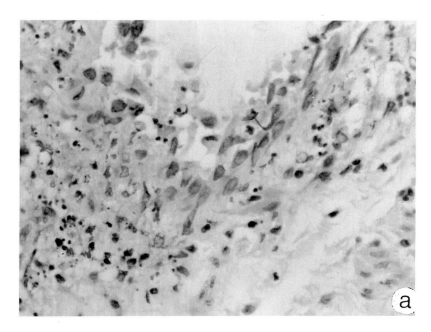

Fig. 5.13a Herpes simplex viral inclusions in bronchial epithelium associated with polymorphs and focal necrosis.

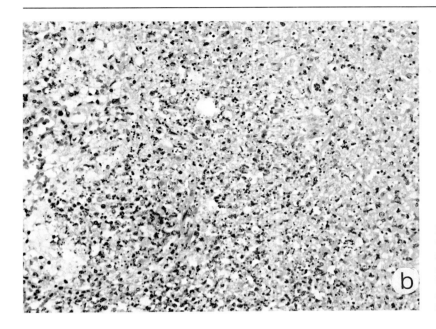

Fig. 5.13b Parenchymal necrosis with abundant polymorphs and pyknotic debris. Intranuclear inclusions are noted in some cells with occasional binucleate forms. Haematoxylin and eosin stain: (a) high power; (b) medium power. Autopsy sections.

Table 5.5 Histopathological differentiation of CMV and HSV pneumonitis

	CMV	HSV
Viral inclusions	Intranuclear	Intranuclear
	Haematoxyphilic	Eosinophilic
	'Owl's eye'	Dense nuclear membrane
	Halo	May be multiple
	Single	Multinucleation unusual
	Intracytoplasmic	No cytoplasmic inclusions
	Eosinophilic	
	Granular	
Infected cells	Alveolar epithelial	Bronchial epithelial
	Endothelial	Metaplastic squames
	Alveolar macrophage	Alveolar epithelial (rare)
Distribution	Interstitial pneumonitis	Bronchopneumonic pattern
	DAD	DAD
Inflammatory cells	Mononuclear with prominent polymorphs	Polymorphs with suppuration
	Microabscesses	Microabscesses
	Necrosis uncommon	Abundant necrosis with pyknosis

CMV, cytomegalovirus; DAD, diffuse alveolar damage; HSV, herpes simplex virus.

Mycobacterial infections

Mycobacterial infections occur in lung allografts but are uncommon compared with other opportunistic infections. Both typical and atypical myco-bacterial infections have been described[37] and patients with obliterative bronchiolitis appear to be the most at risk. Special stains for mycobacteria should be employed when granulomatous inflammation is seen in any biopsy or autopsy material from lung transplant patients or where clear or foamy macrophages are in abundance. Patients transplanted for pulmonary sarcoidosis may have recurrence of granulomatous inflammation in the new graft.[3] This is only compatible with recurrent primary disease if infection has been meticulously excluded. The presence of necrosis would also exclude recurrence of sarcoidosis.

Toxoplasmosis

Toxoplasmosis has also been reported both as a primary (donor-transmitted) and recrudescent infection but the organisms have not been seen on biopsy material.[38]

Epstein–Barr virus

This virus can cause an acute pneumonitis both in immunosuppressed and non-immunosuppressed patients. In the latter it has been described as showing dense interstitial cellular infiltration of the lungs with atypia and necrosis. A perivascular distribution of the infiltrate may also be a prominent feature. This process must be distinguished from both the infiltrates of acute rejection and those of post-transplant lymphoproliferative disease (LPD). This may be difficult with small biopsy material. Unfortunately studies of clonality and viral serology may not be helpful in making these distinctions.[39]

Other differential diagnoses of acute rejection

These include inflammation and scarring of a previous biopsy site, recurrent primary disease including sarcoidosis, Langerhans' cell histiocytosis and cryptogenic fibrosing alveolitis.[3,9] Bronchus-associated lymphoid tissue (BALT) can easily be confused with lymphocytic bronchitis or bronchiolitis.[9] It may be quite prominent in the first few months following transplantation with identifiable follicular and parafollicular areas. The striking vascularity with high endothelial venules helps differentiate this from rejection. Hyperplasia of the BALT may cause stenosis or obstruction as seen in non-immunosuppressed patients with diffuse panbronchiolitis. In lung allograft recipients the BALT becomes depleted over time and may be another factor in the increased susceptibility to infection.[40]

Bronchoalveolar lavage

Bronchoalveolar lavage (BAL) is useful for the diagnosis of opportunistic infections in immunosuppressed hosts. Cytology of the lavage fluid has the advantage of high specificity for non-bacterial agents and a fast turn-around time.[41] Fungus and herpes simplex virus are often readily identified in the lavage but not in the accompanying transbronchial biopsy. In the diagnosis of complications in lung grafts lavage and parenchymal biopsies should be regarded as complementary to each other. The significance of organisms identified in the aspirate can often be determined by the pathological findings in the biopsy. It is very unusual to find CMV-infected cells in the cytological preparation in the absence of pneumonitis in the biopsy. Immunohistochemical and *in situ* hybridization techniques applied to the cytological specimens may enhance sensitivity but are usually confirmatory rather than providing further diagnosis.

Bronchoalveolar lavage allows the examination of the immunocompetent cells infiltrating the graft itself.[42,43] Cellular profiles of lavage fluid have been described in rejection and infection but are disappointingly non-specific.[44] There is no correlation with the accompanying biopsy rejection grade where paired samples have been examined. In other studies the cytological profiles have not been compared with a biopsy 'gold standard'.[43] The allograft lavage fluids are more cellular than those of a control group of non-transplant patients and increase further during episodes of illness due to rejection or infection of the lung. Lymphocyte counts greater than 15% of the total cell count are supportive of a diagnosis of rejection but are found in only 23% of biopsy-confirmed rejection episodes.[44] Neutrophils are non-specifically increased in both rejection and infection. The cellular profiles return to those of healthy lung transplant recipients after treatment of rejection or infection. Donor lymphocytes in the graft are largely replaced by recipient cells in the first 6 weeks after transplantation but macrophages and lymphocytes of donor origin may persist for as long as 32 weeks.[42] The replacement of donor by recipient lymphocytes may explain the lower incidence of acute rejection in the first few months after transplantation. Cytolytic activity of the lavage fluid against donor spleen cells has been demonstrated with some predictive value for rejection. Spontaneous proliferation and interleukin 2 responsiveness is also observed but these functional studies have the same considerable overlap as the absolute cell counts and differentials.[42] During acute rejection episodes a predominance of CD4 cells directed against donor-specific class 2 antigens has been demonstrated in the lavage, contrasting with the preponderance of CD8 cells reactive against class 1 antigens in grafts showing obliterative bronchiolitis.[45] It is likely that the lack of specificity, and particularly the overlap with infection, will limit the usefulness of phenotypic and functional studies and also measurements of cytokines in the lavage fluid.

Lymphoproliferative disorders

Lung allograft recipients, in common with other graft recipients, are at increased risk of developing lymphoproliferative disease; this appears to be related to Epstein–Barr virus infection.[39] Heart–lung and lung grafts are maintained with a relatively high level of immunosuppression compared with other solid organ grafts. Lympho-

proliferative disease (LPD) has been described as occurring soon after transplantation and shows a predilection for the graft.[6,46] It may be an unexpected finding on biopsy material or more commonly presents as new radiological or CT shadows, in the presence of non-specific clinical signs. In transbronchial, open biopsy or autopsy material the disease is manifest by sheet-like cellular infiltrates composed of lymphoid cells with varying proportions of mature and blast-like components.[6,39,47] These infiltrates often have a perivascular distribution strikingly similar to both EBV (Fig. 5.14) pneumonitis and acute lung rejection.[9] In lymphoproliferative disease there is involvement of the full thickness of the vessel wall with more conspicuous endothelial infiltration and sometimes fibrinoid necrosis. Coagulative necrosis of the tumour cells is common and the finding of monotonous 'ghost' cells on transbronchial or fine needle. Aspiration biopsy material should alert the histopathologist to the possibility of LPD, which is usually of B cell origin and can be monoclonal or polyclonal.[39] Some cases probably represent a florid viral pneumonitis on the background of

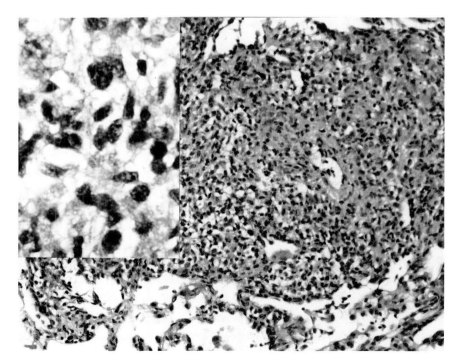

Fig. 5.14 Lymphoproliferative disease showing sheet-like infiltration from a perivascular position and transmural infiltration of the vessel. Atypia and necrosis help distinguish this from acute rejection. Haematoxylin and eosin stain; medium power. Transbronchial biopsy.

exogenous immunosuppression, particularly those cases with polymorphic polyclonal cellular infiltrates. However monomorphic monoclonal proliferations at the other end of the spectrum, have the features of malignant non-Hodgkin's lymphoma. Polyclonal and monoclonal lesions can exist in the same patient concurrently so the classi-fication of LPD in the individual patients has been extremely difficult.[39,48] Post-transplant lympho-proliferation must be distinguished from non-specific lymphoid hyperplasia. This is difficult even in lymph node material and can be virtually impossible in small biopsies. Having identified a probable case of LPD it is essential to try to define its position on the clinico-pathological spectrum of disease. The benign polyclonal hyperplasias often respond to a simple reduction in immuno-suppression and antiviral treatment with acyclovir.[46,47] Chemotherapy is usually considered inappropriate for this variant as it does not behave as a malignant lymphoma in most cases. Those cases that do behave as malignant lymphomas may be acyclovir resistant and are commonly solid tumours. Fatal cases may involve dissemination outside the lung to numerous sites including the central nervous system. Recent evidence has shown the OKT3 monoclonal antibody immunosuppres-sion increases the incidence of LPD, possibly by producing cytokines which favour the oncogenic potential of EBV.[49]

Lung transplant autopsy

Early deaths of heart–lung and lung recipients are usually related to primary donor organ failure or perioperative complications (Table 5.6).[1,8] The former is often related to the combination of pre-servation injury, donor infection and/or aspir-ation and may be pathologically manifest as adult respiratory distress syndrome. This may also be the pathological appearance of some perioperative complications, particularly excessive bleeding and patients with sepsis. Deaths due to acute pulmon-ary rejection are very rare owing to intensive post-operative surveillance of early complications.[1,3,8] Other causes of early death include airway dehiscence, which is unusual in the case of com-bined heart–lung grafts and is associated with ischaemia and/or fungal infection in single or double lung grafts. Recurrent thromboemboli contributed to the early death of a patient given a combined heart–lung transplant for thrombo-

Table 5.6 Causes of death in heart–lung and lung recipients

Early	Late
Donor organ failure	Obliterative bronchiolitis
preservation injury	Aspergillosis
infection	Disseminated herpes
aspiration	simplex
	Pneumocystis carinii
Perioperative complications	pneumonia
bleeding	Lymphoproliferative
Anastomotic dehiscence	disease
Bacterial pneumonia	Renal failure
CMV pneumonia[a]	Myocardial infarction due
Multiorgan failure	to coronary occlusive
Bowel perforation	disease[b]

[a] Primary mismatch.
[b] Combined grafts only.

embolic pulmonary hypertension. Heart–lung and lung recipients are prone to develop gastrointest-inal complications including small and large bowel ischaemia which may require resection. Often no specific cause is identified but opportunistic infection should always be excluded. Cystic fibro-sis patients can develop meconium ileus equivalent with intestinal obstruction but this usually res-ponds to conservative management and should not cause fatal complications. Dehiscence of the aortic anastomosis in combined grafts is usually associ-ated with infection and may cause unheralded fatal haemorrhage. All these early deaths are likely to be within the transplant centre but later deaths may present to pathologists in other institutions.

The commonest cause of late mortality in heart–lung and lung transplantation patients is chronic rejection.[1,3,5,8] Obliterative bronchiolitis and its accompanying bronchiectasis is associated with an inexorable decline in lung function with an increase in respiratory infections. Death is usually due to pneumonia from bacterial pathogens, commonly Gram-negative or *Aspergillus*. At autopsy the ap-pearances are usually obviously pneumonic with heavy, patchily consolidated lungs associated with an infected tracheobronchial tree. Intrapulmonary abscess formation is not uncommon. The terminal pulmonary infections may be associated with pleural disease such as effusion or empyema, but dense fibrosis has often obliterated the pleural cavities within a few months of transplantation. This fibrosis, together with extensive mediastinal fibrosis, can make these autopsies technically very difficult. Cut slices of the lungs, ideally after inflation fixation, may show the cord-like fibrosed bronchioles in areas not severely affected by pneumonia. Bronchiectasis is easily seen in these specimens but can be demonstrated by specimen bronchography if required.[8] In these late deaths all the anastomotic suture lines should be intact and well healed over. Occasionally small diverticula are seen in relation to the tracheal and bronchial anastomosis.

The presence of pulmonary nodules at autopsy raises three main possibilities. (1) The lungs may be involved with lymphoproliferative disease. Such nodules are variable in size and may involve all transplanted lobes. Necrosis may be evident macroscopically. Enlargement of lymph nodes, both intra- and extra-thoracic, should be sought. The spleen may be enlarged and there may be deposits of lymphoma elsewhere including liver, kidneys, adrenals and brain. (2) The necrotizing bronchopneumonia of herpes simplex infection can also produce nodules. These are generally more necrotic to the naked eye than those of lymphoma and can often be seen to be centred on a small airway with purulent exudate. (3) *Aspergillus* infection can result in pulmonary nodules either as colonization of previous cavities or consolidation of bronchopneumonia and local pulmonary infarction. In patients who have had transbronchial biopsies performed a few days before death the sites of these biopsies may be identified as haemorrhagic subpleural nodules. An unsuspected *Pneumocystis carinii* pneumonia has been seen in a late death occurring beyond the periods of prophylaxis.

Macroscopic abnormalities of the pulmonary vessels are uncommon compared with the airway changes.[24] Occasionally thickening of the proximal vessels with yellow lipid-rich concentric plaques can be identified. Evidence of thrombosis and embolism is rare but should be sought all cases, particularly those transplanted for apparently primary pulmonary hypertension.

Combined heart–lung grafts can suffer coronary occlusive disease which may be severe enough to cause fatal myocardial infarction. This is usually seen in association with severely restricting obliterative bronchiolitis. It remains to be seen whether coronary occlusive disease will become a significant clinically problem in heart–lung recipients surviving into the medium and long term.

It is imperative for the assessment of innovative therapies such as cardiac and pulmonary trans-

plantation that a high autopsy rate is maintained. The originating transplant centre should always be informed where possible of autopsy findings so that a profile of the natural history of these grafts can be established.

Recurrent disease

Many of the conditions leading to pulmonary transplantation are of unknown aetiology and the likelihood of recurrence in the graft is not known. Certainly sarcoidosis appears to have a high recurrence rate. The diagnosis can only be made after meticulous exclusion of opportunistic infection, which is a common cause of granulomatous inflammation in these patients. The presence of necrosis within the granulomas excludes the diagnosis of recurrent sarcoidosis and has been seen in a patient transplanted for clinico-pathologically diagnosed sarcoidosis who succumbed to *Mycobacterium tuberculosis* infection 3 years later. The early pre-granulomatous stages of sarcoidosis which show lymphocytic alveolitis and perivascular mononuclear cell infiltration cannot be distinguished reliably from rejection.[9] Similarly the early changes of cryptogenic fibrosing alveolitis which are non-specific cannot be distinguished from rejection. It would appear on present evidence that this condition does not recur in the graft. At autopsy the appearance of obliterative bronchiolitis and any associated organizing pneumonia is quite distinct from the end-stage appearances of cryptogenic fibrosing alveolitis. In particular microcystic or honeycomb change is not a feature of long-term lung allografts. Other diseases which could be expected to recur include Langerhans' cell histiocytosis, the pulmonary vasculitides and pulmonary hypertension, especially thromboembolic. Lung transplant recipients are discouraged from smoking and it is unlikely that in the lifespan of a graft significant emphysema could develop. End-stage obliterative bronchiolitis and bronchiectasis in many ways resembles the appearances of advanced cystic fibrosis. It has been shown that the epithelial defects that result in raised airway potential difference in cystic fibrosis do not apparently recur in the transplanted lung.[50] In practical terms it is essential to be aware of the primary disease of patients when interpreting transbronchial biopsies from the graft as the differential diagnosis of rejection may be enlarged.

Graft-Versus-Host Disease

The transplanted lung contains abundant lymphoid tissue mainly in association with the airways. Peripheral blood and bronchoalveolar lavage fluid have shown lymphocyte chimerism post-operatively until the recipient's own cells have replaced those in the graft. This does not appear to have clinical significance in the majority of patients. Some however exhibit skin rashes, bone marrow failure, diarrhoea and increased susceptibility to infection which would be consistent with graft-versus-host disease. The histological features however lack specificity and infections, particularly CMV, may produce similar appearances. The inclusions of the latter may not be easy to find if ganciclovir prophylaxis or treatment has been given. One patient with persistent chimerism developed colonic perforation with the histological features of pseudomembraneous colitis. No toxins or organisms were demonstrated. Extensive invasive *Apergillus* bronchopneumonia was found at autopsy and lymphoid tissue was noted to be depleted throughout the body. It may be that this is a case of graft-versus-host disease contributing to death.

References

1. Kriett JM, Kaye MP. The Registrar of the International Society for heart and lung transplantation: eighth official report - 1991. *J Heart Lung Transplant* 1991; **10**, 491–498.
2. Stewart S, Ciulli F, Wells F *et al*. The pathology of unused donor lungs. *Transplant Proc* 1993; **25**, 1167–1168.
3. Stewart S. Pathology of lung transplantation. *Semin Diagn Pathol* 1992; **9**, 210–219.
4. Higenbottam TW, Stewart S, Penketh AR *et al*. Transbronchial lung biopsy for the diagnosis of rejection in heart–lung transplant patients. *Transplantation* 1988; **46**, 532–539.
5. Scott JP, Higenbottam TW, Sharples L *et al*. Risk factors of obliterative bronchiolitis in heart–lung transplant recipients. *Transplantation* 1991; **51**, 813–817.
6. Yousem SA, Randhawa P, Locker J *et al*. Post-transplant lymphoproliferative disorders in heart–lung transplant recipients. *Hum Pathol* 1989; **20**, 361–369.
7. Prop J, Wildevuur CRH, Nieuwenhuis P. Lung allograft rejection in the rat. III. Corresponding morphological rejection phase in various rat strain combinations. *Transplantation* 1985; **40**, 132–136.
8. Tazelaar HD, Yousem SA. The pathology of combined heart–lung transplantation: an autopsy study. *Hum Pathol* 1988; **19**, 1403–1416.
9. Yousem SA, Berry GJ, Brunt EM *et al*. A working formulation for the standardisation of nomenclature in the diagnosis of heart and lung rejection: lung rejection study group. *J Heart Transplant* 1990; **9**, 593–601.
10. Millet B, Higenbottam TW, Flower CDR *et al*. The radiographic appearances of infection and acute rejection of the lung after heart–lung transplantation. *Am Rev Respir Dis* 1989; **140**, 62–67.
11. Trulock EP, Ettinger NA, Brunt EM *et al*. The role of transbronchial lung biopsy in the treatment of lung transplant recipients. *Chest* 1992; **102**, 1049–1054.
12. Hunt JB, Stewart S, Cary N *et al*. Evaluation of the International Society for Heart Transplantation grading of pulmonary rejection in 100 consecutive biopsies. *Transplant Int* 1992; **5** (suppl 1), S249–S251.
13. Maurer JR, Tullis E, Grossman RF *et al*. Infectious complications following isolated lung transplantation. *Chest* 1992; **101**, 1056–1059.
14. Dummer SJ, Montero CG, Griffith BP *et al*. Infections in heart–lung transplant recipients. *Transplantation* 1986; **41**, 325–329.
15. Yousem SA, Burke CM, Billingham ME. Pathologic pulmonary alterations in long-term heart–lung transplantation. *Hum Pathol* 1985; **16**, 911–923.
16. Yousem SA, Dauber JA, Keenan R *et al*. Does histologic acute rejection in lung allografts predict the development of bronchiolitis obliterans? *Transplantation* 1991; **52**, 306–309.
17. Yousem SA, Curley JM, Dauber JA *et al*. HLA class II antigen expression in human heart–lung allografts. *Transplantation* **49**, 991–995.
18. Harjula AJL, Baldwin JC, Glanville AR *et al*. Human leukocyte antigen compatibility in heart–lung transplantation. *J Heart Transplant* 1987; **6**, 162–166.
19. Epler GR, Colby TV. The spectrum of bronchiolitis obliterans. *Chest* 1983; **83**, 161–162.
20. Smyth RL, Scott JP, Borysiewicz LK *et al*. Cytomegalovirus infection in heart–lung transplant recipients: risk factors, clinical associations and response to treatment. *J Infect Dis* 1991; **164**, 1045–1050.
21. Duncan SR, Paradis IL, Yousem SA *et al*. Sequelae of cytomegalovirus pulmonary infections in lung allograft recipients. *Am Rev Respir Dis* 1992; **146**, 1419–1425.
22. Yousem SA, Paradis IL, Dauber JH *et al*. Efficacy of transbronchial lung biopsy in the diagnosis of bronchiolitis obliterans. *Transplantation* 1989; **47**, 893–895.
23. Abernathy EC, Hruban RH, Baumgartner WA *et al*. The two forms of bronchiolitis obliterans in heart–lung transplant recipients. *Hum Pathol* 1991; **22**, 1102–1110.

24. Yousem SA, Paradis IL, Dauber JH *et al.* Pulmonary arteriosclerosis in long-term human heart–lung transplant recipients. *Transplantation* 1989; **47**, 564–569.

25. Stewart S, Higenbottam TW, Hutter JA *et al.* Histopathology of transbronchial biopsies in heart–lung transplantation. *Transplant Proc XX* (suppl) 1988; 764–766.

26. Niedobiteck G, Finn T, Herbst H *et al.* Detection of cytomegalovirus by *in situ* hybridisation and histochemistry using a monoclonal antibody CCH2: 1 comparison of methods. *J Clin Pathol* 1988; **41** 1005–1009.

27. Weiss LM, Movahed LA, Berry GJ *et al.* *In situ* hybridization studies for viral nucleic acids in heart and lung allograft biopsies. *Am J Clin Pathol* 1990; **93**, 675–679.

28. Smyth RL, Higenbottam TW, Scott JP *et al.* Herpes simplex virus infection in heart–lung transplant recipients. *Transplantation* 1990; **49**, 735–739.

29. Gryzan S, Paradis IL, Zeevi A *et al.* Unexpectedly high incidence of *Pneumocystis carinii* infection after heart–lung transplantation: implications for lung defence and allograft survival. *Am Rev Respir Dis* 1988; **137**, 1268–1274.

30. Travis WD, Pittaluga S, Lipschik GY *et al.* Atypical pathologic manifestation of *Pneumocystis carinii* pneumonia in the acquired immune deficiency syndrome. *Am J Surg Pathol* 1990; **14**, 615–625.

31. Tazelaar HD. Perivascular inflammation in pulmonary infections: implications for the diagnosis of lung rejection. *J Heart Lung Transplant* 1991; **10**, 437–441.

32. Pennington JE. *Aspergillus* lung disease. *Med Clin North Am* 1980; **64**, 475–491.

33. Kramer MR, Denning DW, Marshall SE *et al.* Ulcerative tracheobronchitis after lung transplantation. *Am Rev Respir Dis* 1991; **144**, 552–556.

34. Tazelaar HD, Baird AM, Mill M *et al.* Bronchocentric mycosis occuring in transplant recipients. *Chest* 1989; **96**, 92–95.

35. Hines DW, Hauber MH, Yaremko L *et al.* Pseudomembranous tracheobronchitis caused by *Aspergillus. Am Rev Respir Dis* 1991; **143**, 1408–1411.

36. Boon AP, O'Brien D, Adams DH. 10 year review of invasive aspergillosis detected at necropsy. *J Clin Pathol* 1991; **44**, 452–454.

37. Trulock EP, Bolman RM, Genton R. Pulmonary disease caused by *Mycobacterium chelonae* in a heart–lung transplant recipient with obliterative bronchiolitis. *Am Rev Respir Dis* 1989; **140**, 802–805.

38. Wreghitt TG, Hakim M, Gray JJ *et al.* Toxoplasmosis in heart and lung transplant recipients. *J Clin Pathol* 1989; **42**, 194–199.

39. Swerdlow S. Post-transplant lymphoproliferative disorders: a morphologic, phenotypic and genotypic spectrum of disease. *Histopathology* 1992; **20**, 373–385.

40. Hruban RH, Beschorner WE, Baumgartner WA *et al.* Depletion of bronchus-associated lymphoid tissue with lung allograft rejection. *Am J Pathol* 1988; **132**, 6–11.

41. Walts AE, Marcheysky AM, Morgan M. Pulmonary cytology in lung transplant recipients. *Diagn Cytopathol* 1991; **7**, 353–358.

42. Zeevi A, Fung JJ, Paradis II *et al.* Lymphocytes of bronchoalveolar lavage from heart–lung transplantation recipients. *Heart Transplantation* 1985; **4**, 417–421.

43. Paradis IL, Marrairi M, Zeevi A *et al.* HLA phenotype of lung lavage cells following heart–lung transplantation. *Heart Transplant* 1985; **4**, 422–425.

44. Clelland CA, Higenbottam TW, Monk JA *et al.* Bronchoalveolar lavage lymphocytes in relation to transbronchial lung biopsy in heart–lung transplants. *Transplant Proc* 1990; **22**, 1479.

45. Reinsmoen NL, Bolman RM, Savile K *et al.* Differentiation of Class I- and Class II-directed donor-specific alloreactivity in bronchoalveolar lavage lymphocytes from lung transplant recipients. *Transplantation* 1992; **53**, 181–189.

46. Nalesnik NA, Locker J, Jaffe R *et al.* Experience with post-transplant lymphoproliferative disorders in solid organ transplant recipients. *Clin Transplant* 1992; **6**, 249–252.

47. Rhandhawa PS, Yousem SA, Paradis IL *et al.* The clinical spectrum, pathology and clonal analysis of Epstein–Barr virus associated lymphoproliferative disorders in heart–lung transplant recipients. *Am J Clin Pathol* 1989; **92**, 177–185.

48. Hanto DW. Polyclonal and monoclonal post-transplant lymphoproliferative disease (LPD). *Clin Transplant* 1992; **6**, 227–234.

49. Goldman M, Gerard C, Abramowicz D *et al.* Induction of interleukin-10 by the OKT3 monoclonal antibody: possible relevance to post-transplant lymphoproliferative disorders. *Clin Transplant* 1992; **6**, 265–268.

50. Wood A, Higenbottam T, Jackson M *et al.* Airway mucosal bioelectric potential difference in cystic fibrosis after lung transplantation. *Am Rev Respir Dis* 1989; **140**, 1645–1649.

6

The clinical and radiological approach to interstitial lung disease: useful points for pathologists

G D Phillips, D M Hansell, M N Sheppard, R M du Bois

Introduction

The histopathological diagnosis of interstitial lung disease requires a close relationship between the pathologist, the clinician and the radiologist, which oftens guides interpretation of a biopsy specimen. Without their input mistakes can occur such as the omission of subtle features which may make a real difference to outlook and therapeutic approach. 'We find only what we look for' is very true for all practising branches of morphology. It is essential for any pathologist looking at such biopsies to be familiar with both the clinical and radiological features of the interstitial lung diseases so that he/she can provide the clinician with useful interpretation of the features seen on biopsy other than to make vague references to 'non-specific changes' or miss focal lesions. With this information the pathologist is in a much stronger position to assist in reaching a specific diagnosis and staging disease activity.

Interstitial lung disease

According to the Oxford English Dictionary 'interstitial' refers to the 'fine connective tissue lying between the cells of other tissue'. In a strict etymological sense, therefore, the term applies to disease processes which are seen to involve only the interstitial compartment of the lung. In practice, however, thoracic physicians use the term somewhat more loosely to describe a group of diseases whose aetiology, pathology, course, prognosis and response to treatment are often very different, but which have in common certain clinical features and

the presence of widespread, persistent abnormalities on chest radiography.[1] Clinically, they are commonly characterized by chronic/subacute onset and the main symptom is gradual and progressive breathlessness. If cough is present it is usually non-productive. By convention, infective disorders are excluded. On physical examination, bibasilar end-inspiratory dry crackles may be heard but examination may be normal even in the face of extensive radiographic abnormality. Physiologically there is usually a restrictive ventilatory defect with reduction in compliance and in gas transfer. Although histological examination of the lung in many of these conditions reveals a disease process predominantly involving the pulmonary interstitium, in many instances the pathological process also involves the alveoli and the respiratory and terminal bronchioles.[1] Thus, although it is a term in very common usage, 'interstitial lung disease' can be seen to be an inaccurate one, perhaps better replaced by 'diffuse lung disease'.

This chapter gives an insight into the clinical problem confronting the clinician and an appreciation of the factors which help to decide whether a biopsy is needed and, if so, the optimum method for obtaining one. It also aims to provide an understanding of what information is needed from the clinician by the histopathologist and vice versa so that the most meaningful interpretation of specimens can be obtained.

The clinician's problem

There are over 180 causes of diffuse lung disease so that, at first sight, establishing a diagnosis may seem an almost impossible task.[2] However, by a systematic use of the information obtained from a full history and examination, blood tests, lung function tests and chest radiography, it is usually feasible to narrow the possibilities down to a smaller number of more likely differential diagnoses. The information thus obtained can then be used to guide the physician to select the most appropriate further investigative procedures. These include thin section, high-resolution computed tomography (HRCT), bronchoalveolar lavage, endobronchial biopsy, transbronchial biopsy, various forms of needle biopsy, drill (trephine) biopsy and open or thoracoscopic lung biopsy. With the proper selection of the most appropriate of these further investigative procedures, the clinician can often achieve a definitive diagnosis.

Stage 1: Narrowing the possibilities

History

A careful history may provide useful diagnostic evidence. Although progressive breathlessness is common, sudden breathlessness may indicate a pneumothorax (Langerhans' cell histiocytosis,[3] lymphangioleiomyomatosis[4] or alveolar haemorrhage (pulmonary vasculitides). Wheeze indicates airway involvement and so is less common in diffuse lung disease, but may occur in Langerhans' cell histiocytosis where a significant airway component may be present.[5] Dry cough is particularly common in fibrosing alveolitis[6] but may also occur where there is either significant airways inflammation (Sjögren's syndrome) or a significant bronchocentric component (extrinsic allergic alveolitis, sarcoidosis, Langerhans' cell histiocytosis). Sputum production is not, however, a typical feature of diffuse lung disease except in some cases of bronchioloalveolar cell carcinoma[7] and when it is due to recurrent infection in end-stage disease. Pleural pain may suggest connective tissue disease, Churg–Strauss syndrome and asbestos exposure, but should point away from the diagnosis of cryptogenic fibrosing alveolitis or extrinsic allergic alveolitis.

Langerhans' cell histiocytosis almost never occurs in non-smokers[8] whereas extrinsic allergic alveolitis[9] and sarcoidosis[10] are less common in smokers. Smoking also increases the risk of lung malignancy, and bronchioloalveolar cell carcin-

oma as well as lymphangitis carcinomatosa may present with diffuse pulmonary shadowing.

A history of extrapulmonary disease may provide important clues as to the nature of the lung pathology. For instance, rheumatoid arthritis may be associated with bronchiectasis,[11] rheumatoid nodules,[12] pleural effusions[13] and thickening, pulmonary vasculitis,[17] obliterative and follicular bronchiolitis,[15] lymphocytic interstitial pneumonia,[16] fibrosing alveolitis[17] and progressive upper lobe fibrosis and cavitation.[18] Systemic sclerosis, systemic lupus erythematosus (SLE), Sjögren's syndrome and polymyositis may also be associated with a variety of pulmonary complications. Moreover, drugs used to treat these conditions may cause diffuse lung disease in their own right as well as producing immunosuppression which may make the patient vulnerable to opportunistic infections. Hence, a thorough drug history is a vital part of assessing the patient with diffuse lung disease.

Previous malignancy may predispose to diffuse pulmonary shadowing by virtue of immunosuppression, side-effects of the chemotherapeutic agents used, and recurrence of the malignant process in the lung in a diffuse pattern, e.g. lymphoma. Some lung diseases occur in association with diseases in which there may be a family history, e.g. neurofibromatosis.[19] A detailed history of occupations and hobbies may also be revealing. Exposure to inorganic dusts such as coal dust,[20] asbestos[21] or beryllium[22] or to organic dusts such as experienced by farmers[23] and bird fanciers[24] may be forthcoming, all of which may produce pulmonary disease.

Examination

The physical examination may reveal features of connective tissue diseases, vasculitides or malignancy which may provide important clues to the nature of the underlying pulmonary pathology. Examination of the respiratory system may be less rewarding. However, digital clubbing is present in 70% of patients with fibrosing alveolitis[25] but is rarely seen in fibrosing alveolitis associated with scleroderma,[26] and is more likely to be present in chronic berylliosis[27] than in chronic sarcoidosis, two conditions which may be difficult to differentiate. Basal inspiratory fine crackles are almost always present in fibrosing alveolitis.[28]

Blood tests

There are only a few investigations which may be helpful in pointing towards a specific diagnosis. A raised blood eosinophil count (>1500/mm³) will suggest pulmonary eosinophilia.[29] The relative concentration of IgE may help discriminate the cause. Very high eosinophil counts with only mild elevations of IgE are suggestive of cryptogenic pulmonary eosinophilia, whilst parallel elevation of the two indicates allergic, drug-induced and helminthic causes.[30] Detection of autoantibodies may indicate the presence of an underlying connective tissue disease and the pattern of these autoantibodies may point towards the specific disease involved.[31] The presence of anti-neutrophil cytoplasmic antibodies (ANCA) may support the diagnosis of a pulmonary vasculitis such as Wegener's granulomatosis.[32] In extrinsic allergic alveolitis, precipitating antibodies directed against the causative organic allergen should be present.[33] Although the serum angiotensin-converting-enzyme level may be elevated in patients with sarcoidosis,[34] it is neither a sensitive nor a specific finding.[35]

Lung function tests

The typical finding in diffuse lung disease is a restrictive ventilatory defect with reduced gas transfer and hypoxaemia at rest or on exercise.[36] The $PaCO_2$ is usually normal or low. If, in addition, an obstructive ventilatory component is present, this suggests either coexistent disease such as emphysema,[37] or the presence of a disease which is bronchocentric, such as advanced sarcoidosis, Langerhans' cell histiocytosis or lymphangioleiomyomatosis.

Chest radiography

Typically, lung size is reduced in patients with fibrosing lung disease and is well maintained until a later stage in those with granulomatous pathologies.[38] As a general guide, the pulmonary opacities in granulomatous lung diseases such as sarcoidosis tend to predominate in the mid zones of the chest radiograph,[39] whereas the changes of fibrosing alveolitis (occurring both alone or in association with the rheumatological conditions) are concentrated in the lower zones, and are more prominent in the periphery of the lung.[40] Occupational lung diseases such as silicosis have an upper zone predilection, as does sarcoidosis and chronic extrinsic allergic alveolitis. There are also a number of other conditions which tend to produce predominant upper zone disease including bronchopulmonary aspergillosis, late ankylosing spondylitis, tuberculosis and, less commonly, late Langerhans' cell histiocytosis.[41]

When the radiographic pattern is reticular, nodular or reticulonodular, the size and shape of individual opacities may help to refine the differential diagnosis. Although the International Labour Office (ILO) has formulated a more formal grading system of such opacities primarily for use in occupational lung disease, a simple scheme is easier to remember.[42] Thus, the smallest nodules, <1 mm in size, are seen in miliary tuberculosis, idiopathic pulmonary haemosiderosis and alveolar microlithiasis ('sandstorm' effect), whereas opacities up to 5 mm in diameter are found in sarcoidosis, extrinsic allergic alveolitis and silicosis. Opacities > 5mm occur in metastatic disease, Wegener's granulomatosis, rheumatoid arthritis and lymphoma. Cavitation is a feature of Wegener's granulomatosis, rheumatoid nodules (particularly in Caplan's syndrome) and some neoplasms, particularly squamous cell carcinoma.[41]

Some diffuse lung diseases produce confluent air space consolidation; these include pulmonary alveolar proteinosis, alveolar haemorrhage, pulmonary eosinophilia, cryptogenic organizing pneumonitis and opportunistic infections such as *Pneumocystis carinii* and cytomegalovirus.[41] Co-existent pleural disease may suggest a particular connective tissue disease, lymphangtis carcinomatosa, a drug-related process or asbestos.[41]

Stage 2: Further investigation

High-resolution computed tomography in interstitial lung disease

It is generally accepted that a chest radiograph may be normal in between 10 and 15% of patients with chronic interstitial lung disease.[43,44] High-resolution computed tomography (HRCT) is now frequently used to increase the confidence of an otherwise uncertain clinical and radiographic diagnosis. What distinguishes 'high-resolution' from standard computed tomography is a narrow scan collimation of 1–2 mm thickness. Most modern CT scanners are capable of high-quality HRCT images.

Usually, but not infallibly, HRCT will show whether or not there is significant interstitial pathology. It has been estimated that HRCT will detect otherwise radiographically cryptic disease in 20% of all patients with diffuse interstitial lung disease.[45] Large studies which have included a variety of diffuse interstitial lung disease have confirmed the improved diagnostic rate of HRCT over chest radiography.[46-48] When there are obvious features of diffuse lung disease, HRCT will show which area of lung will produce diagnostic material and which is the most appropriate biopsy technique (transbronchial, percutaneous, thoracoscopic or open-lung biopsy).

The HRCT images closely resemble macroscopic appearances of pathological specimens of inflated lung[49-51] and this has led to attempts to describe HRCT patterns of disease in anatomical terms[52] rather than employing the vague, and often imprecise, terminology used for the description of diffuse shadowing on chest radiography. There is no doubt that the fine morphological detail available from HRCT images (structures as small as 0.2 mm are visible in the high-contrast environment of the lung) is responsible for the increased confidence with which specific histopathological diagnoses can be made. The precise anatomical appearance and distribution of disease on HRCT may be suggestive or pathognomic of a particular diagnosis. An irregular linear pattern representing thickened interlobular septa may be seen in lymphangitis carcinomatosa;[53] a peripherally distributed reticular pattern is seen in established fibrosing alveolitis (Fig. 6.1) whether occurring alone, as part of a rheumatological disease or as a consequence of exposure to asbestos;[54] a thin-walled cystic pattern is seen in lymphangioleiomyomatosis[55] (Fig. 6.2) and Langerhans' cell histiocytosis.[56] Thickening and beading of the bronchovascular bundles is typical of sarcoidosis[57] (Fig. 6.3). Nodular changes and thickening of the interlobular septa may be seen in silicosis (Fig. 6.4) and coal workers' pneumoconiosis, Langerhans' cell histiocytosis and extrinsic allergic alveolitis.[58] A homogeneous ground-glass pattern may be present in extrinsic allergic alveolitis and desquamative interstitial pneumonia, whilst a consolidative appearance is seen in alveolar proteinosis, cryptogenic eosinophilic pneumonia and cryptogenic organizing pneumonitis.[58]

Another major use of CT is in assessing disease stage; HRCT patterns correlate with histopathological indicators of disease activity in a number of diffuse lung diseases. This has been a particularly fruitful area of research in fibrosing alveolitis.[59-61] In this condition it has been shown that the HRTC pattern predicts the appearances of lung biopsy specimens and this provides prognostic information: a predominantly cellular specimen predicts a better response to treatment than a biopsy which shows mainly fibrotic changes. In the case of fibrosing alveolitis, the distinction between reversible and irreversible disease is made by whether a ground-glass pattern (representing increased cellularity in the interstitium and/or air spaces) or

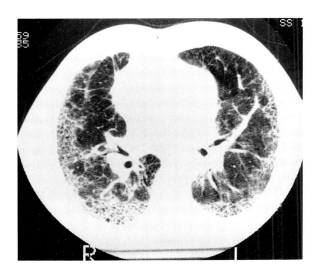

Fig. 6.1 High-resolution computed tomography throughout the lower lobes of lung showing the typical subpleural distribution of a reticular pattern in established fibrosing alveolitis.

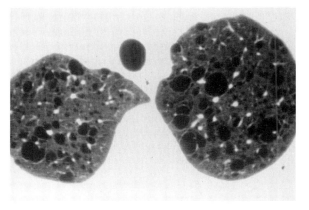

Fig. 6.2 Thin-walled cystic air spaces in the upper lobes of a patient with lymphangioleiomyomatosis.

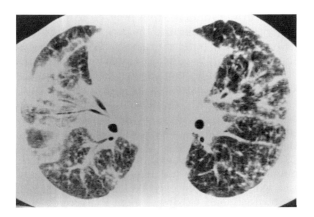

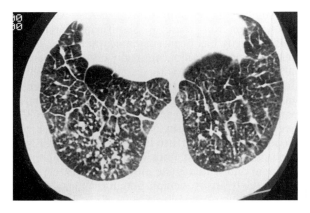

Fig. 6.3 Pulmonary sarcoidosis: widespread nodular opacities, some of which are concentrated around the bronchovascular bundles, giving a 'beaded' appearance.

Fig. 6.4 Scattered nodules in a case of silicosis. There is particularly prominent thickening of the interlobular septa in this case.

a reticular pattern (representing established fibrosis in the interstitium of the disrupted lung) predominates. Wells and co-workers[60] have shown that the presence of a predominantly reticular appearance on CT can correctly predict the finding of more fibrotic appearances on histological examination of open-lung biopsy samples. Thus, in certain patients, this allows staging of the disease process without recourse to open-lung biopsy. The majority of patients with fibrosing alveolitis show a predominantly reticular pattern on HRCT but identifying the 15–20% of patients with a predominantly ground-glass pattern is crucial because of the very different nature of these patients' response to treatment and their prognosis.[62,63] It is important, however, to stress that a ground-glass pattern, whether seen on a chest radiograph or HRCT, is diagnostically an entirely non-specific pattern which can be seen in several diseases. Such judgements about disease activity from the HRCT alone can, with some caveats, be applied to other conditions including sarcoidosis[64] and extrinsic allergic alveolitis.[65] Because high-resolution CT is readily available with newer machines and is a reliable means of determining disease activity in some conditions, it may be used as a means of monitoring patients. The added information about the likely histological diagnosis, the precise extent of the disease, the optimal site for biopsy when indicated, and in some cases the state of disease activity, have established the usefulness of HRCT as an essential tool in the investigation of interstitial lung disease.

Bronchoalveolar lavage

This procedure involves wedging the fibreoptic bronchoscope into a subsegmental bronchus and instilling 20–50 ml aliquots of sterile isotonic saline solution, total instilled volumes usually being 100–300 ml.[66] Using this technique, an assessment can be made of the total and differential counts of cells present on the epithelial lining surface of the lung, the identification of specific cellular constituents and the presence of specific non-cellular components, some of which may be used to refute or confirm a diagnosis, or help assess disease activity. The procedure is a useful adjunct to the other techniques used and in some situations bronchoalveolar lavage (BAL) can be diagnostic. In Langerhans' cell histiocytosis, the characteristic Langerhans' cells may be identified by their ultrastructural appearance on electron microscopy (Birbeck or 'X' bodies)[67] and by immunohistochemistry (S-100 staining);[68] in pulmonary alveolar proteinosis characteristic globules of periodic acid Schiff (PAS)-positive lipoproteinaceous material are present[69] and electron microscopy may demonstrate the eosinophilic myelin-like multilamellar structures which are invariably a feature of the disorder;[70] the presence of bizarre multinucleate giant cells may help to confirm a diagnosis of hard metal disease;[71] and large numbers of iron-laden macrophages may suggest repeated intrapulmonary haemorrhage due to a vasculitis or idiopathic pulmonary haemosiderosis.[72]

In patients with occupational lung disease, the

presence of asbestos bodies in lavage fluid can confirm exposure to this dust,[73] whilst *in vitro* transformation of BAL lymphocytes by beryllium may confirm this as a cause of granulomatous disease.[74] In such patients, examination of the cellular components of the BAL for beryllium using laser analysis by mass spectroscopy can also provide additional confirmation.[75] Energy dispersive analysis by X-rays of lavaged cell pellets can identify the presence of other minerals in the cells indicative of occupational exposure.[76]

Normal BAL fluid contains $10-15 \times 10^6$ cells per 100 ml; details of the normal cellular composition are given in Table 6.1. Changes in the total and percentage counts of individual cells in patients with diffuse lung disease have been used to define two broad categories of chronic inflammatory response; one involving lymphocytes and macro-

Table 6.1 Bronchioloalveolar lavage cellular content

Normal total = $10-15 \times 10^6$ cells
>80% macrophages
<14% lymphocytes (60–70% T cells, helper: suppressor ratio 1.6:1)
<4% neutrophils
<3% eosinophils
<1% basophils
<0.55% mast cells

Smokers' total = 40×10^6 cells
>90% macrophages
<5% lymphocytes
<4% neutrophils

---

phages, and the other neutrophils and macrophages.[66] Whilst changes in differential cell counts are seldom diagnostic, they can often be quite suggestive of a particular disease process or group of processes. A high lymphocyte count occurs in the granulomatous disorders, including sarcoidosis, berylliosis, extrinsic allergic alveolitis and tuberculosis,[77-79] and also in some drug-related conditions such as methotrexate and gold hypersensitivity.[80] In extrinsic allergic alveolitis, lymphocyte counts are particularly high and may reach 60–70% of the total. However, a similar though usually slightly less marked lymphocytosis may be present in asymptomatic but normal individuals exposed to the antigens responsible for farmer's lung and bird fancier's lung.[81] Whilst the predominant lymphocyte type in both sarcoidosis and hypersensitivity pneumonitis is the T cell, CD4+ helper T cells predominate in sarcoidosis

and CD8+ suppressor/cytotoxic T cells in extrinsic allergic alveolitis.[66] Tuberculosis is less constantly associated with a BAL lymphocytosis (55–60% of cases) and the helper:suppressor cell ratio is reduced.[82]

An increase in neutrophil numbers is a feature of the diffuse fibrosing lung disorders,[83] including cryptogenic fibrosing alveolitis, fibrosing alveolitis associated with the rheumatological diseases, asbestosis and the adult respiratory distress syndrome. Although it would therefore appear at first glance that lymphocytes are a feature of granulomatous and neutrophils of fibrosing lung disorders, experience has shown that the situation is, in fact, not so simple. Thus, lymphocytes may be present in the earlier and more inflammatory phase of fibrosing alveolitis when they tend to predict a better response to treatment,[84] but are not always present in granulomatous disorders. On the other hand, neutrophils may be present in BAL from patients recently exposed to the allergen, eliciting an extrinsic allergic alveolitis, and may be a feature of sarcoidosis once it has become chronic and fibrotic.[85] Thus, although BAL may provide diagnostic evidence in a few specific conditions, its main clinical value is as a pointer towards diagnosis and as an indicator of disease activity and the likely response to therapy.

Lung biopsy

There are two main reasons for wanting to obtain biopsy material: first, to establish a diagnosis which less invasive techniques have failed to yield; and second, to stage the disease process. The need to accomplish either or both of these two objectives will vary from case to case. Establishing the diagnosis beyond doubt and staging the disease is important when that individual may be faced with the possibility of many years of potent treatment which has potentially serious side-effects. A clear indication of the balance between potential side-effects and likelihood of improvement is required, particularly in the case of fibrosing alveolitis which has a 50% five-year mortality,[6] but is also important in other inflammatory and granulomatous diseases. At the present time, biopsy remains the 'gold standard' investigation in patients with fibrosing alveolitis who are younger than 55 years of age.[86]

There are five types of biopsy procedure: endobronchial (bronchial and transbronchial), fine needle, cutting needle, transcutaneous drill and open-lung biopsy (including the thoracoscopic approach). The complication rates of the various non-surgical biopsy techniques are given in Table 6.2. In practice, however, fine needle and cutting needle procedures are never used in the investigation of the diffuse lung diseases; the former because they are really only suitable for providing specimens for cytological examination which is rarely helpful in this group of disorders, and the latter because analysis of morbidity and mortality data suggests that the risks are unacceptably high for the interstitial lung diseases compared with the investigation of mass lesions.[87]

The transcutaneous drill biopsy was first described in 1935 by Kirschner[88] and then modified in the UK by Steel,[89] who used a high-speed air drill attached to the trephine. Pneumothorax and haemorrhage rates are similar to other transcutaneous techniques (Table 6.2). Steel has described his experience with a series of 119 consecutive biopsies

Endoscopic techniques
These are most suitable for granulomatous and bronchocentric disorders (Table 6.3). The complication rate is low (Table 6.2), and they are relatively simple and safe procedures. In sarcoidosis, tuberculosis, amyloidosis and metastatic carcinoma of the breast, endobronchial biopsy may provide diagnostic material. However, in interstitial lung disease, it is usual to proceed to transbronchial biopsy (TBB) with its slightly higher complication rate. Studies have shown a positive diagnosis being obtained in only 38% of cases in the series of Wall and co-workers.[93] The two conditions which are particularly likely to reveal themselves on transbronchial biopsy, however, are sarcoidosis and lymphangitis carcinomatosa. In the former, one biopsy gives a positive result in 40% of instances, but this figure rises to 80–90% with four biopsies.[94] The positive rate for lymphangitis carcinomatosa is 50–75%.[95] In fibrosing

Table 6.2 Complication rates from non-surgical lung biopsy procedures

Technique	Pneumothorax (%)	Tube drainage (%)	Haemorrhage (%)	Mortality (%)
Transbronchial	3 (0–5)	1 (0–3)	4 (1–12)	<0.1
Fine needle	28 (8–57)	8 (0–21)	11 (1–26)	<0.1
Cutting needle	26 (17–35)	13 (4–22)	16 (5–31)	<0.5
Trephine drill	36 (18–65)	12 (3–31)	7 (4–10)	<0.6

Range given in parentheses

in patients with diffuse lung disease, diagnostically significant material being obtained in 101 (85%) instances.[89] Several similarly large subsequent series have reported positive diagnostic yields of 75% or more in patients having diffuse lung disease.[90,91] Despite this, the technique has never gained widespread acceptance, perhaps because of concern over the potential instability of the attachment of the trephine to the drill and the rather disconcerting nature of the procedure. In addition, others have failed to duplicate the high diagnostic yield and therefore consider the complication rate unacceptable.[92]

In the modern era, therefore, the choice of biopsy technique lies between endobronchial/transbronchial procedures and open-lung biopsy, and is determined by consideration of the most likely diagnosis.

Table 6.3 Choice of biopsy procedure in diffuse lung disease

Endobronchial/ transbronchial	Open-lung/thoracoscopic
Sarcoidosis	Cryptogenic fibrosing alveolitis
Extrinsic allergic alveolitis	Fibrosing alveolitis in rheumatological disease
Tuberculosis	Pulmonary vasculitis
	Histiocytosis X
Lymphangitis carcinomatosa	Lymphangioleiomyomatosis
	Drug-induced disease
Berylliosis	Other diffuse lung disease, e.g. fungal opportunistic infection

alveolitis, however, the results are often not as good; for instance Mitchell and co-workers obtained positive histology in only 19 (40%) of 48 patients with this disorder in their series.[96] In diseases which are characterized by diffuse interstitial fibrosis, the biopsy specimens obtained at transbronchial biopsy are of insufficient size to establish a firm diagnosis and not sufficiently representative to stage the disease process. In addition, there are the potential problems of crush artefact and sampling error. Under these circumstances, an open-lung biopsy is the procedure of choice.

Open-lung biopsy
This is often regarded as the 'gold standard' of biopsy procedures. Wound problems and other minor complications occur in 2–10% of cases.[97,98] In their series of 502 patients with 'diffuse infiltrative' lung disease, Gaensler and Carrington reported no pneumothoraces and a mortality rate of <0.3% at 10 days.[99] Mortality in most series is probably under 1%.[87] Although tissue is always obtained at open-lung biopsy (in contrast to transbronchial biopsy), the diagnosis is not always obtained. Analysis of the American series of Gaensler and Carrington[99] and of the UK series of Venn and co-workers[100] suggests that a positive result may be achieved in approximately 90% of cases. However, in some reports this figure is as low as 70%[97,101] and even in the American series careful analysis suggests that 13% might be considered to be inadequately classified.[99] The best results in future studies are likely to be obtained by using CT to indicate the best area(s) for biopsy. Experience suggests that the lingula and right middle lobe are often involved in incidental infective/inflammatory changes, and this may limit their usefulness as areas to biopsy.[6] Moreover, the most severely affected regions of lung are often the least useful for providing diagnostic material, and it is in this context that CT often proves to be a powerful tool. Open-lung biopsy is the only method which can be used to obtain large enough specimens to enable the detailed scoring systems for grading histological appearances in fibrosing alveolitis to be used.[102]

Although open-lung biopsy may be performed via a 'mini thoracotomy' incision which produces significantly less morbidity than a standard thoracotomy approach, recent developments in microchip technology have led to the advent of video-assisted thoracoscopic lung biopsy.[103] This pro-

cedure requires an even smaller incision, is more rapid and results in shorter hospital stays. Thus far it would appear to be safe and the requirement for analgesia in the immediate postoperative period appears to be reduced.

It is our policy to perform an open-lung biopsy procedure in two situations. First, if the diagnosis is in doubt, biopsy is necessary. Second, in fibrosing alveolitis, even if other tests such as CT have provided a clear diagnosis, we would still recommend biopsy in the younger patient, in order to establish unequivocally the relative degrees of cellularity and fibrosis in an individual who may otherwise be faced with many years of potent treatment with potential side effects. CT can usually provide this information but can underestimate the amount of fine fibrosis in some instances when a ground-glass pattern is predominant. In the older patient with a predominantly reticular pattern on CT, biopsy provides no further information and is not done in our unit.

Transbronchial and open-lung biopsy

Mary N Sheppard

Transbronchial biopsy

This is often carried out in the investigation of both diffuse and well-circumscribed peripheral lung lesions. The procedure is performed by passing forceps beyond the visible bronchi and out of direct vision and is best done with fluoroscopic guidance. The bronchoscope is wedged in the appropriate subsegmental bronchus and a forceps passed at the bifurcation, out into the lung parenchyma. To biopsy diffuse pulmonary parenchymal disease, particularly interstitial disease, the lateral segment of the right lower lobe is recommended because the forceps will pass to the lung periphery with ease. Up to six biopsies are taken in the investigation of diffuse lung disease.[104] Each is usually 1–2 mm in diameter. The overall diagnostic rate in diffuse lung disease ranges from 38–64%,[105,106] but it is 67–80% in conditions with specific histological features such as lymphangitis carcinomatosa, metastatic malignant disease and lymphoma.[105,106]

Transbronchial biopsies are handled in the same way as bronchial biopsies (see Chapter 2). Some of the special investigations that can be carried out for both transbronchial and open-lung biopsies are given in Table 6.4.

Table 6.4   Special investigations for both transbronchial and open-lung biopsies

Elastic van Gieson for elastic tissue, collagen, blood
  vessels
Ziehl–Neelsen for mycobacteria
Grocott for fungi and *Pneumocystis carinii*
Gram stain for bacteria, i.e. *Actinomyces* or *Nocardia*
Congo red for amyloid
Perls Prussian blue reaction for iron
Polarizing light filters for birefringent material
Immunocytochemistry with specific antibodies for
  infective organisms, tumours

Assessment of transbronchial biopsy

The number of specimens should be checked. The adequacy of each piece should be assessed by checking the tissues present. There should be bronchial wall, peribronchial tissues including bronchial arteries, bronchioli with accompanying pulmonary arterial branches, lung parenchyma with alveoli and interstitium as well as capillaries. It is rare to obtain pulmonary veins and interlobular septae with overlying pleura. The biopsy should then be approached in the same way as an open-lung biopsy.

Artefacts

If previous bronchial biopsies have been performed, airway inflammation, ulceration and granulation tissue may occur so this must be checked from the history. In transbronchial biopsies the lung parenchyma is always collapsed (atelectasis). This can give the erroneous impression of solid fibrotic lung and may be misinterpreted as interstitial pneumonia or interstitial fibrosis because the apposition of alveolar walls produces thickening and hypercellularity (Fig. 6.5). A trichrome stain to demonstrate collagen will show the absence of scarring. If the alveoli are fully expanded, the lumen should be assessed for oedema, alveolar lipoproteinosis or the foamy exudate of *Pneumocystis carinii*. The parenchyma can be compressed by forceps (Fig. 6.5) which also makes interpretation difficult or impossible. Fresh red blood cells filling alveoli is common operative trauma and

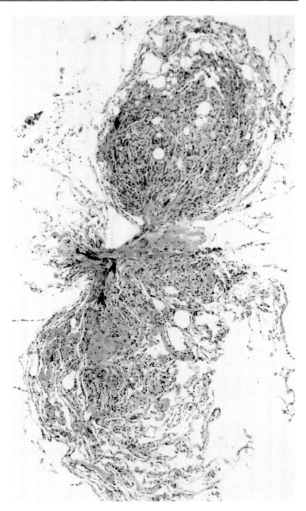

Fig. 6.5   Section of transbronchial biopsy showing lung parenchyma with squeeze artefact, nuclear crushing and atelectasis of alveoli. Air bubbles are also seen.

should never be used to diagnose haemorrhage (Fig. 6.6). In addition large holes will be seen in the parenchyma which are air bubbles introduced during biopsy (Figs 6.5 and 6.6). They are confused with lipid droplets but lack the macrophage and giant cell reaction around them. There is slight thickening of the alveolar walls directly leading from the adventitia of bronchi so assessment of interstitial thickening and fibrosis on small pieces of bronchial wall with attached alveoli is to be avoided. Transbronchial biopsy rarely if ever yields sufficient tissue for a definitive diagnosis of cryptogenic fibrosing alveolitis or interstitial pneu-

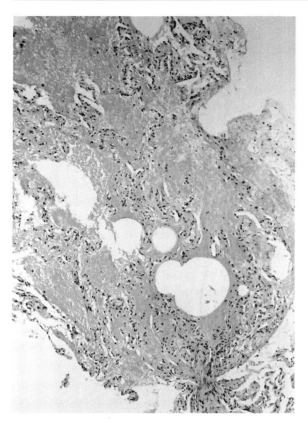

Fig. 6.6 Transbronchial biopsy in which there is squeeze artefact, air bubbles and red blood cells within alveoli.

monia. An open-lung biopsy is required for this. However, the discovery of specific lesions such as granulomas on transbronchial biopsy will point the way towards infection, particularly tuberculosis or fungi, sarcoidosis or extrinsic allergic alveolitis as the cause of interstitial shadowing. Pulmonary blood vessels may also be distorted by squeeze arte-fact and give the impression of fibrous scarring. A trichrome will highlight the outline of muscular pulmonary arteries with internal and external elastic lamina in such cases.

Transbronchial biopsy in immunosuppressed patients

Fibreoptic biopsy is well established as a diagnostic tool for a large range of pulmonary infections[104] and is widely used in immunosuppressed pa-tients.[107] This is becoming very important because of the increasing numbers of patients being treated for malignancies, undergoing transplantation and

having AIDS. The diagnostic yield in opportunistic infections is in the region of 76–88%[108] but in the case of *Pneumocystis carinii* pneumonia, trans-bronchial biopsy may not add to the detection rate provided by brushings and lavage.[108] Many groups now consider that the results from lavage alone are sufficiently good to justify omission of routine transbronchial lung biopsy. This is because the lavage procedure samples many alveoli, far more than are represented in a transbronchial biopsy specimen.[109] However, lavage alone will not iden-tify diffuse alveolar damage, lymphocytic inter-stitial pneumonia, chronic interstitial pneumonia/fibrosis or lymphoma. These diagnoses will be missed unless transbronchial or open-lung biopsy is performed. It has been reported that up to 91% of infections can be diagnosed by a combination of transbronchial biopsy and bronchoalveolar lavage.[108] In addition, the significance of cyto-megaloviral change in cells from lavage and the detection of *Candida* spp. and *Aspergillus* spp. may not indicate significant lung infection and it is the detection of invasive organisms and cytological changes in pneumocytes on transbronchial or open-lung biopsy that confirms lung disease.[109] In patients with AIDS there is an increased risk of complications with transbronchial biopsy, especi-ally haemorrhage and pneumothoraces if the patient suffers from *Pneumocystis carinii* pneu-monia.[110] The endobronchial lesions of Kaposi's sarcoma can also be seen[111] but patients usually have multiple lesions at other sites, particularly the skin. Since bronchial and transbronchial biopsy can produce severe haemorrhage, they are gen-erally avoided. Because of the high success rate with lavage and transbronchial biopsy, open-lung biopsy is rarely done nowadays and only when these less invasive procedures fail to provide a diagnosis.[112] Open-lung biopsy may be required in particular to diagnose interstitial pneumonia, lym-phocytic interstitial pneumonia and lymphoma.

Open-lung biopsy

Open-lung biopsy provides a high diagnostic yield in patients with diffuse pulmonary disease.[113] The specimen is large enough to provide adequate material for histology as well as culture and there is a low morbidity.[114] The specimen, generally measuring 5cm x 3cm x 2cm, is received fresh in the laboratory. Some pathologists recommend frozen section in order to give an early broad outline of the pathological process going on, i.e. infection, vasculitis, fibrosis, diffuse alveolar damage, tumour, etc. Special stains can then be ordered together with the routine sections the next day (see Table 6.4). In desperately ill patients, frozen section is often required to establish rapidly if the patient has a treatable condition, such as infection, so that therapy can commence immediately. Finally, the frozen section can be used to assess the adequacy of the tissue obtained and to determine whether it is representative of the disease process, which is particularly important when the imaging shows a focal or nodular process. Other pathologists avoid doing frozen sections on open-lung biopsies, believing that it is the paraffin sections which give the best picture and that frozen sections can be difficult to interpret. Pieces can be taken for viral and bacterial culture, electron microscopy, immunofluorescence and immunocytochemical or biochemical studies, imprints, etc. (see Table 6.4) The imprints can be fixed immediately in ethanol as with cytological specimens or they can be air-dried, in which form they are suitable for staining for infectious agents.

The remaining tissue can then be fixed in formalin. Careful inflation of the lung biopsy is essential if the lung architecture is to be fully assessed. The lung parenchyma must be well inflated in order to avoid atelectasis. This can be done using a small-gauge needle attached to a small (5cc or tuberculin-type) syringe filled with buffered formalin. The needle is put into the parenchyma and the specimen inflated until the pleural surface is uniformly smooth and all areas of the biopsy inflated. Little damage to the lung is done with this technique. Patchy atelectatic portions of lung involving single lobules are common in biopsies which have not been uniformly inflated.

Thoracoscopic microthoracotomy

The pioneering work of Donnelly *et al.* has paved the way for the use of video-directed thoracoscopy to undertake lung biopsy.[115] This practice is gaining widespread acceptance because of surgical audit with shorter in-patient stays and patient demands for the latest keyhole technology with minimal scarring. The specimens are adequate and representative in the case of interstitial lung disease and the endoscopic staples provide secure closure in patients with chronic fibrosing lung conditions.[116]

Artefacts

Focal atelectasis, recent biopsy haemorrhage and crush artefact can all be seen in open-lung biopsy. Patients who have had positive pressure ventilation may have dilatation and distension of bronchioles and alveolar ducts. Prolonged manipulation may lead to margination of neutrophils in capillaries (especially prominent in the pleura) which will mimic capillaritis and forceps compression can also lead to lymphatic dilatation.

General abnormalities

One must remember when looking at an open-lung biopsy that there are changes that are non-specific. Biopsies from lobar tips, particularly the lingula or right middle lobe, can show non-specific interstitial fibrosis with honeycombing and inflammatory changes. This is particularly important since these areas are easily biopsied – and all surgeons should be advised to avoid them in interstitial lung disease. Apical subpleural fibrosis with a chronic inflammatory infiltrate is very common, particularly in tall thin individuals. Previously these changes were linked to tuberculosis, but are now considered to be the result of an imbalance in perfusion/ventilation with ischaemia in the apical regions of the lung and are non-specific. These areas show fibrosis in the pleura and subpleural parenchyma with thickened elastic fibres. Calcification and ossification are common. Scattered non-specific scars with hyperplasia of type 2 cells around them can be an incidental finding in the lung parenchyma. They may represent old infarcts or healed granulomas. Often the pulmonary arteries in the area of scarring exhibit medial hypertrophy with intimal fibrosis which is indicative of endarteritis obliterans and not pulmonary hypertension. Intimal thickening in both arteries and veins is also a normal ageing process.

Scattered subpleural blebs are air-filled spaces less than 1 cm in diameter (Fig. 6.7). These are associated with parenchymal fibrosis and are common in adult patients with a history of pneumothoraces. Often the spaces are lined by epithelium and contain dense collagen in the wall as well as smooth muscle hyperplasia with many blood vessels. They can go on to form bulla (> 1 cm in diameter) which can expand and compromise the lung.[117] Often the overlying pleura shows a florid mesothelial proliferation with many macrophages, some forming multinucleate giant cells and eosino-

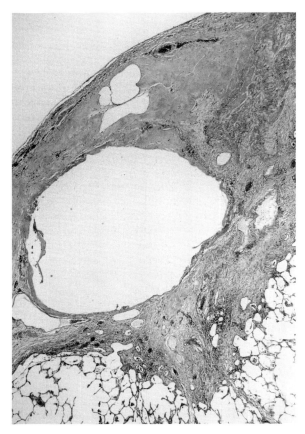

Fig. 6.7 Section from apex of lung showing a subpleural bulla with surrounding fibrosis and overlying thickening of the pleura.

phils, when air is introduced into the pleural space and a reactive eosinophilic pleuritis. Because of the presence of the eosinophils and macrophages as well as the history of pneumothoraces, histiocytosis X must be ruled out. Interstitial emphysema with or without a giant cell reaction may be associated with the pneumothorax.

Both airway and parenchymal changes are common in smokers. The larger airways show evidence of chronic bronchitis (see Chapter 2) while the bronchioles also show extension of goblet cells into the epithelium, which is not a normal finding. There are chronic inflammatory cells in the bronchiolar wall accompanied by peribronchiolar fibrosis, together with accumulation of pigmented macrophages in the interstitium and in the adjacent air spaces with mild interstitial fibrosis. The pigment, which is in the form of brown

granules, results from the ingestion of debris from cigarette smoke and the alveolar macrophages have prominent lysosomes. The lysosomes stain with PAS while the granules are Perl's Prussian blue-positive (indicating the presence of haemosiderin). Black dense anthracotic pigment is also prominent in smokers around bronchioles, often between the bronchiole and the accompanying artery. Centrilobular emphysema is very common in smokers in the upper lobes and histology is not useful in assessing the severity of disease. In all urban-dwelling adults one may find short needle-like birefringent material (silica or silicates) associated with anthracotic pigment. It is often very prominent in smokers who do not have an occupational history. Therefore the presence of birefringent material alone is not diagnostic of silicosis.

Corpora amylacea, similar to those found in the prostate, are eosinophilic rounded lamellated proteinaceous bodies 5–15 µm in diameter, which stain positively with PAS and are common in older patients. They are found within alveoli and are of no pathological significance. Blue bodies are calcified lysosomal debris 15–40 µm in diameter and are smaller than corpora amylacea, which are also found in alveoli. They consist of calcium carbonate with a rim of iron and accumulate within macrophages and outside them, in conditions where macrophages increase. They are of no diagnostic significance on their own.

An occasional non-necrotizing granuloma or cholesterol granuloma or single giant cell containing cholesterol clefts may be incidental findings in an open-lung biopsy, but in most cases they are significant findings in the context of interstitial lung disease.

The lung is a major reservoir for megakaryocytes. These can be found within the alveolar capillaries and may be confused with malignant cells or virally infected cells. Although they are believed to be prominent in sepsis and diffuse alveolar damage, their identification has no specific pathological significance.

Bone marrow emboli are common in autopsy material but can also be seen in biopsy which may be a consequence of cutting into ribs during the procedure. Small pulmonary thrombi can form *in situ* when there is local lung damage and are not by themselves indicative of deep vein thrombosis and embolization.

Intraparenchymal lymph nodes can be an accidental finding in the lung.

Assessment of open-lung biopsy

The open lung should always be examined at low power to assess the anatomical location of any abnormality. The following patterns can be useful

1. *Bronchial/bronchiolar*, i.e. obliterative bronchiolitis, viral infection, extrinsic allergic alveolitis, bronchiectasis, cystic fibrosis, inhalation pneumonia, follicular bronchiolitis, bronchocentric granulomatosis, allergic bronchopulmonary aspergillosis, histiocytosis X, sarcoidosis.
2. *Vascular*, i.e. foreign material, tumour, vasculitis, thromboemboli, primary pulmonary hypertension, lymphoma.
3. *Lymphatic*, i.e. sarcoidosis, tumour.
4. *Septal*, i.e. venous congestion, pulmonary veno-occlusive disease, interstitial emphysema.
5. *Subpleural*, i.e cryptogenic fibrosing alveolitis/interstitial pneumonia, asbestosis.
6. *Interstitium*, i.e. interstitial pneumonia/cryptogenic fibrosing alveolitis, lymphocytic interstitial pneumonia, amyloid, histiocytosis X, lymphangioleiomyomatosis, calcification, granulomas in sarcoidosis and extrinsic allergic alveolitis, asbestos bodies in asbestosis.
7. *Alveolus*, i.e. bacterial pneumonia, eosinophilic pneumonia, lipid pneumonia, desquamative interstitial pneumonia, alveolar lipoproteinosis, oedema, *Pneumocystis carinii*, cryptogenic organizing pneumonia, radiation and paraquat fibrosis, microlithiasis, haemosiderosis.
8. *Alveolar lining* i.e. diffuse alveolar damage, hyaline membrane disease, viral inclusions, foamy pneumocytes in amiodarone toxicity, giant cells in hard metal disease.
9. *Pleural*, i.e. connective tissue disease, tumour.
10. *Haphazard*, i.e. infectious granulomas, hyalinizing granulomas, tuberculosis, silicosis, nodular amyloid.

Infection

With the widespread access to travel and migration, the pathologist is now more likely to see exotic infections in the lung presenting as nodules or interstitial infiltrates. In addition with increased immunosuppression and the emergence of AIDS, multiple and repeated lung infection can occur.

Viral infection

Often one see evidence of lymphocytic interstitial pneumonia or diffuse alveolar damage as a general response to viral infection particularly in malnourished or immunosuppressed individuals. A neutrophilic response with necrosis may also be seen, so this reaction is not confined to bacterial infections.

Cytomegalovirus
A diffuse interstitial pneumonia or a miliary nodular pattern can develop. The virus infects pneumocytes which become enlarged with centrally located amphophilic to basophilic intranuclear inclusions with a definite clear halo giving the classical 'owl's eye' appearance. The cells can desquamate off for cytology. Cytomegalovirus (CMV) often co-exists with *Pneumocystis carinii* in the lung.

Adenovirus
Adenovirus usually affects the bronchi and bronchioles with basophilic/eosinophilic intranuclear inclusions with a cuff of lymphocytes in the airways. Smudge cells with amphophilic to basophilic smudgy inclusions without halos may also be seen.

Herpes simplex
The pattern shown by herpes simplex infection can be of necrotizing bronchopneumonia or of nodules in the lung. Multinucleated cells with moulding of nuclei can be seen but these can be rare. Intranuclear ground-glass inclusions or central eosinophilic inclusion with a halo are present in lining bronchial/bronchiolar epithelium. It is often seen in patients who develop adult respiratory distress syndrome (ARDS) and are on ventilators.[118]

Measles
The dominant feature of measles infection is syncytial proliferation of type 2 pneumocytes with clustered nuclei and the presence of intranuclear and less commonly intracytoplasmic inclusions. These can also desquamate off into the alveolar lumen.

Respiratory syncytial virus
Respiratory syncytial virus causes an acute and chronic cellular bronchiolitis with loss of ciliated cells. The metaplastic cells become enlarged and can be multinucleated (hence the name of syncytial virus). Eosinophilic cytoplasmic inclusions may occasionally be seen.

Influenza

Influenza infection usually produces swelling of bronchial and bronchiolar epithelial cells with vacuolation of cytoplasm and nuclear degeneration. Intracellular inclusions are rare but apical decapitation of ciliated epithelial cells (termed ciliocytophthoria) has been suggested as an indication of viral disease but is now considered non-specific. A lymphocytic infiltrate around the airways is prominent.

Parainfluenza

Parainfluenza can give rise to similar appearances but there is a lack of inclusions. In immunosuppressed patients, parainfluenza virus may cause a giant cell pneumonia which is similar to measles.

Chickenpox (varicella)

Chickenpox (varicella) pneumonia is rare and occurs usually in the immunosuppressed patient. Eosinophilic intranuclear viral inclusions may be seen in bronchiolar or alveolar epithelium. Healing results in circumscribed fibrous nodules that measure up to 5 mm in diameter and usually calcify. These are typically seen on chest X-ray.

Specific antibodies are available for immunolocalization of virus in infected cells and have proven very useful in routinely fixed material.

Bacterial infection

Specific bacterial infections which can be identified on transbronchial or open-lung biopsy include actinomycosis (Fig. 6.8), nocardiasis or tuberculosis. Miliary tuberculosis is seen most commonly now in the elderly and these patients can often present with non-specific findings, not pointing to the diagnosis. The miliary tubercle is 1–2 mm in diameter and can be necrotizing or non-necrotizing. The granulomas are found along airways and in alveolar septae. The granulomas may enlarge by confluence and these are more likely to have necrotic centres. Blood vessels can be involved in the vicinity of the granulomas. In immunosuppressed individuals, there may be a histiocytic response often in the interstitium with no granulomas and large numbers of acid-fast bacilli fill these histiocytes (Fig. 6.9a and b).

Fungal infection

The most common fungal infections are those due to *Candida*, *Aspergillus* and *Histoplasma* spp. Fungal spores and hyphae can be seen in haematoxylin and eosin (HOE) sections but periodic acid Schiff (PAS) and silver stains such as Grocott's are best for visualization.

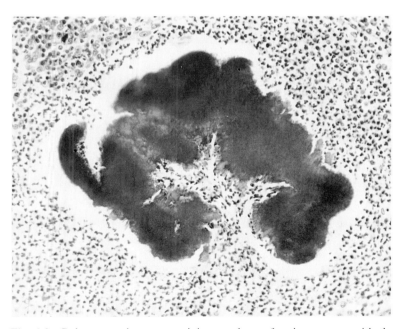

Fig. 6.8 Pulmonary abscess containing a colony of actinomycoses with the Splendore–Hoeppli phenomenon at the periphery.

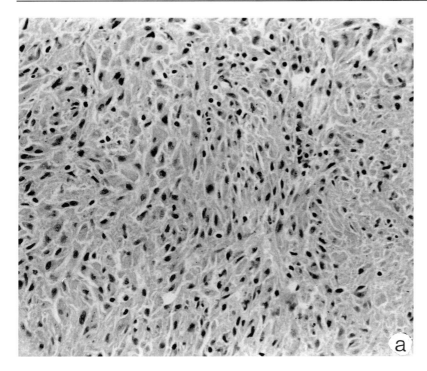

Fig. 6.9a Sheets of histiocytes within the lung of an AIDS patient.

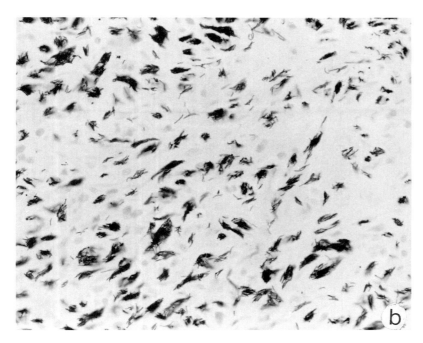

Fig. 6.9b The same section stained with Ziehl–Neelsen showing large numbers of mycobacteria within the histiocytes.

Fungi can cause airway plugging, acute ulceration, abscess formation and granulomatous reactions with and without necrosis. Fungi should be looked for in extracellular debris, necrotic material and within histiocytes and multinucleated histiocytes.

Aspergillus

Aspergillus spp. produce 3–5 µm septate hyphae with branching at 45° and no narrowing at branching points. *Aspergillus fumigatus* is the most common cause of pulmonary disease and can easily be detected on H&E section when the hyphae can be intensely haematoxyphilic. The hyphae grow in waves extending out from the centre of the plug of hyphae present in blood vessels and alveoli. The hyphae at the growing edge are often coated with intensely eosinophilic material representing immune-complex deposition (Splendore–Hoeppli phenomenon). Widespread haemorrhagic infarction is common because of vessel involvement with thrombosis. Crystals of calcium oxalate can be seen, particularly in association with *Aspergillus niger* infection. These are birefringent, which can be useful in diagnosis.

Candida

Candida is a 3–4 µm diameter yeast with multiple buds and narrow points of attachments. The yeasts can have elongated bodies, giving the impression of pseudohyphae.

Mucormycosis

Mucormycosis spp. produce 10–35 µm large irregular hyphae, non-septate with branching at 90°.

Blastomycosis

Blastomycosis is a 8–15 µm thick walled yeast with single buds and no narrowing. Multiple nuclei are seen.

Coccidioides

Coccidioides are 20–60 µm spherules with 3–5 µm endospores.

Cryptococcus

Cryptococcus is a 5–20 µm yeast with single buds and narrow attachment. It has a gelatinous capsule which stains with PAS or mucicarmine stains.

Histoplasma

Histoplasma is a 1.5–3.5 µm ovoid yeast often found within histiocytes. It is best visualized with silver stains. The granulomas calcify rapidly.

Protozoan infection

Pneumocystis

Pneumocystis infection produces 5–8 µm cysts and 2–12 µm trophozoites. Silver stains aid diagnosis demonstrating the cyst wall and the focal area of cyst wall thickening, distinguishing it from red blood cells which can be similar in shape and size. The trophozoites stain with cresyl violet and toluidine blue but results can be variable and a known *Pneumocystis* case must be used as the control. Antibodies to *Pneumocystis* are available for diagnosis in routine formalin-fixed tissue.

Toxoplasmosis

Toxoplasmosis infection causes an interstitial pneumonia. Cysts are generally 6 µm in diameter, and filled with merozoites, lie free in the intercellular tissues and provoke little inflammatory reaction. When the merozoites are released from the cyst they are called trophozoites and these can invade and proliferate within host cells to form pseudocysts. The identification of the cysts or pseudocysts can be difficult in pulmonary tissue because they are so few.

Helminthic infection

Schistosomiasis

In schistosomiasis the ova of the parasite lodge in blood vessels and can induce granulomatous and eosinophilic responses.

Paragonimiasis

Infestation by the lung fluke is endemic in the Middle East. Paragonimiasis causes an eosinophilic pneumonia. The worms lie singly in the connective tissue of the lung.

Echinococcus (hydatid disease)

This disease takes the form of cysts within the pulmonary parenchyma formed by a laminated capsule and an inner germinal layer which give rise to brood capsules and from which arise the scolices. Free brood capsules and scolices make up the grains seen within the hydatid fluid. In a sterile cyst, often the germinal layer and brood capsules with scolices are absent. The cyst can rupture into a bronchus, causing bronchocentric granulomatosis.

Nematodes

Strongyloides
This can be seen in immunosuppressed patients, especially those with AIDS in which large numbers of the parasite can be seen in airways and in alveoli. They induce a chronic inflammatory response.

Ascaris, Ancyclostoma and *Filaria*
The larval and microfilarial forms of these nematodes can cause fleeting eosinophilic pneumonia. A careful search should therefore be made of any biopsy showing large numbers of eosinophils.

Dirofilaria spp.
The canine heartworm can induce a necrotizing granuloma in the lung parenchyma which often presents as a coin lesion but can be multiple and bilateral. The diagnosis is made by identifying the dead worm within a thrombosed artery in the central area of necrosis.

Bronchial/bronchiolar lesions

Extrinsic allergic alveolitis

This condition results from an immunological reaction to inhaled allergens. Three main categories of antigens are associated with the disease: thermophilic actinomyces found in mouldy hay; moulds; and animal proteins found in pigeons, parrot and parakeets. Both acute and chronic forms of the disease occur. Although the clinical history, radiological features and presence of precipitating antibodies are usually sufficient to establish the diagnosis, a transbronchial or open-lung biopsy may be required in questionable cases. The pathology localizes on the respiratory bronchioles and consists of aggregates of lymphocytes, plasma cells, histiocytes, some neutrophils and eosinophils which infiltrate the bronchiolar wall and extend out into the surrounding alveolar interstitium. This bronchiolar concentration of inflammation is very useful in drawing the pathologist's attention to the possibility of this condition and emphasizes the need to look at a low power view of the open-lung biopsy. The presence of usually single, scattered, non-necrotizing loosely formed granulomas with epithelioid cells and multinucleate giant cells typically containing cholesterol clefts, in the bronchiolar wall and also

in the interstitium surrounding bronchioles (Fig. 6.10) is another important feature of this condition. Buds of granulation tissue within alveoli indicating organizing pneumonia can be a prominent finding also. The final histological feature to look for is a mild lymphocytic alveolitis. In the chronic stages of the disease, interstitial fibrosis is prominent with the other pathological features being obscured, which may make the underlying cause of the scarring uncertain. Occasionally the bronchioles become occluded by the granulomas resulting in obliterative bronchiolitis, but this is a rare complication. Obviously in all cases of granulomatous inflammation, special stains for mycobacteria and fungi should be done. The other differential diagnosis is of course sarcoidosis, but

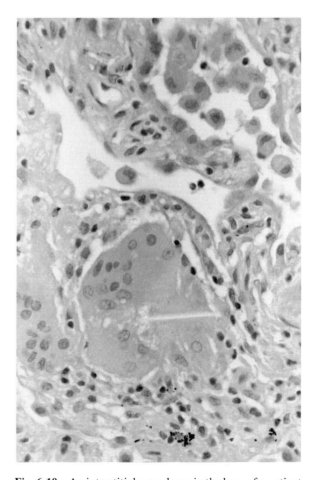

Fig. 6.10 An interstitial granuloma in the lung of a patient with extrinsic allergic alveolitis showing a multinucleate giant cell containing a cholesterol cleft.

the granulomas are usually more well defined and distributed with the lymphatics in perivascular and subpleural locations as well as being peribronchiolar.

In some open-lung biopsies showing all the features of extrinsic allergic alveolitis, no antigen can be identified on detailed work-up of the patient.[119] On the other hand, clinically proven cases may have a non-specific interstitial infiltrate in up to 20% of patients.

Follicular bronchitis/bronchiolitis

Bienenstock and co-workers[120] were the first to detail the normal anatomical distribution of lymphoid tissue in the lung along the bronchial tree and interlobular septa and apply the term 'bronchial associated lymphoid tissue' (BALT) to these lymphoid aggregates. Such aggregates are very prominent in the rabbit but in other species, particularly in man, their presence and significance has been debated. Recent studies have shown that BALT is present in the human fetal lung and its presence related to exposure to infection.[21] In adults BALT is not found in healthy lungs and even in inflammatory conditions BALT is seen in only a few cases.[122] What initiates the formation of BALT in the human lung remains to be defined and its role in normal lung immunology remains unclear.

In follicular bronchitis/bronchiolitis there is proliferation of BALT-containing lymphoid follicles with germinal centres infiltrating the bronchial and bronchiolar wall.[123] An open-lung biopsy shows that the bronchioles are distorted by lymphoid follicles of variable size which impinge upon the epithelium. Often it is difficult to identify the bronchioles, which appear as slit-like spaces, and the lymphoid aggregates merge with each other so that the normal lung architecture is lost. The bronchiolar epithelium is attenuated over the follicles with individual lymphocytes infiltrating the epithelium to form a lymphoepithelial lesion which is an important feature of BALT. The BALT contains a germinal centre with a central zone of B cells and a more loosely arranged mixture of B and T cells at the periphery. The intraepithelial lymphocytes are mainly B cells.

Follicular bronchiolitis is seen in chronic airway infection such as bronchiectasis, in autoimmune disease such as rheumatoid arthritis, systemic lupus erythematosis, Sjögren's syndrome and in immune deficiency states. It has also been reported in association with peripheral eosinophilia. Many cases are idiopathic.[123-125] The stimuli for its formation may be viral or a hypersensitivity/allergic reaction. Radiographic changes show bilateral interstitial infiltrates, primarily reticulonodular.

The differential diagnosis includes lymphoid interstitial pneumonia in which it is the interstitium which is expanded by lymphocytes, these mainly being T lymphocytes (Fig. 6.11). In extrinsic allergic alveolitis there is peribronchiolar inflammation and granulomas are identified, which are not seen in follicular bronchiolitis. An important distinction from low-grade B cell lymphoma (BALToma) must also be made and immunocytochemistry is important in showing the monotypic B cells and light chain restriction in malignant lesions. Lymphoepithelial lesions can be found in both benign and malignant conditions and reactive follicles can be seen in lymphomas, so care has to be exercised in the interpretation of both open-lung biopsies and especially transbronchial biopsies when one is dealing with a lymphoid infiltrate.

Obliterative bronchiolitis

Obliterative bronchiolitis refers to chronic inflammation with fibrous destruction of the bronchiolar wall so that all that may be left is a fibrous scar next to a pulmonary arterial branch (Fig. 6.12). This is known as the disappearing bronchiole syndrome and can be missed on both transbronchial and open-lung biopsy. It is best seen with connective tissue stains, which will show the fibrosis with destruction of the bronchiolar wall next to a pulmonary arterial branch. It can follow viral infection, or mycoplasma infection or toxic gas exposure, can occur after bone marrow transplantation or lung transplantation, or can be associated with connective tissue disorders, especially rheumatoid arthritis.[126] The lesions are usually confined to the bronchioles but larger airways can also become involved. It is important to remember that the lesions are usually focal and a biopsy may show normal bronchioles. Multiple levels and complete sampling of the biopsy is required in order to avoid missing a single lesion which may be present. The pulmonary parenchyma is usually not involved in the inflammatory process. However, a typical feature is accumulation of foamy macrophages in the alveoli around the obstructed bronchioles and also foamy macrophages in bronchioles which are distal to focal lesions. These can be the only findings in a biopsy which may hint at, but

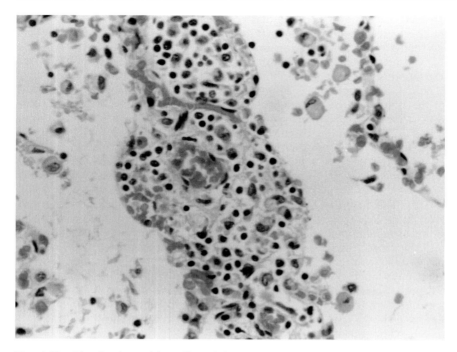

Fig. 6.11 Alveolar interstitium distended by lymphocytes in lymphocytic interstitial pneumonia.

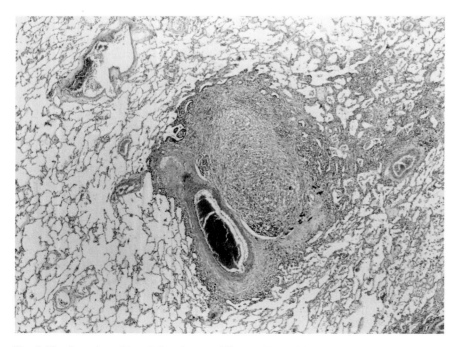

Fig. 6.12 Open-lung biopsy showing an obliterated bronchiole filled with inflammatory cells and fibrous tissue next to a pulmonary artery.

not be diagnostic of, obliterative bronchiolitis. Some authors indicate a high specificity and sensitivity of detecting obliterative bronchiolitis on transbronchial biopsy in lung transplant patients[127] while others give a less optimistic figure of only a 14% detection rate.[128]

Histiocytosis X

This condition affects young adults, predominantly males, and the majority of patients are cigarette smokers. Histiocytosis X tends to form nodular aggregates around terminal bronchioles and it is the presence of the characteristic cellular infiltrate of Langerhans' cells admixed with other inflammatory cells which is the diagnostic landmark (Fig. 6.13). Lesions can also be found beneath the pleura. Only interstitial nodular infiltrates of Langerhans' cells are regarded as diagnostic. On open lung biopsy, cellular, intermediate, fibrotic and cavitating lesions can be found, often all in the same biopsy. The majority of lesions are adjacent to or centred on bronchioles. On low-power examination, discrete focal nodular infiltrates are usually seen to be distributed evenly throughout the specimen with intervening normal lung parenchyma. The lesions often have a stellate border with finger-like extensions into adjacent alveolar interstitium. The majority of lesions have a diameter of 1–3 mm but some can reach 1–2 cm. Intra-alveolar collections of pigmented macrophages are often found adjacent to the nodules. Central cavitation can be seen which represents infiltration of bronchiolar wall with weakening and dilatation. The cellular lesions contain more than 50% of Langerhans' cells, with their distinctive folded grooved clear nuclei and eosinophilic cytoplasm (Fig. 6.13) admixed with lymphocytes, plasma cells and eosinophils. These are seen within the interstitium and can be missed with the interstitial expansion by inflammatory cells being regarded as interstitial pneumonia. One has to be highly sensitive to the possibility of this condition in the lung which, because of the variable histology, can be easily mistaken for other inflammatory conditions. The clinical history and radiological findings are essential to tie into the histology. The intermediate lesions have up to 50%

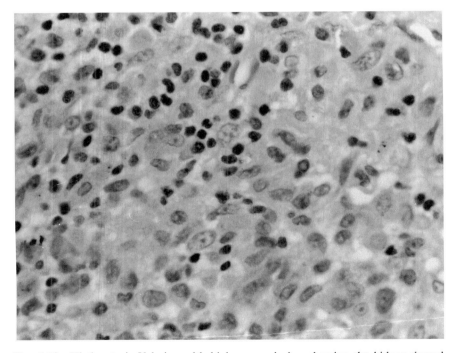

Fig. 6.13 Histiocytosis X lesion with high-powered view showing the kidney-shaped nuclei and granular cytoplasm of Langerhans' cells admixed with other inflammatory cells.

of Langerhans' cells with admixed fibrosis. Fibrotic lesions contain less than 10% of Langerhans' cells and consist of mainly acellular scars which retain the stellate configuration of the cellular lesions. This can be a useful feature in distinguishing it from other causes of pulmonary fibrosis. Associated with the fibrosis, dilatation of airways with honeycombing may occur with vascular thickening, mimicking fibrosing alveolitis or usual interstitial pneumonia. With the accumulation of macrophages in alveoli surrounding the lesions, one of the differential diagnoses is desquamative interstitial pneumonia. Reactive eosinophilic pleuritis may also be present in association with pneumothoraces, which are common complications in this condition.

A recent study by Travis *et al.*[129] showed that while open-lung biopsy usually gives the diagnosis, transbronchial biopsy can also be successful. Four out of ten transbronchial biopsies yielded diagnostic lesions which were interstitial nodular infiltrates of Langerhans' cells. S-100 immunostaining may be useful in confirming the presence of Langerhans' cells, since they can be mistaken for simple macrophages at light microscopy. The finding of the characteristic Birbeck granules at electron microscopy may also be useful but it must be remembered that the normal lung can contain Langerhans' cells which do increase in number with inflammation and fibrosis. In order to distinguish reactive Langerhans' cells from the large numbers found in histiocytosis X, it has been suggested that the finding of over 75 S-100 positive cells per ten high-powered fields in a nodular aggregate is diagnostic of histiocytosis X.[130] Since other cells can be positive for S-100 (chondrocytes, myoepithelial cells, Schwann cells) caution is advised in the interpretation of positivity. Positivity is diagnostically meaningful only in the appropriate histological setting already described.

Sarcoidosis

The diagnostic yield from transbronchial biopsy in sarcoidosis is generally high, rising from 57% in stage 1 (hilar adenopathy) to 91% in stage 3 disease (parenchymal disease alone).[131] The well-defined non-necrotizing granulomas are found in the bronchial wall, along lymphatic routes, blood vessels, interlobular septae and pleura. Usually the intervening lung appears normal, although it has been postulated that a lymphocytic alveolitis precedes granuloma formation. The granulomas can be seen

on occasion obstructing bronchioles and they can also infiltrate blood vessels. Eventually the granulomas undergo fibrosis, which leads to interstitial fibrosis and eventual honeycomb lung. The changes are usually bilateral and more prominent in the upper lobes. Conchoidal bodies also known as Schaumann bodies, are commonly found in sarcoid granulomas, within the cytoplasm of the multinucleated giant cells. They are larger than blue bodies, up to 80 µm in diameter, and have an irregular laminated densely calcified appearance. Often Schaumann bodies are all that are left to indicate the presence of previous granulomatous disease, but they can be found in any healed granulomatous process. The differential diagnosis includes granulomatous infection, berylliosis, extrinsic allergic alveolitis, aspiration pneumonia, intravenous drug abuse, granulomatous reaction to tumours, Wegener's granulomatosis, rheumatoid nodules and bronchocentric granulomatosis.

Alveolar lining/lumen

Diffuse alveolar damage

Diffuse alveolar damage occurs in the clinical setting of adult respiratory distress syndrome (ARDS) and up to 50% of cases are fatal.[132] The lungs become heavy and beefy-red at autopsy. This syndrome is due to a variety of insults which mainly affect the endothelial and epithelial cells of the pulmonary parenchyma.[133] The main cell lining the alveoli are the type 1 pneumocyte with the type 2 cell being much less common. The type 2 cell is the progenitor of the type 1 pneumocyte and regenerates rapidly in response to injury. Damage can result from many insults and the features of diffuse alveolar damage are very similar, regardless of the cause.[133]

The changes include oedema with fibrinous exudate as one of the earliest features of damage at the light microscopic level. Death of lining type 1 cells with detachment and hyaline membrane formation follows. Fibrin thrombi are present in alveolar capillaries. These are the findings of the early exudate phase of diffuse alveolar damage and are most prominent at 2–6 days following injury. Reactive type 2 cell hyperplasia occurs within 48–72 hours and appears as rows of cuboidal cells lining alveolar spaces. During this reparative process interstitial fibrosis becomes obvious also. With cytotoxic drugs such as bleomycin and busulphan, a characteristic feature is the presence of atypical type 2 cells. Interstitial and intra-alveolar organization may be encountered 6–7 days after the insult. This results in fibrosis of the interstitium and obliteration of alveoli, alveolar ducts and bronchioles. Radiation damage and paraquat poisoning can show prominent interstitial and intra-alveolar fibrotic changes. With extensive epithelial damage to bronchi, bronchioles and alveoli, reactive squamous metaplasia of lining cells can be prominent and this proliferating epithelium can simulate squamous cell carcinoma in an open-lung biopsy. Squamous metaplasia is particularly prominent at the edge of pulmonary infarcts. Bronchiolization of alveoli can occur with damage in which ciliated cuboidal/columnar epithelium can be found lining alveoli. This can be the result of ingrowth of epithelium from bronchioli or *in situ* metaplasia. Clues to the cause of diffuse alveolar damage may be observed with viral infections. However, the changes are general and will not point to a specific cause in most cases.

Katzenstein *et al.*[134] described eight patients with acute interstitial pneumonia or Hamman–Rich syndrome with rapid onset of respiratory symptoms resembling a viral illness in previously healthy young adults. The histological features on open-lung biopsy resembled those of diffuse alveolar damage in the reparative phase with variable numbers of chronic inflammatory cells and fibroblasts in the interstitium but little collagen deposition. There was proliferation of type 2 cells with remnants of hyaline membranes also. No cause for the lung disease could be elicited in these patients, the majority of whom succumbed rapidly to the disease.[134]

When a severely ill patient who is likely to be ventilated has an open-lung biopsy, the features are often of diffuse alveolar damage with no obvious pointer to the underlying aetiology; these features have been described as diagnostically non-specific. Our clinicians, however, find this category useful because they feel that they are not missing treatable conditions and the degree of fibrosis if present may also point to the long-term outlook for these patients.

Eosinophilic pneumonia

Pulmonary eosinophilia, the association of lung shadows with peripheral blood eosinophilia, was first described by Loeffler in 1932.[135] It occurs most commonly with allergic bronchopulmonary asperillosis (ABPA),[136] but also in association with nematode and filarial infection, drugs (particularly nitrofurantoin), bacterial infections and asthma. It can also be associated with a systemic vasculitis as the Churg–Strauss syndrome, but this condition is rare.[136] In up to 20% of patients it occurs as a primary event, known as cryptogenic pulmonary eosinophilia[137] or chronic eosinophilic pneumonia.[138] The latter is a better term from the pathologist's viewpoint since earlier definitions included blood eosinophilia, which may not be present in all cases.

Microscopically, the ground-glass radiographic appearance is attributed to the flooding of air spaces by many eosinophils and macrophages, often with an abundant eosinophilic exudate. Many of the macrophages are multinucleate. There is type 2 cell hyperplasia and expansion of the interstitium by inflammatory cells, mainly eosinophils, but some lymphocytes and plasma cells are also seen. The macrophages within the alveoli are pinker than usual due to the ingestion of eosinophil granules. Clumps of eosinophilic cells are seen undergoing necrosis within alveoli with basophilic debris in the centre of these clumps and many Charcot–Leyden crystals. Such lesions are known as eosinophilic abscesses. Some, but not all, of these foci are surrounded by a rim of plump epithelioid cells and Langhans'-type giant cells. Rarely small sarcoid-like granulomas can be seen away from areas of necrosis. Organizing pneumonia with Masson bodies within alveoli and bronchioles can be a prominent feature and may even obscure the eosinophils. The differential diagnoses include other granulomatous lung diseases including tuberculosis, sarcoidosis, desquamative interstitial pneumonia, histiocytosis X or rarely Hodgkin's disease. Eosinophils are not seen in tuberculosis or sarcoidosis and only a rare eosinophil is seen in desquamative interstitial pneumonia. However, in eosinophilic pneumonia macrophages can sometimes predominate so the picture may overlap with this condition. Histiocytosis X usually has fewer eosinophils with nodular collections of Langerhans' cells near bronchioles and many pigmented macrophages.

Response to steroids is rapid with dramatic clearing of radiological changes[139] in both acute and chronic forms. Recurrences can occur, particularly in patients with ABPA, which can lead to permanent lung damage. However this is rare in cryptogenic cases.[139]

Alveolar lumen

Neutrophils are seen within the alveolar lumen in bacterial pneumonia and within alveoli surrounding bronchioles in bronchopneumonia. Intra-alveolar macrophages are a feature of desquamative interstitial pneumonia. Intra-alveolar foamy macrophages are seen in obstructive pneumonia and obliterative bronchiolitis, where the vacuolization is often fine. Multinucleate giant cell macrophages, often containing cholesterol crystal clefts, may also be seen in obstructive pneumonia, as well as mucin accumulation within alveoli and bronchioles. In exogenous lipid pneumonia the cells can be distended in a 'Swiss-cheese' pattern. Patients on the cardiac drug amiodarone can have accumulation of fine foamy macrophages in the alveolar lumen because the drug blocks lysosomal activity in the macrophages, allowing the accumulation of phospholipids.

Intra-alveolar macrophages filled with haemosiderin as demonstrated with Perl's stain, can be found with any bleeding disorder, Goodpasture's syndrome, pulmonary vasculitis, veno-occlusive disease, chronic venous congestion, connective tissue disease and lymphangioleiomyomatosis. Therefore the presence of haemosiderin-laden macrophages will not point to a specific diagnosis, but only indicate previous haemorrhage, whatever the cause. Small numbers of haemosiderin-positive macrophages can be seen in smokers, so large numbers are required to indicate haemorrhage. Prolonged chronic bleeding leads to interstitial fibrosis with accumulation of haemosiderin and calcium in blood vessel walls, alveolar septa and interstitial macrophages.

Idiopathic pulmonary haemosiderosis

Idiopathic pulmonary haemosiderosis is a diagnosis of exclusion and requires a detailed clinical history. Great care must be taken, especially with transbronchial biopsy in which fresh blood may fill alveoli, to avoid reaching a diagnosis of pulmonary haemorrhage because the trauma of biopsy can lead to this appearance. Idiopathic pulmonary haemosiderosis occurs in childhood and young adulthood. The disease is characterized by recurrent haemoptysis with iron deficiency anaemia and can be fatal, especially in children under 10 years of age.[140] The cause of the disease is unknown but there is an association with coeliac disease, dermatitis herpetiformis, autoimmune haemolytic anaemia and cow's milk sensitivity. High-resolu-

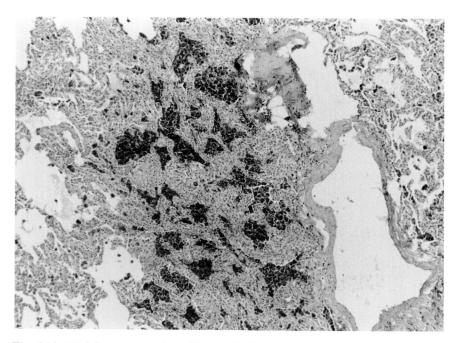

Fig. 6.14 Nodular accumulation of haemosiderin-laden macrophages within alveoli in idiopathic pulmonary haemosiderosis. Perl's stain.

tion computed tomography shows a nodular pattern which correlates with nodular accumulation of haemosiderin-laden macrophages within alveoli (Fig. 6.14).[141] Pulmonary calcification is often found with chronic haemorrhage, especially in children. Basement membranes and elastic fibres are common sites of iron and calcium deposition and the walls of blood vessels and alveoli are particularly affected. The iron-impregnated elastic fibres tend to fragment and induce a foreign body giant cell reaction which can be very prominent in some lungs.

Alveolar lipoproteinosis

The alveoli are filled with eosinophilic granular debris often containing cholesterol clefts in alveolar lipoproteinosis (Fig. 6.15). The material can be surrounded by multinucleate giant cells also containing cholesterol clefts, but it is usually free of other cells. The material is PAS-positive and at electron microscopy contains numerous electron-dense lamellar membranes similar to surfactant. The material will stain for surfactant antibodies. The major differential diagnosis is *Pneumocystis carinii* pneumonia, which should have a more cystic appearance. Most cases are idiopathic but it can be seen in association with silicosis and infections, particularly in immunosuppressed individuals and especially with nocardiosis. It can also occur in children with haematological malignancies. Treatment involves bronchoalveolar lavage with removal of the material, which has a milky appearance.

Microlithiasis

Microlithiasis is a rare condition mainly affecting adults. It is characterized by a distinctive micronodular calcification within the alveoli. The microliths are large, up to 300 μm in diameter, and have a wavy concentrically laminate structure which fills the alveoli. There is usually no inflammatory reaction in the alveolar wall. The radiological appearance is classical with fine calcified opacities throughout the lungs, so that biopsy is rarely required.

Cryptogenic organizing pneumonia

Although changes of organizing pneumonia have long been recognized in postmortem lungs,[142,143]

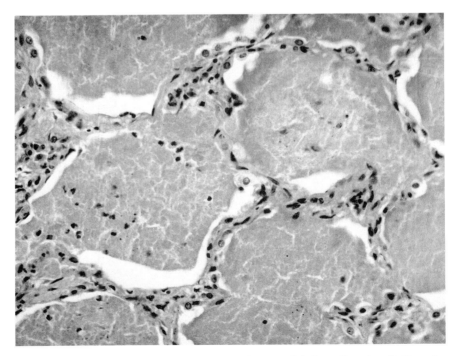

Fig. 6.15 Alveolar lipoproteinosis with granular material distending alveoli. Note the lack of an inflammatory reaction.

only in the 1980s have they been reported in association with a clinical syndrome unrelated to previous infection.[144-146] This entity has been defined as cryptogenic or idiopathic organizing pneumonia (COP)[145] or as bronchiolitis obliterans organizing pneumonia (BOOP).[146] Organizing pneumonia is characterized by the presence of small buds of granulation tissue (also called Masson bodies or bouton conjunctive) in alveoli and alveolar ducts, and infiltration of alveolar walls with chronic inflammatory cells with preservation of alveolar architecture.[145] Bronchioli can also contain buds of granulation tissue, which is the feature emphasized in BOOP, but the buds are mainly in alveoli with involvement of bronchioli being a secondary phenomenon. In addition there is preservation of the bronchiolar wall architecture with the Masson bodies attached at one point only in the wall and extending out to fill the lumen. There is never fibrous obliteration of the bronchiolar wall. It is now accepted that idiopathic BOOP and COP are the same condition and represent a distinct clinical entity which responds impressively to corticosteroids.[144] COP has been reported in association with rheumatoid arthritis[149] and as a seasonal disease with biochemical cholestasis.[150]

As histological changes may be non-specific and seen in other conditions, it is essential to stick to stringent clinical criteria in diagnosing COP/BOOP.[151] Diagnosis on transbronchial biopsy has been reported.[145,150,152] However, as the process is often focal with variation in the radiological changes,[145,153] strict criteria should be used to define organizing pneumonia histologically on transbronchial biopsy. In a recent study we looked at 11 cases of clinically diagnosed COP. The essential pathological feature on transbronchial biopsy was the presence of buds of granulation tissue (Masson bodies), indicating organizing of a persistent exudate by fibroblasts and capillaries within alveoli, with preservation of the alveolar architecture. These buds could be seen as rounded bodies (Fig. 6.16a) or as streaming sheaths extending from one alveolus to another through the pores of Kohn (Fig. 6.16b). The elastic van Gieson stain shows that the Masson bodies stain very faintly for collagen. In addition, both acute and chronic inflammatory cells are present in the interstitium. These features, combined with the clinical history, were sufficient to come to a diagnosis in seven of the 11 cases on transbronchial biopsy. Five of the six cases coming to open-lung biopsy confirmed COP with Masson bodies within

alveoli with extension into bronchioles. However these changes were focal in three cases, and very few Masson bodies were observed in one case. One other case which had the features on transbronchial biopsy lacked them in the open-lung biopsy. Therefore transbronchial biopsy can yield diagnostic material in the majority of patients with COP while open-lung biopsy, which is considered the gold standard for interstitial lung disease, may yield negative results because of sampling error and the rapid evolution and changing pattern of the disease.

Interstitium

Cryptogenic fibrosing alveolitis/interstitial pneumonia

The interstitium is the area between opposing alveoli and consists of alveolar capillaries with a few lymphocytes, elastic fibres and fibroblasts with little collagen. Liebow and Carrington classified interstitial reactions into several morphological categories with the general label of interstitial pneumonia.[154] In the UK Scadding used the term 'fibrosing alveolitis' to describe this disorder of the lower respiratory tract characterized by inflammation and fibrosis of the pulmonary interstitium and peripheral air spaces.[155] This leads to derangement of the alveolar wall and loss of functional alveolar capillary units resulting in a restrictive pattern on lung function tests. The term has been used to describe a large number of pulmonary diseases including those caused by inhaled dusts such as asbestos or the administration of drugs such as amiodarone, gold, busulphan and bleomycin, or those associated with collagen vascular diseases such as rheumatoid arthritis and systemic sclerosis. However, in 75% of cases no causative agent or associated condition can be found and these are referred to as cryptogenic fibrosing alveolitis or idiopathic interstitial fibrosis. Its prevalence in the UK is five per 100 000 of the population and mortality is increasing with 1200–1400 people dying each year from the condition.

Many studies, particularly in animals exposed to toxic agents such as bleomycin or asbestos, support the concept that direct injury to capillary endothelial cells, shared basement membrane and pneumocyte type 1 epithelial cells, leads to interstitial fibrosis.[156] Furthermore in the adult respiratory distress syndrome, acute lung injury

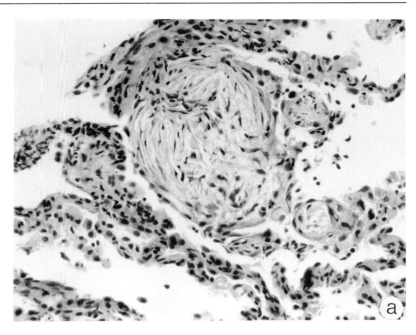

Fig. 6.16a Alveolus containing a Masson body consisting of granulation tissue.

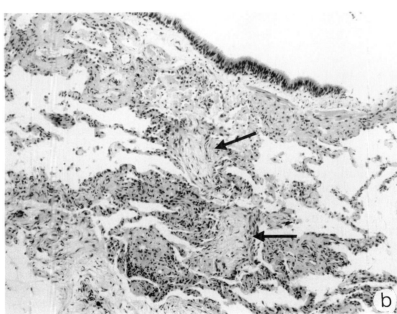

Fig. 6.16b Transbronchial biopsy showing a stream of granulation tissue extending between two alveoli (arrow).

results in similar damage and survivors often develop diffuse interstitial fibrosis. Injury is believed to be due either to inhalation of injurious agents or to blood-borne agents. Many consider that viral infections play a role but this has never been proven. In general the initiating agent(s) in fibrosing alveolitis remain unknown.

Interstitial chronic inflammatory cells and fibrosis occur in the chronic interstitial pneumonias/fibrosing alveolitis diseases which affect the lower lobes of the lung most severely. A trichrome stain to demonstrate the normal architecture and collagen deposition in the interstitium is essential for judging the extent of interstitial fibrosis and architecture destruction. In an open-lung biopsy the inflammation and fibrosis is usually subpleural, most marked in the lower lobes and can often be focal (Fig. 6.17). In what is called usual interstitial pneumonia (UIP), there is a mixture of chronic inflammatory cells, mainly lymphocytes, and plasma cells in the interstitium with fibrosis and expansion of the interstitium and accumulation of macrophages in the alveolar lumen. There is also marked type 2 cell hyperplasia. In the interstitium there can also be hyperplasia of intrapulmonary lymphoid follicles that are particularly prominent in connective tissue diseases such as rheumatoid arthritis. Another feature is intimal proliferation and thickening of the media of small pulmonary arteries in the areas of interstitial fibrosis. There can also be smooth muscle hyperplasia around terminal and respiratory bronchioles resulting in 'muscular cirrhosis' which becomes most marked in the later stages of the disease. However, even in the same biopsy there can be extreme variation in the histological features, ranging from increased interstitial cellularity to dense fibrous scarring. A very cellular biopsy with little fibrosis can indicate a better prognosis and response to therapy.[156] An important negative finding is that the pleura is not involved in the inflammatory process.

If there is pleural thickening and inflammation in association with interstitial pneumonia/fibrosis, the patient probably has a connective tissue disease such as rheumatoid arthritis or systemic lupus erythematosus. The pathologist should look out for necrobiotic nodules with palisading histiocytes in rheumatic lung disease as well as a vasculitis, haemorrhage or secondary amyloidosis. Interstitial pneumonia/fibrosis can also occur in systemic sclerosis as well as pulmonary vascular changes with intimal onion-skin-like fibrosis and medial hypertrophy with mucinous foci which can cause pulmonary hypertension.

'Honeycomb' lung or cyst formation occurs with severe fibrosis and can be seen in any condition causing interstitial fibrosis. The cysts are due to either distension of residual bronchioles or the bronchiolization of spaces formed by the interstitial expansion. Often these spaces can become acutely infected or mucus accumulation occur with the formation of multinucleate macrophage giant cells containing cholesterol clefts within these

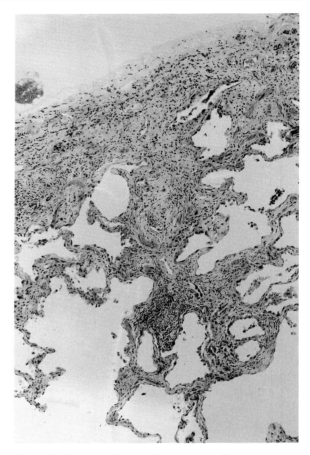

Fig. 6.17   Subpleural interstitial chronic inflammation and fibrosis in cryptogenic fibrosing alveolitis/usual interstitial pneumonia.

spaces. Sometimes these multinucleate giant cells can accumulate in the interstitium and may lead to the erroneous impression that the interstitial fibrosis is due to extrinsic allergic alveolitis.

Desquamative interstitial pneumonia

In contrast to usual interstitial pneumonia/filnosing alveolitis, which has a poor prognosis, desquamative interstitial pneumonia is characterized by the accumulation of macrophages within alveoli with minimal inflammation or fibrosis of the interstitium. It was originally labelled as desquamative in the erroneous belief that the cells were pneumocytes. The appearance must be uniform throughout the biopsy before the above label is put on it since the condition responds to steroids while UIP response is more variable. Blue bodies may be prominent in the macrophages and alveoli. This condition is seen in younger patients than those presenting with usual interstitial pneumonia/filnosing alveolitis and can also occur in children.

Giant cell interstitial pneumonia

As described by Liebow and Charrington,[154] numerous multinucleated giant cells fill the alveoli and the interstitium contains small numbers of lymphocytes, plasma cells and macrophages. The giant cells in the alveoli are derived from macrophages and macrophages can be multinucleate in many conditions, so this is not unique to giant cell pneumonia. The most important feature is that the lining pneumocytes are also multinucleate. It is now established that most cases are caused by exposure to hard metals containing cobalt and tungsten. The main differential diagnosis is measles pneumonia.

Pulmonary vessels

Wegener's granulomatosis

This vasculitis is typically associated with lesions in the upper airways, lungs and kidneys. Rare cases are limited to the lung. Open-lung biopsy usually confirms the diagnosis and there are few cases diagnosed on transbronchial biopsy alone.[157] In Wegener's granulomatosis the chief features are areas of geographic basophilic necrosis surrounded by palisading histiocytes, multinucleate giant cells, mixed neutrophilic and granulomatous inflammation and vasculitis. In the areas of mixed inflammation there are often scattered multinucleate giant cells without well-formed granulomas. The presence of well-formed granulomas is against the diagnosis of Wegener's. The blood vessels away from areas of inflammation and necrosis should be examined. Look for inflammation of the vessel walls composed of histiocytes and lymphocytes associated with destruction of the wall as demonstrated by fragmentation of the internal and external elastic lamina. An elastic stain is essential in the analysis of these cases.[157] Giant cells can be a prominent part of the infiltrate in the vessel walls. Involvement of muscular arteries, veins and capillaries with a mixed neutrophilic and granulomatous inflammation and areas of necrosis are the chief features essential to the diagnosis of Wegener's granulomatosis. There can be consider-

able variability in the intensity and extent of lesions at any one time. Special stains must be done to eliminate infection. The presence in the serum of anti-neutrophil cytoplasmic antibodies (ANCA), which give a granular immunofluorescence in neutrophils, has proven very useful in the diagnosis of this condition, especially in the active phase.[158] As a result the use of open-lung biopsy has been reduced in cases that clinically are obviously positive.

Transbronchial biopsy rarely shows all the features of Wegener's granulomatosis. In only four of 57 biopsies was vasculitis identified and granulomatous inflammation was noted in three of these four.[158]

Capillaritis with neutrophilic infiltration of capillaries and interstitium, fibrinoid necrosis of capillaries and extravasation of red blood cells and haemorrhage may also be seen. Capillaritis with haemorrhage can be seen in other systemic vasculitic conditions such as Behçet's disease, connective tissue diseases, Goodpasture and Henoch–Schönlein purpura. Patients respond well to steroids combined with cyclophosphamide.

Allergic angiitis and granulomatosis (Churg–Strauss syndrome)

Patients often have a history of asthma and peripheral eosinophilia. Microscopically the lesions resemble chronic eosinophilic pneumonia. However a number of features are present which distinguish the two conditions. Large foci of necrosis with a prominent granulomatous reaction around the necrotic foci and an eosinophilic vasculitis, often with a giant cell reaction, are present. Involvement of small muscular arteries, arterioles, and venules is typical of the vasculitis associated with the Churg–Strauss syndrome. Like Wegener's granulomatosis, the prognosis for Churg–Strauss syndrome has been dramatically altered by the use of cyclophosphamide combined with steroids.

Necrotizing sarcoid granulomatosis

Nodular lesions are typically seen in the lung with numerous sarcoid-like granulomas in association with areas of necrosis and vasculitis affecting arteries and veins. Granulomas may also be seen in the bronchiolar walls. The relationship of this condition, which is extremely rare, to sarcoidosis is controversial. Extrathoracic involvement is rare and other features of sarcoidosis can be lacking.

The syndrome has an excellent prognosis, responding well to steroids or spontaneously resolving.

Lymphomatoid granulomatosis

Atypical lymphoid cells infiltrating pulmonary vessel walls should suggest lymphomatoid granulomatosis. Another useful feature is that the elastic lamina is preserved in this condition. Most cases are malignant T cell lymphomas.

Lymphangioleiomyomatosis

This is a rare disease confined to women in the reproductive years of life which affects lymphatics and lymph nodes of the mediastinum, retroperitoneum and lung. It can be associated with tuberous sclerosis. The pulmonary disease is characterized by diffuse or focal proliferation of immature smooth muscle cells which involve terminal bronchioles, arterioles, venules and lymphatics and diffusely thicken the surrounding alveolar septa. Airway obstruction gives rise to characteristic cysts. The differential diagnosis includes interstitial fibrosis, with which it is often confused. The use of antibodies to smooth muscle actin may aid diagnosis.[159] It has been claimed that transbronchial biopsy combined with immunocytochemistry using antibodies to melanoma-related antigen HMB45 can also provide a diagnosis in the appropriate clinical setting.[160] Diffuse haemorrhage is common because of the obstruction of veins that occurs and may obscure the histology. The relevant clinical history with cyst formation, pneumothoraces and chylous effusions in young females will often point to the diagnosis.

Amyloidosis

The lung can be involved in secondary amyloidosis with diffuse interstitial deposition. Pulmonary blood vessels are also involved. Usually the pulmonary involvement is an incidental finding at autopsy and rarely gives rise to symptoms. The interstitial thickening can be mistaken for fibrosis but the lack of an inflammatory infiltrate and the deep pink colour of the infiltrate should point to the diagnosis. The amyloid stains for Congo red with apple-green birefringence on polarization. It is usually of the AA type and is digested by potassium permanganate treatment.

Amyloid primary to the lung parenchyma usually takes the form of multiple nodules replacing the pulmonary parenchyma and shows extensive plasma cell infiltration and multinucleate giant cells around the nodules of amyloid. It is very similar to amyloid involving the airways and is of the AL type. It has no association with amyloid elsewhere or previous lung disease.

Malignancy

Both primary and metastatic tumours can diffusely involve the lung in the form of lymphangitis carcinomatosa or they can permeate veins; such involvement can be detected by both transbronchial and open-lung biopsy. Lymphomas, both B and T cell in type, can diffusely involve the lung, with the T cell lymphoma having an angiocentric pattern in many cases and leukaemia can be seen in the interstitium and pulmonary blood vessels.[133] In addition bronchioloalveolar cell carcinoma (BAC) can mimic interstitial lung disease with diffuse involvement of one or several lobes of the lung. Extreme care should be taken in diagnosing bronchioloalveolar carcinoma on a transbronchial biopsy because: (1) tissue is limited; (2) any tumour, particularly adenocarcinomas (both primary and metastatic), can grow along the alveoli at the periphery and mimic BAC; and (3) reactive type 2 pneumocytes can look bizarre and malignant in interstitial lung disease and mimic BAC, especially if the patient has had radiation or chemotherapy. It is only the mucin-producing type that can be diagnosed with confidence on a transbronchial biopsy in the presence of normal lung parenchyma (Fig. 6.18). Thus one must pay careful attention to the lung parenchyma. However, metastatic colonic adenocarcinoma can mimic exactly mucin-positive bronchioloalveolar cell carcinoma.

Summary

Both transbronchial and open-lung biopsy are commonly used for interstitial lung diseases. Detailed clinical and radiological information are required in order that the biopsies can be interpreted in a meaningful and useful manner for the clinician. Thus clinico-pathological conferences are essential to make the most use of the tissue provided by biopsy, which often involves using a general anaesthetic in an already-ill patient. It is the duty of the pathologist to use all the techniques at his disposal to maximize the histological find-

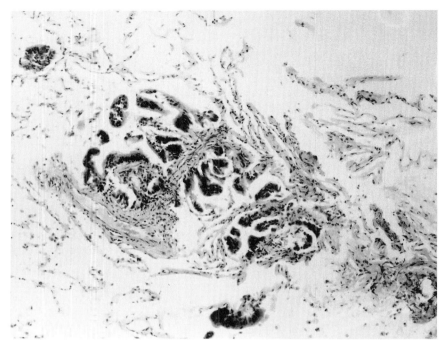

Fig. 6.18 Open-lung biopsy showing a mucin-positive bronchioloalveolar cell carcinoma growing along the alveolar walls.

ings and be alert to often subtle features which may point to a helpful diagnosis.

References

1. Mortenson RL, Bogin RM, King TE. Interstitial lung diseases. In: *Recent Advances in Respiratory Medicine*, vol. 5, Mitchell DM (ed.). Edinburgh: Churchill, Livingstone, 1991 pp. 163–184.
2. Turner-Warwick M. Infiltrative and interstitial lung disease. In: *Respiratory Medicine*, Brewis RAL, Gibson GJ, Geddes DM (eds). London: Balliere Tindall, 1990, pp. 1078–1087.
3. Lacronique J, Roth C, Battesti J-P, Basset F, Chretien J. Chest radiological features of pulmonary histiocytosis X: a report based on 50 adult cases. *Thorax* 1982; **37**, 104–113.
4. Stovin PGJ, Lum LC, Flower CDR, Darke CS, Belley M. The lungs in lymphangiomyomatosis and tuberous sclerosis. *Thorax* 1975; **30**, 497–509.
5. Corrin B, Liebow AA, Friedman PJ. Pulmonary lymphangiomyomatosis. *Am J Pathol* 1975; **79**, 348–367.
6. du Bois RM. Cryptogenic fibrosing alveolitis. In: *Respiratory Medicine*, Brewis RAL, Gibson GJ, Geddes DM (eds). London: Balliere Tindall, 1990, pp. 1088–1103.
7. Spiro SG, Lopez-Vidrieto MT, Charman J, Das I, Reid L. Bronchorrhoea in a case of alveolar call carcinoma. *J Clin Pathol* 1975; **28**, 60–65.
8. Hance AJ, Basset P, Sammon G *et al.* Smoking and interstitial lung disease. The effect of cigarette smoking on the incidence of pulmonary histiocytosis X and sarcoidosis. *Ann NY Acad Sci* 1986; **465**, 643–656.
9. Warren CPW. Extrinsic allergic alveolitis: a disease commoner in non-smokers. *Thorax* 1977; **32**, 567–569.
10. Douglas JG, Middleton WG, Gaddie J, Petrie GR, Choo-Kang YFJ, Prescott RJ, Crompton GK. Sarcoidosis: a disorder commoner in non-smokers? *Thorax* 1986; **41**, 787–791.
11. Walker WC. Pulmonary infections and rheumatoid arthritis. *Q J Med* 1967; **36**, 239–251.
12. Caplan A. Certain unusual radiological appearances in the chest of coal miners suffering from rheumatoid arthritis. *Thorax* 1953; **8**, 29–37.
13. Walker WC, Wright V. Rheumatoid pleuritis. *Ann Rheum Dis* 1967; **26**, 467–474.
14. Kay JM, Banik S. Unexplained pulmonary hypertension with pulmonary arteritis in rheumatoid disease. *Br J Dis Chest* 1977; **71**, 53–59.
15. Geddes DM, Corrin B, Brewerton DA. Progressive airway obliteration in adults and its association with rheumatoid disease. *Q J Med* 1977; **46**, 427–444.

16. Strimlan CV, Rosenow EC, Divertie MB, Harrison EG. Pulmonary manifestations of Sjögren's syndrome. *Chest* 1976; **70**, 354–361.
17. Stack BHR, Grant IWB. Rheumatoid interstitial lung disease. *Br J Dis Chest* 1965; **59**, 202–211.
18. Petrie GR, Bloomfield P, Grant IWB, Crompton GK. Upper lobe fibrosis and cavitation in rheumatoid disease. *Br J Dis Chest* 1980; **74**, 263–267.
19. Massato D, Katz S, Matthews MJ, Higgins G. Von Recklinghausen's neurofibromatosis associated with cystic lung disease. *Am J Med* 1975; **38**, 233–240.
20. Fletcher CM. Pneumoconiosis of coal miners. *Br Med J* 1948; **1**, 1015–1065.
21. Lee DHK, Selikoff IJ. Historical background to the asbestos problem. *Environ Res* 18, 300–314.
22. Hardy HL, Tabershaw IR. Delayed chemical pneumonitis occurring in workers exposed to beryllium compounds. *J Industr Hyg Toxicol* 1946; **28**, 197.
23. Seal RME, Hapley EJ, Thomas GO, Meek JC, Hayes M. The pathology of the acute and chronic stages of farmers' lung. *Thorax* 1968; **23**, 469–500.
24. Pearsall HR, Morgan EH, Tesluk H, Beggs D. Parakeet dander pneumonitis. Acute psittacokerato-pneumoconiosis. *Bull Mason Clin* 1960; **14**, 127–137.
25. McKusick VA, Fisher AM. Congenital cystic disease of the lung with progressive pulmonary fibrosis and carcinomatosis. *Ann Inst Med* 1958; **48**, 774–790.
26. Alton E, Turner-Warwick M. Lung involvement in scleroderma. In: *Systemic Sclerosis: Scleroderma*, Jayson MIV, Black CM (eds). New York: J Wiley & Sons Ltd, 1988, pp. 181–205.
27. Stoeckle JD, Hardy HL, Weber AL. Chronic beryllium disease. *Am J Med* 1969; **46**, 545–561.
28. DeRemee RA, Harrison EG, Anderson HA. The concept of classic interstitial pneumonitis – fibrosis (CIP-F) as a clinicopathologic syndrome. *Chest* 1972; **61**, 213–220.
29. Crofton JW, Livingstone JL, Oswald NC, Roberts ATM. Pulmonary eosinophilia. *Thorax* 1952; **7**, 1–35.
30. Newman-Taylor AJ. Pulmonary eosinophilia. *Med Int* 1991; **90**, 3769–3774.
31. Turner Warwick M, Doniach D. Autoantibody studies in interstitial pulmonary fibrosis. *Br Med J* 1965; **1**, 886–891.
32. Van der Woude FJ, Rasmussen N, Lobatto S *et al.* Autoantibodies against neutrophils and monocytes: tool for diagnosis and marker of disease activity in Wegener's granulomatosis. *Lancet* 1985; **1**, 425–429.
33. Do Pico GA, Reddan WG, Chmelik F, Peters ME, Reed CE, Rankin J. The value of precipitating antibodies in screening for hypersensitivity pneumonitis. *Am Rev Respir Dis* 1976; **113**, 451–455.
34. Lieberman J. Elevation of serum angiotensin converting enzyme (ACE) levels in sarcoidosis. *Am J Med* 1975; **59**, 365–372.
35. Turton CWG, Grundy E, Firth G, Mitchell D, Rigden BG, Turner-Warwick M. Value of measuring serum angiotensin converting enzyme and serum lysozyme in the management of sarcoidosis. *Thorax* 1979; **34**, 57–62.
36. Austrian R, McClement JH, Renzett AD. Clinical and physiological features of some types of pulmonary diseases with impairment of alveolar capillary diffusion. The syndrome of alveolar capillary block. *Am J Med* 1951; **11**, 667–685.
37. Wiggins J, Strickland B, Turner-Warwick M. Combined cryptogenic fibrosing alveolitis and emphysema: the value of high resolution computed tomography in assessment. *Respir Med* 1990; **84**, 365–369.
38. Kerr IH. Radiology of interstitial lung disease. In: *Seminars in Respiratory Medicine*, vol. 6, no. 1, Petty TL, Cherniak RM (eds). New York: Thieme Stratton Inc., 1984, 80–91.
39. Smellie H, Hoyle C. The natural history of pulmonary sarcoidosis. *Q J Med* 1960; **29**, 539–559.
40. Turner-Warwick M. Cryptogenic fibrosing alveolitis. *Br J Hosp Med* 1972; **7**, 697–704.
41. Armstrong, P. Radiological signs of diseases of the lungs. In: *Imaging of Diseases of the Chest*, Armstrong P, Wilson AG, Dee P (eds). Chicago: Yearbook Medical Publishers Inc. 1990, pp. 75–151.
42. White FE, Simpson W. Radiographic images. In: *Respiratory Medicine*, Brewis RAL, Gibson GJ, Geddes DM (eds). London: Ballière Tindall, 1990, pp. 244–287.
43. Epler GR, McLoud TC, Gaensler EA, Mikus JP, Carrington CB. Normal chest roentgenograms in chronic diffuse infiltrative lung disease. *New Engl J Med* 1978; **298**, 935–939.
44. Gaensler EA, Carrington CB. Open lung biopsy for chronic diffuse infiltrative lung disease: clinical, roentgenographic and physiological correlations in 502 patients. *Ann Thorac Surg* 1980; **30**, 411–426.
45. Muller NL. Clinical value of high resolution CT in chronic diffuse lung disease. *Am J Radiol* 1991; **157**, 1163–1170.
46. Mathieson JR, Mayo JR, Staples CA, Muller NL. Chronic diffuse infiltrative lung disease: comparison of diagnostic accuracy of CT and chest radiography. *Radiology* 1989; **171**, 111–116.
47. Padley SPG, Hansell DM, Flower CDR, Jennings P. Comparative accuracy of high resolution computed tomography and chest radiology in the diagnosis of chronic diffuse infiltrative lung disease. *Clin Radiol* 1991; **44**, 227–231.
48. Grenier P, Valeyre D, Cluzel P, Brauner MW, Lenoir S, Chastang C. Chronic diffuse interstitial

lung disease: diagnostic value of chest radiography and high-resolution CT. *Radiology* 1991; **179**, 123–132.

49. Hruban RH, Meziane MA, Zerhouni EA, Khouri NF, Fishman EK, Wheeler PS, Dumler S, Hutchins GM. High resolution computed tomography of inflation-fixed lungs. *Am Rev Respir Dis* 1987; **136**, 935–940.

50. Webb WR, Stein MG, Finkbeiner WE, Im JG, Lynch D, Gamsu G. Normal and diseased isolated lungs: high-resolution CT. *Radiology* 1988; **166**, 81–87.

51. Itoh H, Murata K, Konishi J, Nishimura K, Kitaichi M, Izumi T. Diffuse lung disease: pathologic basis for the high-resolution computed tomography findings. *J Thorac Imag* 1993; **8**, 176–188.

52. Webb WR, Muller NL, Naidich DP. Standardized terms for high resolution lung CT: a proposed glossary. *J Thorac Imag* 1993; **8**, 167–175.

53. Stein MG, Mayo J, Muller N, Aberle DR, Webb MR, Gamsu G. Pulmonary lymphangitic spread of carcinoma: appearance on CT scans. *Radiology* 1987; **162**, 371–375.

54. Hansell DM, Kerr IH. The role of high resolution computed tomography in the diagnosis of interstitial lung disease. *Thorax* 1991; **46**, 77–84.

55. Merchant RN, Pearson MG, Rankin RN, Morgan WKC. Computerized tomography in the diagnosis of lymphangioleiomyomatosis. *Am Rev Respir Dis* 1985; **131**, 295–297.

56. Moore ADA, Godwin JD, Muller NL, Naidich DP, Hammar SP, Buschmann DL, Takasugi JE, de Carvalho CRR. Pulmonary histiocytosis X: comparison of radiographic and CT findings. *Radiology* 1989; **172**, 249–254.

57. Muller NL, Kullnig P, Miller RR. The CT findings of pulmonary sarcoidosis: analysis of 25 patients. *Am J Radiol* 1989; **152**, 1179–1182.

58. Muller NL, Miller RR. State of the art: Computed tomography of chronic diffuse infiltrative lung disease: Part 2. *Am Rev Respir Dis* 1990; **142**, 1440–1448.

59. Muller NL, Staples CA, Miller RR, Vedal S, Thurlbeck WM, Ostrow DN. Disease activity in idiopathic pulmonary fibrosis: CT and pathologic correlation. *Radiology* 1987; **165**, 731–734.

60. Wells, AU, Hansell DM, Corrin B *et al*. High resolution computed tomography assessment of disease activity in the fibrosing alveolitis of systemic sclerosis: a histopathological correlation. *Thorax* 1992; **47**, 738–742.

61. Lee JS, Im JG, Ahn JM, Kim YM, Han MC. Fibrosing alveolitis: prognostic implication of ground-glass attenuation at high-resolution CT. *Radiology* 1992; **184**, 451–454.

62. Vedal S, Welsh EV, Miller RR, Muller NL. Desquamative interstitial pneumonia: computed tomographic findings before and after treatment with corticosteroids. *Chest* 1988; **93**, 215–217.

63. Wells AU, Hansell DM, Rubens MB, Cullinan P, Black CM, du Bois RM. The predictive value of appearances on thin section computed tomography in fibrosing alveolitis. *Am Rev Respir Dis* 1993; **148**, 1076–1082.

64. Brauner MW, Lenoir S, Grenier P, Cluzel P, Battesti JP, Valeyre D. Pulmonary sarcoidosis: CT assessment of lesion reversibility. *Radiology* 1992; **182**, 349–354.

65. Silver SF, Muller NL, Miller RR, Lefcoe MS. Hypersensitivity pneumonitis: Evaluation with CT. *Radiology* 1989; **173**, 441–445.

66. Daniele RP, Elias JA, Epstein PE, Rossman MD. Bronchoalveolar lavage: role in the pathogenesis, diagnosis and management of interstitial lung disease. *Ann Intern Med* 1985; **102**, 93–108.

67. Basset F, Soler P, Jaurand MC, Bignon J. Ultrastructural examination of bronchalveolar lavage for diagnosis of pulmonary histiocytosis X: Preliminary report on four cases. *Thorax* 1977; **32**, 303–306.

68. Mierau GW, Favara BE. S100 protein immunohistochemistry and electron microscopy in the diagnosis of Langerhans cell proliferative disorders: A comparative assessment. *Ultrastruct Pathol* 1986; **10**, 303–309.

69. Martin RJ, Coalson JJ, Rogers RM, Horton FO, Manus LE. Pulmonary alveolar proteinosis: the diagnosis by segmental lavage. *Am Rev Respir Dis* 1980; **121**, 819–825.

70. Stack BHR, St. J. Thomas J. Pulmonary infiltrations. In: *Respiratory Medicine*, Brewis RAL, Gibson GJ, Geddes DM (eds). London: Ballière Tindall, 1990, pp. 1183–1197.

71. Parkes WR. *Occupational Lung Disorders*, 2nd edn. London: Butterworths, pp. 464–467.

72. Morgan PGM, Turner-Warwick M. Pulmonary haemosiderosis and pulmonary haemorrhage. *Br J Dis Chest* 1981; **75**, 225–242.

73. Jaurand MC, Gaudichet A, Atassi K, Sebastien P, Bignon J. Relationship between the number of asbestos fibres and the cellular and enzymatic content of bronchoalveolar fluid in asbestos-exposed subjects. *Bull Eur Physiopathol Respir* 1980; **16**, 595–606.

74. Williams WR, Jones-Williams W. Development of beryllium lymphocyte transformation tests in chronic beryllium disease. *Int Arch Allergy Appl Immunol* 1982; **67**, 175–180.

75. Jones-Williams W, Wallach ER. Laser probe mass spectrometry (LAMMS) analysis of beryllium, sarcoidosis and other granulomatous diseases. *Sarcoidosis* 1991; **6**, 111–117.

76. White EW, Denney PJ, Irving SM. Quantitative microprobe analysis of microcrystalline powders. In: *The Electron Microprobe*. New York: J Wiley & Sons, 1966, pp. 791–804.

77. Dauber JH, Rossman MD, Daniele RP. Broncho-alveolar cell populations in acute sarcoidosis: observations in smoking and non-smoking patients. *J Lab Clin Med* 1979; **94**, 862–871.

78. Daniele RP, Dauber JH, Altose MD, Rowlands DT, Gorenberg DJ. Lymphocyte studies in asymptomatic smokers. *Am Rev Respir Dis* 1977; **116**, 997–1005.

79. Hunninghake GW, Crystal RG. Pulmonary sarcoidosis: A disorder mediated by excess helper T-lymphocyte activity at sites of disease activity. *New Engl J Med* 1981; **305**, 429–434.

80. Weinberger SE, Kelman J, Elson NA, Young RC, Reynolds HY, Fulmer JD. Bronchoalveolar lavage in interstitial lung disease. *Ann Intern Med* 1978; **89**, 459–466.

81. Schatz M, Patterson R, Fink J, Moore VL. Pigeon Breeders' disease ii: pigeon antigen induced proliferation of lymphocytes from symptomatic and asymptomatic subjects. *Clin Allergy* 1976; **6**, 7–17.

82. Ainslie GM, Solomon JA, Bateman ED. Lymphocytes and lymphocyte subset numbers in blood and in bronchoalveolar lavage and pleural fluid in various forms of human pulmonary tuberculosis at presentation and during recovery. *Thorax* 1992; **47**, 513–518.

83. Haslam PL, Turton CWG, Heard B, Lukoszek A, Collins JV, Salsbury AJ, Turner-Warwick M. Bronchoalveolar lavage in pulmonary fibrosis: comparison of cells obtained with lung biopsy and clinical features. *Thorax* 1980; **35**, 9–18.

84. Rudd RM, Haslam PL, Turner-Warwick M. Cryptogenic fibrosing alveolitis. Relationships of pulmonary physiology and bronchoalveolar lavage to response to treatment and prognosis. *Am Rev Respir Dis* 1981; **124**, 1–8.

85. Studdy PR, Rudd RM, Gellert AR, Uthaya Kumar S, Sinha G, Geddes DM. Bronchoalveolar lavage in diffuse pulmonary shadowing. *Br J Dis Chest* 1984; **78**, 46–54.

86. Scadding JG. Lung biopsy in the diagnosis of diffuse lung disease. *Br Med J* 1970; **2**, 557–564.

87. Lane DJ. Lung biopsy. In: *Respiratory Medicine*, Brewis RAL, Gibson GJ, Geddes DM (eds). London: Ballière Tindall, 1990, pp. 329–338.

88. Kirschner. Die Probebohrung. *Schweiz Med Wochenschr* 1935; **65**, 28–30.

89. Steel SJ, Winstanely DP. Trephine biopsy of the lung and pleura. *Thorax* 1969; **24**, 576–584.

90. King EG, Bachynski JE, Mielke B. Percutaneous trephine lung biopsy: evolving role. *Chest* 1976; **70**, 212–216.

91. Castillo G, Ahmad M, van Ordstrand HS, McCormack LJ. Trephine drill biopsy of the lung. *J Am Med Assoc* 1974; **228**, 189–191.

92. Gaensler EA. Open and closed lung biopsy. In: *Diagnostic Techniques in Pulmonary Disease*, Sackner MA (ed.). New York: Marcel Dekker, 1981.

93. Wall CP, Gaensler EA, Carrington CB, Hayes JA. Comparison of transbronchial and open biopsies in chronic infiltrative lung diseases. *Am Rev Respir Dis* 1981; **123**, 280–285.

94. Gilman MJ, Wang KP. Transbronchial lung biopsy in sarcoidosis. *Am Rev Respir Dis* 1980; **122**, 721–724.

95. Haponik EF, Summer WR, Terry PB, Wang KP. Clinical decision making with transbronchial lung biopsies. *Am Rev Respir Dis* 1982: **125**, 524–529.

96. Mitchell DM, Emerson CJ, Collins JV, Stableforth DE. Transbronchial lung biopsy with the fibreoptic bronchoscope. Analysis of results in 433 patients. *Br J Dis Chest* 1981; **75**, 258–262.

97. Ray JF, Lawton BR, Myers WO, Toyama WM, Reyes CN, Emanuel DA, Buras JL, Pederson DP, Dovenbarger WV, Wenzel FJ, Sautter RD. Open pulmonary biopsy. *Chest* 1976; **69**, 43–47.

98. Burt ME, Flye MW, Webber BL, Wesley RA. Prospective evaluation of aspiration needle, cutting needle, transbronchial and open lung biopsy in patients with pulmonary infiltrates. *Ann Thorac Surg* 1981; **32**, 146–153.

99. Gaensler EA, Carrington CB. Open lung biopsy for chronic diffuse infiltrative lung disease: clinical, roentgenographic and physiological correlations in 502 patients. *Ann Thorac Surg* 1980; **30**, 411–426.

100. Venn GE, Kay PH, Midwood CJ, Goldstraw P. Open biopsy in patients with diffuse pulmonary shadowing. *Thorax* 1985; **40**, 931–935.

101. Graeve AH, Saul VA, Aki BF. Role of different methods of lung biopsy in the diagnosis of lung lesions. *Am J Surg* 1980; **140**, 742–746.

102. Cherniak RM, Colby TV, Flint A, Thurlbeck WM, Waldron J, Ackerson L, King TE. Quantitative assessment of lung pathology in idiopathic pulmonary fibrosis. *Am Rev Respir Dis* 1991; **144**, 892–900.

103. Wong PS, Goldstraw, Kaplan D. Early experience with video assisted thoracoscopic lung biopsy. *Thorax* 1993; **48**, 440 (abstract).

104. Fechner RE, Greenburg SD, Wilson RK. Evaluation of transbronchial biopsy of the lung. *Am J Clin Pathol* 1977; **68**, 17–24.

105. Haponik EF, Summer WR, Terry PB, Wang KP. Clinical decision making with transbronchial biopsy. *Am Rev Respir Dis* 1982; **125**, 524–528.

106. Wall CP, Gaensler EA, Carrington CP, Hayes JA. Comparison of transbronchial and open lung biopsies in chronic infiltrative lung disease. *Am Rev Respir Dis* 1981; **123**, 280–288.

107. Griffiths MM, Kocjan G, Miller RF, Godfrey-Faussett P. Diagnosis of pulmonary disease in human immunodeficiency virus infection: role of transbronchial biopsy and bronchoalveolar lavage. *Thorax* 1989; **44**, 554–558.

108. Murray JF, Felton CP, Garay S. *et al*. Pulmonary

complications of the acquired immunodeficiency syndrome. Report of the National Heart Lung and Blood Institute Workshop. *New Engl J Med* 1984; **310**, 1682–1688.

109. Baughman RP. Use of bronchoscopy in the diagnosis of infection in the immunocompromised host. *Thorax* 1994; **49**, 3–7.

110. Milligan SA, Luce JM, Golden J *et al.* Transbronchial biopsy without fluroscopy in patients with diffuse roentographic infiltrates and the acquired immune deficiency syndrome. *Am Rev Respir Dis* 1988; **137**, 486–488.

111. Meduri GU, Stover DE, Lee M *et al.* Pulmonary Kaposi's sarcoma in the acquired immunodeficiency syndrome. *Am J Med* 1986; **81**, 11–13.

112. Fitzgerald W, Bevelaqua FA, Garay SM, Aranda CP. The role of open lung biopsy in patients with the acquired immunodeficiency syndrome. *Chest* 1987; **91**, 659–661.

113. Sattersfield JR, McLaughlin JS. Open lung biopsy in diagnosing pulmonary infiltrates in immunosuppressed patients. *Ann Thorac Surg* 1979; **28**, 359–362.

114. Venn GE, Kay PH, Midwood CJ, Goldstraw P. Open lung biopsy in patients with diffuse pulmonary shadowing. *Thorax* 1985; **40**, 931–935.

115. Donnelly RJ, Page RD, Cowen ME. Endoscope assisted microthoractomy: initial experience. *Thorax* 1992; **47**, 490–493.

116. Carnochan FM, Walker WS, Cameron EWJ. Efficacy of video assisted thoracoscopic lung biopsy – an historical comparison with open lung biopsy. *Thorax* 1994; **49**, 361–363.

117. Soosay GN, Baudouin SV, Hanson PJV *et al.* Symptomatic cysts in otherwise normal lungs of children and adults. *Histopatnol* 1992; **20**, 517–522.

118. Turner JS, Evans TE, Hunter DN *et al.* Adult respiratory distress syndrome. Advances in diagnosis and ventilatory management [clinical conference]. *Br Med J* 1990; **301**, 1087–1089.

119. Coleman A, Colby TV. Histologic diagnosis of extrinsic allergic alveolitis. *Am J Surg Pathol* 1988; **12**, 514–518.

120. Bienenstock J, Johnston N, Perey DYE. Bronchial lymphoid tissue. 1. Morphologic characterstics. *Lab Invest* 1973; **28**, 686–692.

121. Gould SJ, Isaacson PI. Bronchus-associated lymphoid tissue (BALT) in human fetal and infant lung. *J Pathol* 1993; **169**, 229–234.

122. Pabst R. Is BALT a major component of the human lung immune system? *Immunol Today* 1992; **13**, 119–121.

123. Yousem SA, Colby TV, Carrington CB. Follicular bronchitis/bronchiolitis. *Hum Pathol* 1985; **16**, 700–706.

124. Yousem SA, Colby TV, Carrington CB Lung biopsy in rheumatoid arthritis. *Am Rev Respir Dis* 1985; **131**, 770–777.

125. Fortoul TI, Cano-Valle F, Oliva E, Barrios R. Follicular bronchiolitis in association with connective tissue disease. *Lung* 1985; **163**, 305–314.

126. McLoud TC, Epler GR, Colby TV, Gaensler EA, Carrington CB. Bronchiolitis obliterans. *Radiology* 1986; **159**, 1–8.

127. Yousem SA, Paradis I, Griffith BP. Can transbronchial biopsy aid in the diagnosis of bronchiolitis obliterans in lung transplant recipients? *Transplantation* 1994; **57**, 151–153.

128. Kramer MR, Stoehr C, Whang JL *et al.* The diagnosis of obliterative bronchiolitis after heart-lung and lung transplantation. Low yield of transbronchial lung biopsy. *J Heart Lung Transplant* 1993; **12**, 675–681.

129. Travis WD, Borok Z, Roum JH *et al.* Pulmonary Langerhans cell granulomatosis (histiocytosis X). *Am J Surg Pathol* 1993; **17**, 971–986.

130. Webber D, Tron V, Askin F, Churg A: S-100 staining in the diagnosis of eosinophilic granuloma of lung. *Am J Clin Pathol* 1985; **84**, 447–452.

131. Koontz CH, Joyner LR, Nelson RA. Transbronchial lung biopsy via the fibreoptic bronchoscope in sarcoidosis. *Ann Intern Med* 1976; **85**, 64–70.

132. Rinaldo JE, Rogers RM. Adult respiratory distress syndrome. *New Engl J Med* 1986; **315**, 578–579.

133. Doran H, Sheppard MN, Collins PW, Jones L, Newland AC, Van Der Walt JD. Pathology of the lung in leukaemia and lymphoma: a study of 87 autopsies. *Histopathol* 1991; **18**, 211–219.

134. Katzenstein A-LA, Myers JL, Mazur MT. Acute interstitial pneumonia. *Am J Surg Pathol* 1986; **10**, 256–267.

135. Loeffler W. Zur differential-diagnose der lungen infiltrier ungen 11. Uben fluchtige Succedan-Infiltrate (mit eosinophilie). *Beitr Z Klin D Tuberk* 1932; **79**, 368–392.

136. McCarthy DS, Pepys J. Allergic bronchopulmonary aspergillosis. Clinical immunology: (1) Clinical features. *Clin Allergy* 1971; **1**, 261–286.

137. McCarthy D, Pepys J. Cryptogenic pulmonary eosinophilia. *Clin Allergy* 1973; **3**, 339–349.

138. Liebow AA, Carrington CB. The eosinophilic pneumonias. *Medicine* 1969; **48**, 251–285.

139. Capewell S, Chapman BJ, Alexander F, Greening AP, Crompton GK. Corticosteroid treatment and prognosis in pulmonary eosinophilia. *Thorax* 1989; **44**, 925–929.

140. Bush A, Sheppard MN, Warner JO. Idiopathic pulmonary haemosiderosis, clinical findings and response to chloroquine. *Arch Dis Child* 1992; **67**, 625–627.

141. Cheah FK, Sheppard MN, Hansell DM. Computed tomography of diffuse pulmonary haemorrhage with pathological correlation. *Clin Radiol* 1993; **48**, 89–93.

142. Milne LS. Chronic pneumonia (including a discussion of two cases of syphilis of the lung). *Am J Med Sci* 1911; **142**, 408–438.

143. Floyd R. Organisation of pneumonic exudates. *Am J Med Sci* 1922; **163**, 527–548.

144. Grinblat J, Mechlis S, Lewitus Z. Organizing pneumonia-like process: an unusual observation in steroid responsive cases with features of chronic interstitial pneumonia. *Chest* 1981; **80**, 259–263.

145. Davison AG, Heard BE, McAllister WAC, Turner-Warwick M. Cryptogenic organizing pneumonitis. *Q J Med* 1983; **52**, 382–394.

146. Epler GR, Colby T, McLoud TC, Carrington CB, Gaensler EA. Bronchiolitis obliterans organizing pneumonia. *New Engl J Med* 1985; **312**, 152–158.

147. Izumi T, Kitaichi M, Nishimura K, Nagai S. Bronchiolitis obliterans organizing pneumonia – clinical features and differential diagnosis. *Chest* 1992; **102**, 715–719.

148. Editorial. Organizing pneumonia: COP/BOOP and SOP. *Lancet* 1992; **340**, 699–670.

149. Rees JH, Woodhead MA, Sheppard MN, du Bois RM. Rheumatoid arthritis and cryptogenic organising pneumonitis. *Respir Med* 1991; **85**, 243–246.

150. Spiteri MA, Klenerman P, Sheppard MN, Padley S, Clarke TJK, Newman-Taylor A. Seasonal cryptogenic organising pneumonia with biochemical stasis: a new clinical entity. *Lancet* 1993; **540**, 281–284.

151. Bellomo R, Finlay M, McLaughlin P, Tai E. The wider clinical spectrum of cryptogenic organising pneumonitis. *Thorax* 1991; **46**, 554–558.

152. Cordier JF, Loire R, Brune J. Idiopathic bronchiolitis obliterans organizing pneumonia: definition of characteristic clinical profiles in a series of 16 patients. *Chest* 1989; **96**, 999–1004.

153. Miyagawa Y, Nagata N, Shigematsu N. Clinicopathological study of migratory lung infiltrates. *Thorax* 1991; **46**, 233–238.

154. Liebow AA, Carrington CB. The interstitial pneumonias. In: *Frontiers of Pulmonary Radiology*, Simon M, Potchen EJ, LeMay M (eds). New York: Grune and Stratton, 1969, pp. 102–141.

155. Scadding JG. Fibrosing alveolitis. *Br Med J* 1964; **2**, 686.

156. Sheppard MN, Harrison NK. New perspectives on basic mechanisms in lung disease. 1. Lung injury, inflammatory mediators, and fibroblast activation in fibrosing alveolitis. *Thorax* 1992; **47**; 1064–1074.

157. Lombard CM, Duncan SR, Rizk NW, Colby TV. The diagnosis of Wegener's granulomatosis from transbronchial biopsy specimens. *Hum Pathol* 1990; **21**, 838–842.

158. Hoffman GS, Kerr GS, Leavitt RY *et al.* Wegener's granulomatosis: an analysis of 158 patients. *Ann Intern Med* 1992; **116**, 488–498.

159. Matthews TJ, Hornall D, Sheppard MN. Comparison of the use of antibodies to alpha-smooth muscle actin and desmin in pulmonary lymphangioleiomyomatosis. *J Clin Pathol* 1993; **46**, 479–480.

160. Bonetti F, Chiodera PL, Pea M *et al.* Transbronchial biopsy in lymphangiomyomatosis of the lung – Hmb45 for diagnosis. *Am J Surg Pathol* 1993; **17**, 1092–1102.

7

Pulmonary vascular disease

C A Wagenvoort ─────────────────────────

Introduction

The pulmonary circulation differs from the systemic circulation in its functions, its haemodynamics and its vascular morphology. Each of these three factors has a bearing upon the pathology of the lung vessels, which in turn may affect the pulmonary circulation. This explains the considerable differences existing between pulmonary and systemic vascular pathology.

The main function of the pulmonary circulation is gas exchange. Changes in the composition of the respiratory air, such as alveolar hypoxia, affect the lung vessels, eventually causing vascular alterations. There is also a sieve function of the lung vessels; thromboemboli, when arrested in pulmonary arterial branches, produce characteristic lesions which subsequently may cause an increased pulmonary vascular resistance.

Haemodynamic changes are responsible for much pulmonary vascular pathology. This applies especially to pulmonary hypertension, that is a mean pulmonary arterial pressure of over 25 mmHg at rest. However, not only an increased mean pulmonary arterial pressure but also changes in pulse pressure, left atrial pressure, and pulmonary arterial flow, have a distinct and specific relation to the pathology of the lung vessels and may in turn be influenced by these lesions.

In the normal pulmonary circulation, the pressure is low and the flow high as compared to the systemic circulation. Correspondingly, pulmonary arterial branches are wide and have a thin muscular coat; also alveolar capillaries and pulmonary veins are relatively wide. In pulmonary hypertension, the mean pressure may be four or five times as high as

normal, so that adaptation of and damage to the thin arterial walls will leave their imprint on the vasculature.

There is always a strong interaction between pulmonary arteries and the heart. Obstruction of lung vessels and pulmonary hypertension lead to hypertrophy and often to failure of the right ventricle. Pulmonary vascular disease is not only often the cause but may also be the result of heart disease, congenital as well as acquired. The mechanisms involved vary but most often it concerns a left-to-right shunt as in ventricular septal defect, or a pulmonary venous outflow obstruction as in mitral valve disease. Moreover various forms of lung disease may affect the lung vessels.

Lesions in the pulmonary vasculature are exceedingly common in the sense that they may be found, generally in small numbers, in the vast majority of hospital autopsies, when multiple blocks of lung tissue are studied. Usually they have no clinical significance. Extensive vascular alterations, however, are found in the presence of pulmonary hypertension.

Lung biopsies

Autopsies always provide lung tissue for the study of pulmonary vascular disease, in adequate quantity and from different areas. It is far more complicated where lung biopsies are concerned. Lung vessels cannot be studied in transbronchial biopsies in which, if present at all, they are scarce and often distorted by the procedure. An open-lung biopsy is required, but this means that the patient is subjected to an intrathoracic operation with concomitant distress and risk. Particularly in patients with pulmonary hypertension, fatalities have been reported as a result of the procedure, although with modern anaesthesia and good monitoring of the patient this risk appears to be minimal. It is imperative, however, that when it is decided to carry out a biopsy, the benefit for the patient should be optimal. This means that the surgeon should take a piece of lung tissue of adequate size with as little damage to it as possible, while the pathologist should handle and study the specimen with the utmost care.

For the purpose of evaluating pulmonary vascular pathology, the optimal size of a biopsy specimen is approximately the size of that part of the patient's thumb which is distal to a line along the base of the nail. This comparison has the advantage that it can be used independent of the age and stature of the patient.[1] The surgeon must realize that lung tissue damaged by clamps and immediately adjacent to such an area is generally worthless for a judgement.

To increase the chance that focal lesions are represented in the histological slides, the specimen is cut in slices 2–3 mm thick parallel to the surgical cutting plane. Usually three or four pieces are obtained in this way. Blood vessels in collapsed and particularly in compressed lung tissue are usually difficult to assess. The vessel wall or its layers may seem much thicker than they really are, while some lesions may completely escape attention within the compressed tissue. This can be countered by vacuum fixation of the specimen, which provides both removal of air bubbles and expansion of lung tissue, including its vessels. This method is still useful when the specimen has been received in fixative. Histological slides, cut from the embedded specimen, preferably at various levels, are stained by haematoxylin and eosin (H & E), a good elastic stain and a stain for iron pigment. Other stains are rarely essential.

Normal pulmonary vasculature

Pulmonary trunk and main pulmonary arteries are of the elastic type but, unlike the wall of the aorta, in which elastic lamellae are complete and arranged parallel to each other in a regular fashion, in the wall of extrapulmonary lung vessels, elastic laminae are disrupted and often fragmented. In contrast, within the lungs all pulmonary arteries down to a calibre of approximately 1mm have a very regular elastic configuration with intact and parallel laminae. There is a gradual transition from elastic to muscular type of artery as the external diameter decreases from 1mm to 0.5mm.

Muscular pulmonary arteries are generally of a calibre between 500 and 1000 µm and, just as their larger parent arteries, closely follow the bronchial ramifications. There are, however, also many so-called supernumerary muscular arteries which do not fit into a dichotomous branching pattern but which arise more or less perpendicularly from elastic or muscular arteries. Muscular arteries have a media consisting of circularly arranged smooth muscle cells, bounded on either side by an elastic lamina. The thickness of this media is approximately 5% of the external arterial diameter. The endothelium rests almost immediately upon the inner elastic lamina. Between an external diameter of 100 and 50 µm the arterial branches gradually lose their muscular coat. Since smooth muscle cells now follow a spiral rather than a circular course around the vessel, the media appears to be interrupted in a histological section before the arteriole becomes completely non-muscular. Such branches have also lost their association with bronchioles and have become intra-acinar where they break up into networks of alveolar capillaries.

From these networks pulmonary venules arise, initially with a wall consisting of an endothelial layer resting upon a basement membrane. Such venules cannot be distinguished from non-muscular arterioles if not traced in serial sections. When larger veins are formed by merging of venules, smooth muscle cells appear in their walls along with collagen and elastic fibres which are arranged in an irregular way. There is no muscular coat sharply delimited by elastic laminae, as in the muscular arteries. There are no valves in pulmonary veins. The veins are located in the interlobular fibrous septa.

In the perinatal period there are some important differences in comparison to the adult pulmonary vasculature, particularly regarding the arteries. The transition from an elastic to a muscular type of artery takes place at a calibre in the range of 200 µm. During fetal life and at birth, the muscular pulmonary arteries are thick-walled with a media between 15 and 20% of the external diameter. The smallest branches in particular are exceedingly narrow, reflecting the high resistance in this period. Medial thinning takes place rapidly during the first 4 postnatal weeks and more gradually during the rest of the first year of life.

Bronchial arteries have a thicker media than pulmonary arteries of the same calibre but on average the medial thickness is somewhat less than in other systemic arteries. Often bronchial arteries and collateral systemic arteries within pleura and interlobular septa have bundles of longitudinal smooth muscle cells within their intima. These bundles are more numerous and prominent under pathological conditions but may be found in normal individuals, even sometimes in children. Anastomoses are very common between bronchial and pulmonary veins but scarce or absent where arteries are concerned, at least under normal circumstances.

Classification of pulmonary vascular disease

Pathological alterations in lung vessels are particularly common, and may have serious implications, in patients with pulmonary hypertension. However, hypertensive pulmonary vascular disease is not a morphological entity. There are various pathogenetic mechanisms which alone or in combination may produce certain patterns of microscopic lesions, and thus various types of pulmonary vascular disease. The aetiology shows an even greater variation. Agents or stimuli of a completely different nature may elicit the same pathogenetic sequence and thus the same type of disease.

The pathologist is therefore in a position to recognize a morphological entity within the pulmonary vasculature. He may also identify a group of possible aetiological agents. However, he should be careful in diagnosing pulmonary hypertension, not only because establishing pressures is not in his domain, but because pressures and severity of lesions do not always correlate well. In some forms of pulmonary vascular disease very extensive changes are found in the presence of normal or near-normal pressures, while in other instances severe pulmonary hypertension is accompanied by lesions that are scattered over the lung tissue, so that they may be scarce in a random slide. Even so, there are occasional instances in which it may be useful to suggest to the clinician the possibility of an elevated pulmonary arterial pressure.

Vascular disease, whether or not associated with pulmonary hypertension, may be classified according to its pattern of lesions. Moreover, the stage, severity and possible reversibility of the condition can be assessed, potentially with important consequences for diagnosis, prognosis and treatment.

It should be realized that in the early stages of a disease, vascular lesions may be scarce or inconspicuous, and sometimes unreliable for identification of a pattern. Moreover, a combination of more than one pattern may occur, for instance when any form of pulmonary vascular disease is complicated by thrombosis or chronic congestion of the lungs. However, in the vast majority of cases, the pathologist who is familiar with the various forms of pulmonary vascular disease will be able to make a definite diagnosis on a lung biopsy specimen, which will be of help to the clinician and the patient. The various forms of pulmonary vascular disease that can be recognized on morphological grounds are listed in Table 7.1.

Table 7.1 Morphological forms of pulmonary vascular disease

Classification of pulmonary vascular disease
Plexogenic arteriopathy
Thrombotic arteriopathy
Hypoxic arteriopathy
Congestive vasculopathy
Pulmonary veno-occlusive disease
Persistent fetal circulation syndrome
Misalignment of lung vessels
Pulmonary vascular medial defects
Capillary haemangiomatosis
Vasculopathy in lung fibrosis
Non-thrombotic embolism
Vasculopathy in diminished pulse pressure

Terminology

Considerable confusion has arisen by loose or incorrect use of terminology or by the use of differing definitions. Before embarking upon a description of the various forms of pulmonary vascular disease, it may be useful to pay some attention to this problem. Here we will discuss only terms known to have been defined in various ways, or to have been misunderstood, or otherwise to have caused problems.

Arteriole, Venule. There have been multiple attempts to define an 'arteriole', and to a lesser extent a 'venule', but throughout this book we use these words in a rather non-specific way as small artery and small vein, respectively.

Arteriopathy, Vasculopathy. Arteriopathy is used as a general term for a vascular disease limited to (e.g. plexogenic), or predominantly affecting (e.g. hypoxic) the arterial part of the pulmonary vasculature; vasculopathy is used when both arteries and veins are involved.

Medial hypertrophy. By definition, hypertrophy refers to an increase in size of cells, in this case of smooth muscle cells. There is also an increase in number of these cells. Since it is rather cumbersome to speak of both hypertrophy and hyperplasia, common usage of 'medial hypertrophy' for both processes, is followed here.

Intimal fibrosis, concentric-laminar. This type of intimal thickening has an onion-skin arrangement, while the lumen is approximately in the centre. Sometimes the term 'concentric intimal fibrosis' is used but this may be confused with a non-laminar type which happens to be circumferential. Concentric-laminar is therefore preferable.

Intimal fibrosis, eccentric. This is one of the non-laminar forms of intimal thickening which is observed for instance as a result of organization of thrombi or of chronic congestion. It must be realized that occasionally this type of intimal fibrosis may be circumferential. As it is non-laminar, the term concentric should be avoided in these instances.

Plexogenic arteriopathy. As we will see, this is a specific disease of the pulmonary vasculature with a variety of arterial alterations developing sequentially. In the final stages of this condition the most characteristic of these changes is the so-called plexiform lesion. This instigated the name 'plexogenic' for the whole disease,[2] but it is clear that plexiform lesions are not necessarily present. The term 'plexiform' arteriopathy, which is sometimes used, is misleading and confusing, even if applied only to the final stage of the disease.

Thrombotic arteriopathy. This condition is based upon thrombi and their sequelae, irrespective of their origin. Thus the name applies to cases of primary thrombosis as well as to thromboembolism. Of course, the lesions should be generalized enough, or severe enough, to warrant the term 'arteriopathy'.

Pulmonary veno-occlusive disease. Often the lesions in this condition are not limited to the venous part of the pulmonary vasculature. For that reason the term pulmonary vaso-occlusive disease has been proposed.[3] We have kept the current term in view of its general usage.

Primary pulmonary hypertension. Originally, its morphological substrate was generally accepted to be the primary form of plexogenic arteriopathy. Later it became clear that there was more than one

disease of the pulmonary vasculature, not associated with any identifiable cause. Since in many instances the clinician was unable to establish the aetiology, a World Health Organization (WHO) committee[2] suggested the term 'primary pulmonary hypertension' for clinically unexplained pulmonary hypertension. It listed as most common examples: primary plexogenic arteriopathy, silent thromboembolic pulmonary hypertension and pulmonary veno-occlusive disease. It has since been pointed out[4,5] that it is better to speak of 'unexplained' than of 'primary' pulmonary hypertension, since secondary forms of pulmonary hypertension (thromboembolism) fall into this category. It also cannot easily be accepted as a diagnosis, since a whole variety of conditions potentially may remain unexplained. More recently primary pulmonary hypertension was more strictly defined by requiring careful exclusion of secondary causes and by listing criteria for investigative methods.[6] However, even if it were possible to exclude every secondary form, which is unlikely, as a diagnosis the term has limited value since it would include multiple unrelated conditions. When the morphological substrate has been identified by the pathologist, it is imperative to use the name of that condition; if not, then the term unexplained pulmonary hypertension is preferable.

Cor pulmonale. Defined as right ventricular hypertrophy due to pulmonary hypertension, this clinical term is not very satisfactory because it is often used in a loose way. It should be avoided by the pathologist.

Plexogenic arteriopathy

Plexogenic pulmonary arteriopathy is a disease of the muscular pulmonary arteries which in its complete form shows a characteristic pattern of lesions. Some of these lesions are pathognomonic for this condition. The pathogenesis is intriguing and the aetiology is varied so that completely different agents may be responsible for the disease, which in turn influences clinical behaviour, treatment and prognosis.

Aetiology and pathogenesis

The most well-known cause of plexogenic arteriopathy is congenital cardiovascular disease with a left-to-right shunt such as a ventricular septal defect or patent ductus arteriosus, to a lesser extent an atrial septal defect. In some cases of transposition of the great arteries, this arteriopathy may develop in the absence of a septal defect. Rarely an acquired left-to-right shunt may be responsible, as in tetralogy of Fallot when too large a surgical shunt is created.

In a very small percentage of patients with hepatic disease, pulmonary plexogenic arteriopathy develops. This may possibly explain its occurrence in patients with visceral involvement by schistosomiasis, since there is usually pipe-stem fibrosis of the liver. Anorexigenic drugs are known to produce plexogenic arteriopathy in some of those who use them. This was first observed with the drug Aminorex fumarate in Central Europe 25 years ago.[7] There is, however, an increasing number of cases attributed to other anorexigens, particularly fenfluramine derivatives. The Spanish toxic oil syndrome may also be associated with plexogenic arteriopathy[8] and so is seropositivity for human immunodeficiency virus. Finally, there is a rare primary form.

In all these instances, but particularly in primary plexogenic arteriopathy, a hyperreactivity of the peripheral lung vessels which is genetically determined, is likely to play an important part in the pathogenesis of the disease. There is, however, little doubt that the pathogenesis is complicated and multifactorial. Vasoconstriction of muscular pulmonary arteries is almost certainly one factor but it is likely that endothelial injury and possibly sheer stress are also responsible. Immunological factors have been implicated because of the association with some collagen diseases.

Morphology

From biopsy studies in congenital heart disease in man[9,10] and from experimental work, we know that there is a sequence of lesions in the muscular pulmonary arteries and arterioles (Fig. 7.1). Medial hypertrophy of muscular arteries and muscularization of arterioles are the first morphological expressions of the disease. This increased muscularity is roughly proportional to pressure and resistance in the pulmonary circulation. In infants in whom these changes are almost always the only alterations, they can be particularly striking. The average thickness of the media as percentage of the external arterial diameter may increase from a normal 5% to 20% or more. Since medial hypertrophy and arteriolar muscularization

occur in several forms of hypertensive pulmonary vascular disease, their significance for the diagnosis is limited.

Medial hypertrophy of muscular arteries is followed by intimal thickening which begins as a cellular intimal proliferation. Myofibroblasts and smooth muscle cells from the media penetrate through the fenestrations of the internal elastic lamina to form a cellular layer that may become so thick that it virtually occludes the lumen. This change is particularly striking in small muscular arteries close to their point of origin from a parent artery.

Gradually collagen and elastin are deposited within the intimal layer so that the cells are compressed or in the end replaced by fibrotic tissue. This happens in a peculiar way by formation of concentrically arranged layers by which the intimal thickening acquires an onion-skin appearance. This concentric-laminar intimal fibrosis may lead to complete obliteration of the artery. If distinctly present it is pathognomonic for plexogenic arteriopathy. It may occasionally be mimicked by thrombotic intimal fibrosis but this rarely produces confusion. When in doubt, it is important to look for the onion-skin arrangement of the intimal fibrosis in a haematoxylin stain, where it should always be clearly recognizable. Concentric-laminar intimal fibrosis is commonly observed in combination with cellular intimal proliferation, which forms the inner layer. In contrast to thrombotic changes of the intima, the concentric-laminar form rarely contains recanalization channels, which, if present, are very narrow. In arterial branches completely obstructed by concentric-laminar intimal fibrosis, contraction of the obliterated vessel leads to loss of recognizable layers and consequently to loss of its diagnostic significance.

Dilation of the arterial tree may be generalized. In histological slides it is only recognizable when it is pronounced, as often happens when the pulmonary blood flow is greatly enlarged as may occur in atrial septal defect. This dilatation may cause thinning of the muscular coat, thereby masking medial hypertrophy. It can be considered an adaptation rather than a lesion.

Completely different alterations are the so-called dilatation lesions. These are localized dilatations of muscular pulmonary arteries, again with a preference for the proximal parts of arterial branches. Dilatation lesions occur in various forms[11] which should be distinguished since they may differ with regard to clinical implications.

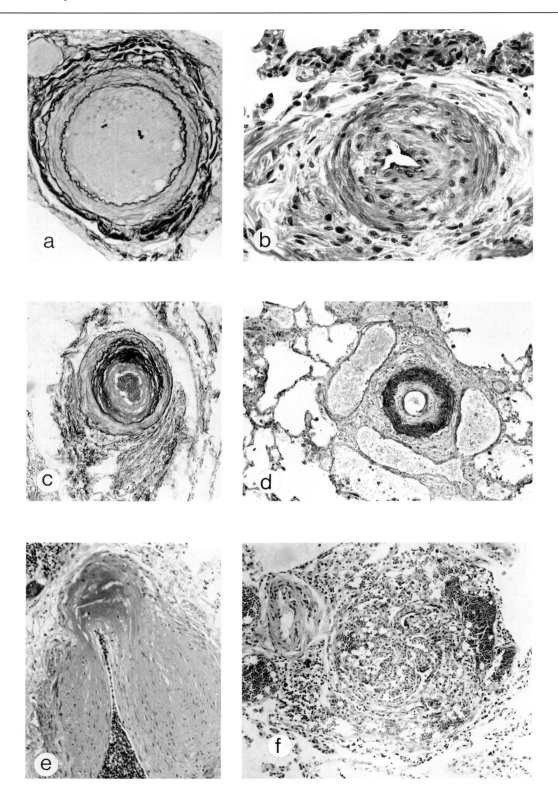

Often branches of a thick-walled muscular artery are dilated over a certain distance. The wall of these 'vein-like branches' is very thin and may have lost its entire muscular coat, in striking contrast to the hypertrophied wall of the parent artery. In a more severe stage of the disease, clusters of dilatation lesions are formed; occasionally these clusters of extremely thin-walled branches may become so large as to mimic angioma's: the so-called angiomatoid lesions.

Another lesion to be observed in the final stages of plexogenic arteriopathy is fibrinoid necrosis of muscular pulmonary arteries. This is not characteristic for the condition as it may occur in a variety of diseases. In plexogenic arteriopathy it is localized particularly in small branches, shortly after their origin from a larger artery and always over a very short distance. Here there may be some peri-arterial, mainly mononuclear, infiltrate but rarely prominent arteritis. Pulmonary arteritis in plexogenic arteriopathy is uncommon and then affects usually somewhat larger muscular pulmonary arteries. Although usually accompanied by fibrinoid necrosis, it is often observed over considerable distances; it is also unrelated to the origin of a branch and regularly leaves much of the muscular coat and of the elastic laminae intact.

The hallmark of plexogenic arteriopathy is the plexiform lesion. This is a complicated structure; its development is almost certainly linked to fibrinoid necrosis and it is found in the same location near the origin of a branch. This alteration consists of a local dilatation of a muscular pulmonary artery. The arterial wall, overall or in part, is damaged by fibrinoid necrosis so that the muscular coat is thin or absent in these areas. Often imbibition of the wall by fibrin can still be recognized. The dilated lumen is filled by a plexus of capillary-like channels, probably resulting from recanalization of the fibrin clot that was accompanying the preceding fibrinoid necrosis. The channels are separated by myofibroblastic cells with dark,

hyperchromatic nuclei. As a result, the plexiform lesion stands out particularly clearly in a haematoxylin-stained slide, more distinctly than in the elastic stain. Distal to the plexus, the capillary-like channels open up into very thin-walled dilated branches. Although the plexiform lesion is not only pathognomonic for plexogenic arteriopathy but, if present, also its most conspicuous and characteristic alteration, there are occasionally problems with its recognition. One of these can be its incomplete representation in a slide when tangentially cut. Incipient lesions also may be difficult to identify.

Complicating lesions are often found, usually sequelae either of thrombosis or of chronic congestion. Thrombotic changes are regularly present and sometimes even numerous in adult patients with secondary or primary plexogenic arteriopathy. They are absent or very scarce in children. They also increase in relation to the duration of the illness.[12] Plexogenic arteriopathy is complicated by chronic congestion whenever there is additional mitral valve disease as is often the case in atrioventricular septal defect, or when there is left heart failure. The combination of both pattern of lesions is not easily recognized as such in the pulmonary arteries but may be detected in the pulmonary veins. Generally, the pulmonary veins are not affected in plexogenic arteriopathy but when there is chronic congestion, their media is thickened and bounded by elastic laminae, known as arterialization.

Reversibility of lesions

Plexogenic arteriopathy is a serious form of hypertensive pulmonary vascular disease. When the cause is recognized and eliminated, regression is possible but only in the early stages of the disease. This happens particularly in patients with congenital heart disease following surgical correction of the defect. It may also occur when anorexigenic drugs are responsible and subsequently discontinued. In the primary form of plexogenic arteriopathy regression cannot be expected since the unknown cause cannot be removed.

Lung biopsies taken during cardiac surgery for congenital heart disease[10] and particularly those taken during pulmonary artery banding and some years later again in the same patient during corrective surgery,[9] have given some insight into the process of regression. Medial hypertrophy of pulmonary arteries and muscularization of arterioles

Fig. 7.1 (facing page) Lesions of plexogenic arteriopathy in muscular pulmonary arteries. (a) Medial hypertrophy; the media is bounded by internal and external elastic laminae; (b) cellular intimal proliferation; (c) concentric-laminar intimal fibrosis, with many elastic fibres; (d) dilatation lesions: an artery with pronounced medial hypertrophy and intimal fibrosis is surrounded by its wide, thin-walled 'vein-like branches'; (e) fibrinoid necrosis in a branch of an artery with intimal fibrosis; there is complete loss of wall structure (top); (f) plexiform lesion with some dilated peripheral branches to the right.

are the earliest changes and usually the only ones in infants since all other changes are uncommon below the age of 2 years. They also appeared to be reversible, sometimes completely, at other times in part. However, when in a preoperative lung biopsy specimen in infants or young children with cardiac defects, medial hypertrophy is very severe, the pathologist should notify the cardiac surgeon because such infants are often prone to bouts of severe pulmonary vasoconstriction.

Cellular intimal proliferation is reversible but concentric-laminar intimal fibrosis only so when it is mild. Severe intimal fibrosis of this type appears to be irreversible. The same situation seems to apply to dilatation lesions. Isolated vein-like branches are almost certainly reversible but this is questionable for clusters and most unlikely for angiomatoid lesions. Fibrinoid necrosis and plexiform lesions carry a very serious prognosis and essentially form a contraindication for corrective surgery if found in an open-lung biopsy specimen. An unusual and disturbing feature of plexogenic arteriopathy is that in the presence of irreversible lesions the condition tends to progress, even if the cause is removed. This means that if prominent concentric-laminar intimal fibrosis is found in a biopsy specimen taken during closure of a cardiac shunt, the patient may develop increasingly severe pulmonary hypertension and end up with fibrinoid necrosis and plexiform lesions. Thus, the condition behaves as an independent and self-perpetuating disease, as soon as a point of no return has been passed.

Thrombotic arteriopathy

In this condition pulmonary arteries of any size may be affected. The lesions belonging to this pattern may be found in the great majority of hospital autopsies, although rarely in such large numbers that the term arteriopathy applies. The finding of isolated thrombotic lesions usually has no clinical significance. Pulmonary hypertension due to thrombotic arteriopathy occurs but is very uncommon.

Aetiology and pathogenesis

As the name implies, the lesions forming the pattern of thrombotic arteriopathy are the sequelae of thrombosis. However, it is generally impossible to distinguish morphologically alterations caused by thromboembolism from those resulting from primary thrombosis. When elastic arteries are involved, particularly segmental, lobar or main arteries, an embolic origin is generally assumed as long as recognizable damage to the arterial wall does not suggest otherwise. This becomes problematic when the lesions are limited to peripheral, muscular arteries, particularly when they are so numerous as to cause pulmonary hypertension. Pulmonary hypertension may develop without any previous symptoms of embolism but cases of silent embolic pulmonary hypertension are well documented. It has been pointed out that the number of these small thromboemboli to be arrested in the pulmonary circulation must be enormous, before elevation of pulmonary arterial pressure can be expected. On the other hand such a process may be developing over several years and, possibly more important, larger emboli may break up into small fragments not only as a result of mechanical causes but probably also during thrombolysis. It is most likely that thrombotic arteriopathy, with or without pulmonary hypertension, may result from thromboembolism as well as from primary thrombosis. Primary thrombosis is certainly the cause in some patients, for instance in sickle cell anaemia[13] or in tetralogy of Fallot. As a cause for thrombotic arteriopathy it may well be far more common than embolism.

Morphology

Gross appearance
Thrombi found at autopsy in major pulmonary arteries are generally embolic in nature, although under special circumstances primary thrombosis, even of the pulmonary trunk, has been observed.[14] Thromboemboli are most often recent and then usually a main or contributory cause of death. They may occlude the pulmonary trunk or one or both main pulmonary arteries and/or one or more intrapulmonary arteries. Fresh thromboemboli are dry, brittle and dark red, sometimes with grey lines of Zahn on the surface. Their form is usually cylindrical with blunt ends but some are V- or Y-shaped so that they may not be completely occlusive when lodging in a pulmonary arterial branch. This also applies to a 'saddle embolus', arrested at a pulmonary arterial bifurcation. Thrombolysis, by which a thrombus may become fragmented or completely solved, is very common, but organization, usually followed by recanalization, may occur, resulting in eccentric plaques of intimal

fibrosis or in the striking 'bands and webs' as remnants of the organized clot. Complete occlusion of large vessels is uncommon.[15]

Histology

There are no essential differences between thrombotic alterations in major elastic, as compared to those in muscular pulmonary arteries. In this survey the emphasis will be on a discussion of the changes in the latter vessels (Fig. 7.2).

In muscular pulmonary arteries fresh thrombi are not common. Even in patients who died of thromboembolic pulmonary hypertension, they may be scarce or absent.[16] This is because they change relatively rapidly into more permanent sequelae. Recent thrombi tend to adhere to the vascular wall and it is from these sites that endothelial cells may multiply to form a lining over the clot, while myofibroblasts from the arterial wall invade and gradually replace it. In this way intimal

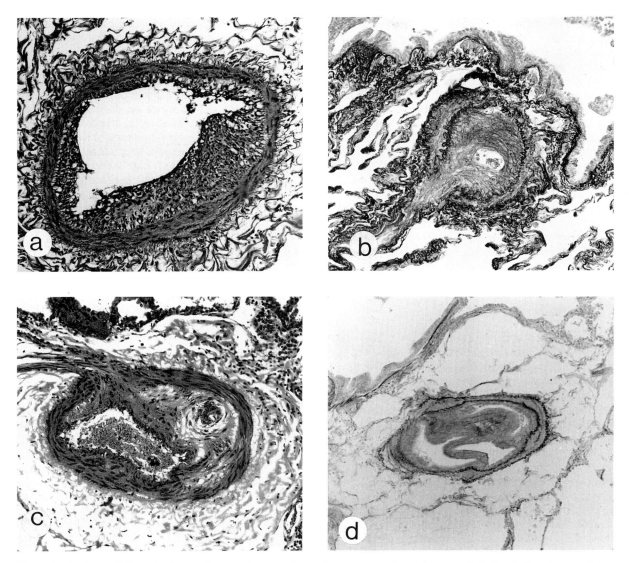

Fig. 7.2 Lesions of thrombotic arteriopathy in muscular pulmonary arteries. (a) Eccentric intimal fibrosis containing longitudinal smooth muscle cells in cross-section; (b) subtotal occlusion of arterial lumen; (c) recanalization: within a patch of intimal fibrosis a new channel is formed; (d) intravascular fibrous septa resulting from recanalization.

fibrosis develops which may completely occlude the arterial lumen but which often, by retraction of the fibrous tissue, results in an eccentric patch. Sometimes this post-thrombotic intimal fibrosis contains some iron pigment. Occasionally there is calcification of a patch. Longitudinal smooth muscle fibres are often embedded in the fibrous tissue.

Simultaneously with the myofibroblasts, or shortly thereafter, capillaries grow into the clot, eventually forming a network therein, which connects with the lumen on either side of an occlusive patch of intimal fibrosis. Some of these capillaries widen at the expense of others, so as to form channels that allow some blood flow through the obstruction. Recanalization channels, although perhaps not strictly pathognomonic for thrombotic arteriopathy, are very characteristic for it. This is particularly the case when the channels become so wide that the remnants of the original organized thrombus stand out as intravascular fibrous septa.

Obstruction of muscular pulmonary arteries by post-thrombotic intimal fibrosis occurs usually over a relatively short distance. This means that in a random histological slide, even in the presence of pulmonary hypertension, arteries may seem normal, while in fact they are narrowed or obliterated at a different level. It is therefore essential to study a biopsy specimen in multiple slides taken at various levels.[17]

Reversibility of lesions

In the earliest stage many thrombi are removed by thrombolysis. Organized thrombi have limited potential for regression. Retraction and shrinkage of a fibrous plaque will restore some blood flow, and so will recanalization. However, even when recanalization channels have become very wide, the vascular resistance is likely to remain considerably increased.

Hypoxic arteriopathy

In chronic hypoxia the pulmonary vasculature may react with changes that are rather discrete so that they are easily overlooked, particularly because they may be overshadowed by complicating alterations. However, it is not a rare arterial disease, since it may occur in a number of widely varying conditions.[18]

Aetiology and pathogenesis

Alveolar hypoxia has long been known to induce pulmonary vasoconstriction and acute pulmonary hypertension. Persistent hypoxic pulmonary hypertension is observed when there is an impediment in the respiratory tract. This may be in the upper tract as occasionally seen in children with enlarged adenoids or tonsils. More often it is the lower respiratory tract, particularly in patients with chronic bronchitis and/or pulmonary emphysema with narrowing or destruction of bronchioles.

Also without airway obstruction, hypoxic pulmonary hypertension may develop as a result of impaired respiratory movements. This may be due to kyphoscoliosis, disturbance of respiratory musculature, or to a central nervous system abnormality, as probably plays a part in Pickwickian syndrome and sleep apnoea syndrome.

Finally, the cause of alveolar hypoxia may be a low oxygen pressure of the respiratory air as exists at high altitude. Residents of areas above 3000 m often have mild to moderate pulmonary hypertension which reverses when they descend to sea level. The changes of hypoxic arteriopathy are found in a number of otherwise normal inhabitants of these altitudes.

Morphology

An unusual feature of hypoxic arteriopathy is that its lesions are found predominantly in the smallest muscular pulmonary arteries and arterioles (Fig. 7.3), leaving the larger ones intact. This applies particularly to an increased muscularity. Intra-acinar arterioles, normally devoid of smooth muscle cells, develop a distinct muscular coat with elastic laminae on either side.[19] Muscular pulmonary arteries associated with bronchi or bronchioli often do not show any medial hypertrophy, or if they do it is usually so mild that there is still a contrast between the muscularization of the larger and the smaller branches.

Intimal fibrosis is not a feature of hypoxic arteriopathy but some thickening of the intima may occur in the form of longitudinal smooth muscle cells.[20] These form bundles or layers within a reduplication of the internal elastic lamina, but only in pulmonary arterioles. Larger muscular arteries are usually not affected. While increased muscularization of arterioles is a constant and widespread feature of hypoxic arteriopathy, the longitudinal muscle bundles or layers are often

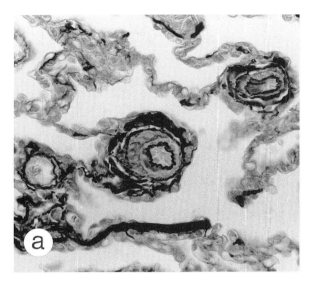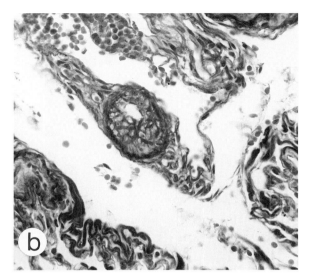

Fig. 7.3 Lesions of hypoxic arteriopathy in arterioles. (a) Intra-acinar arterioles with prominent muscularization resulting in a media with elastic laminae on either side; (b) intra-acinar arteriole narrowed by a layer of longitudinal smooth muscle cells in the intima.

scarce so that they must be searched for. They may even be absent. In pulmonary emphysema they are particularly common.

Similar changes as in the arterioles may be found in pulmonary venules, but here they are far more difficult to recognize. An increased medial thickness in these, normally very thin-walled vessels, generally requires morphometric assessment. Since the intimal muscle bundles are also scarce, the pulmonary venous system is of little diagnostic help.

Of course, muscularization of arterioles is not diagnostic either, as it is found in various forms of hypertensive pulmonary vascular disease. However, the discrepancy between the muscularized arterioles and a not or hardly increased medial thickness of muscular pulmonary arteries may give an indication as to the cause of pulmonary hypertension in these instances. The presence of longitudinal smooth muscle bundles or layers in these small vessels may support this suggestion.

Particularly when alveolar hypoxia is caused by lung disease such as chronic bronchitis, lung fibrosis or emphysema, complicating lesions may disturb the picture so that recognition of hypoxic arteriopathy becomes difficult or impossible. In the presence of chronic inflammatory reactions or fibrosis, lung vessels may become very thick-walled (p.160) so that almost all muscular arteries and

pulmonary veins tend to show medial hypertrophy and intimal fibrosis.

Reversibility of lesions

If alveolar hypoxia can be eliminated, as is possible in children with hypertrophic adenoids or in high-altitude residents, we know that pulmonary hypertension may regress. What happens to the vascular alterations in such instances can only be deduced from experimental work. It appears that the increased muscularity of arterioles is reversible, although this process takes more time than the formation of the muscular layer.

Congestive vasculopathy

One of the most common forms of pulmonary vascular disease is congestive vasculopathy in which arteries as well as veins are involved (Fig. 7.4), while even the lung tissue may show its effect. This pattern of lesions may be associated with pulmonary hypertension, occasionally even with very high pressures in the pulmonary circulation. However, the correlation between these pressures and the severity of the lesions is often poor. In part this is because an intermittent rise in pressure, for instance during effort, may already produce prom-

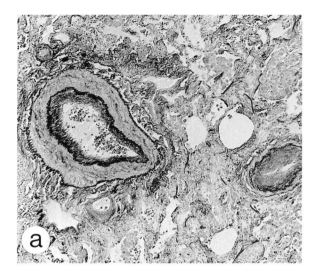 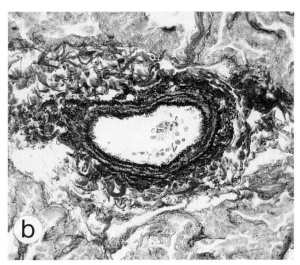

Fig. 7.4   Lesions of congestive vasculopathy. (a) Muscular pulmonary arteries with prominent medial hypertrophy and non-laminar intimal fibrosis; (b) pulmonary vein with a thickened media and with internal and external elastic laminae (arterialization).

inent alterations. Therefore, in most instances patients with this vasculopathy, who may form a significant percentage of hospital autopsy cases, do not have or are not known to have pulmonary hypertension.

Aetiology and pathogenesis

A postcapillary pulmonary hypertension leading to congestive vasculopathy is caused by an impeded blood flow through the pulmonary veins. This impediment is rarely located in the veins themselves, and then most likely in the form of congenital stenoses of the major vessels. It may be in the left atrium as in cor triatriatum, or in the presence of a cardiac myxoma. Far more often the cause of increased left atrial pressure is situated in the mitral valve whenever this is stenotic or incompetent. Left ventricular failure due to hypertension, myocarditis or myocardial infarction is another frequent cause of congestive vasculopathy. Finally, aortic insufficiency and even an aortic coarctation may be at the basis of this pattern of lesions.

The wedge pressure, established during cardiac catheterization, is elevated in these cases as it corresponds with the left atrial pressure. While it is clear that the pulmonary venous pressure is increased and that some rise in pressure will be transmitted over the capillary bed to the pulmonary arteries, it is not well understood why the arterial

pressure is so often far in excess of that in the veins. Constriction of pulmonary arteries and probably interstitial oedema may play a part, but the exact mechanism is not clear.

Morphology

Arteries
The changes in muscular pulmonary arteries are usually conspicuous.[21] The arterial media is thickened, often to a large extent. However, in congestive vasculopathy medial thickness correlates poorly with pulmonary arterial pressure, particularly in the sense that very pronounced medial thickening often occurs in the presence of mild pulmonary hypertension. However, in this condition the thickness of the media is not, as in plexogenic arteriopathy, entirely dependent on the amount of smooth muscle. There is an increased amount of intercellular substance, in part interstitial oedema and mucopolysaccharoids, in part medial fibrosis and elastosis. This fibrosis, which is often easily recognizable, may to some extent explain the lack of correlation of thickness of the media with the pressure. Muscularization of arterioles is regularly present.

Intimal fibrosis of muscular pulmonary arteries is common in congestive vasculopathy and is often severe. This type of intimal fibrosis is usually eccentric but even if circumferential, it does not show the onion-skin appearance characteristic of plexogenic arteriopathy. The lumen may be very narrow, often split-like, but complete obliteration is rarely observed in contrast to thrombotic arteriopathy. Another difference is that the intimal fibrosis as a rule extends over considerable distances, so that in a random histological slice it is present in many arteries. Other arterial lesions are rare. Occasionally fibrinoid necrosis and arteritis have been observed.

Veins

Since the normal pulmonary venous wall is very thin, its thickening in response to an elevated pulmonary venous pressure is usually not spectacular. Even so, in a few cases the venous media may increase to 15 or 20% of its external diameter. But also a modest medial hypertrophy is accentuated by the development of distinct internal and external elastic laminae.[1] By this so-called arterialization, a pulmonary vein may resemble an artery so closely that only the location of the vessel in an interlobular septum away from the bronchial tree may reveal its identity. Intimal fibrosis is often present but is usually mild.

Lung tissue

In congestive vasculopathy the alveolar-capillary wall is generally thickened.[22] In approximately half of the cases there is some, sometimes even pronounced, interstitial fibrosis affecting not only alveolar walls but also interlobular septa and peribronchial and perivascular tissue. It is often associated with deposition of iron pigment in this tissue or in macrophages. Such haemosiderosis can be very prominent and widespread. Pulmonary lymphatics are generally dilated.

Reversibility of lesions

When the cause of chronic pulmonary congestion can be removed, as may happen by correction of mitral stenosis or insufficiency, pulmonary arterial pressure does not always decrease immediately or to a great extent. In other cases, however, there may be a rapid fall in pressure despite the presence of severe medial hypertrophy and severe intimal fibrosis in a lung biopsy specimen taken during cardiac operation. This probably is due to interstitial oedema of the vascular wall, which may greatly increase the thickness of its various components, and can be rapidly removed. There is evidence that intimal fibrosis that has produced

severe narrowing of the arterial lumen can regress to a very thin compact layer with complete restoration of the lumen.[23]

Pulmonary veno-occlusive disease

Pulmonary veno-occlusive disease is a rare condition leading to pulmonary hypertension and regularly to right heart failure and death. It occurs in a wide age range. In children the sex ratio is approximately equal; in adults males are somewhat more affected than females. The clinical diagnosis is exceedingly difficult so that this is one of the forms of hypertensive pulmonary vascular disease that is commonly labelled as unexplained pulmonary hypertension.

Aetiology and pathogenesis

Veno-occlusive disease of the lungs essentially results from a thrombotic vascular occlusion, beginning in, and often limited to, the pulmonary veins and venules. These patients are not known to have disturbances in the clotting mechanism and it seems likely that thrombosis is induced by endothelial damage. However, pulmonary veno-occlusive disease is not an aetiological entity. There are a variety of agents that are held responsible for its development, although the scarcity of the cases so far has hampered an accurate evaluation of the causes. Respiratory viral infections, toxins and chemotherapy are likely candidates, but the disease has also been observed in association with Hodgkin's disease, Raynaud's phenomenon and following renal transplantation, so that immunological factors cannot be dismissed. There are also a few records of the disease in siblings, suggesting the possibility of genetic factors.

Morphology

The term veno-occlusive disease indicates that it is primarily a disease of the pulmonary veins and venules, although arteries may be affected (Fig. 7.5). While the obstruction of these vessels is thrombotic in origin, thrombi that are fresh, or still recognizable as such despite early organization, are scarce or absent. As in thrombotic arteriopathy this undoubtedly is due to the relatively rapid turnover of thrombi into organized, fibrotic tissue. The resulting intimal fibrosis is generally loose, oedematous and paucicellular. Less often, and probably in later stages, collagen and elastic fibres

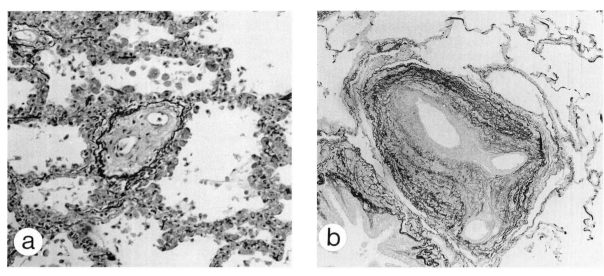

Fig. 7.5 Lesions of pulmonary veno-occlusive disease. (a) Small vein occluded by loose connective tissue with two recanalization channels; there is congestion in adjacent alveolar walls. (b) Muscular artery narrowed by loose connective tissue.

dominate the picture. Particularly small veins are often completely occluded. In larger veins there may be eccentric patches of intimal fibrosis. Recanalization is common and may lead to the formation of intravascular fibrous septa. Often there is also distinct arterialization of veins, which is remarkable because this is not necessarily associated with obstruction of larger venous trunks.

In approximately 50% of cases, the muscular pulmonary arteries are involved in the disease.[23] Of course, it can be expected that an impediment to the blood flow at the pulmonary venous level will have an effect on the pulmonary arteries just as it has in congestive vasculopathy. However, in pulmonary veno-occlusive disease, the intimal fibrosis in pulmonary arteries often has the same loose and oedematous structure as in the veins, may be completely occlusive and tends to show recanalization channels and intravascular fibrous septa. These are features not belonging to the histological picture of congestive vasculopathy, and this strongly suggests that the same agent responsible for the venous alterations has also affected the arteries. Moreover, often the arterial lesions impress as being of more recent date than those in the veins. Notably, fresh thrombi are distinctly more common in arteries than in veins.

A very striking feature of pulmonary veno-occlusive disease is almost always present in the lung tissue itself. It consists of small, round, well-demarcated foci of congestion, or in later stages, of interstitial fibrosis. Since these foci can be immediately recognized at low magnification, it is often their presence which directs the attention of the pathologist to the small veins, usually within or adjacent to these areas, and thus to the correct diagnosis. This is of particular importance because in some cases small venules may be exclusively affected, which easily may escape recognition. Haemosiderin deposition in the lung tissue is common and in some cases very extensive. The pulmonary lymphatics are usually much dilated.

The pattern of lesions in pulmonary veno-occlusive disease is variable. This depends to a great extent on the stage of the condition. Arterial involvement, size of vessels involved, degree of vascular obstruction and of recanalization, variation in extent of interstitial lung fibrosis and haemosiderosis – all these factors may occasionally conspire in making the diagnosis one of the most difficult in pulmonary vascular pathology.

Reversibility of lesions

The course of pulmonary veno-occlusive disease, although most often insidious, generally has a fatal outcome. The most logical treatment seems to be anticoagulant therapy, but with only a few exceptions[25] this has not proved to be successful, almost certainly because the obliterating lesions in the lung vessels were no longer reversible at the time that the diagnosis had been made.

Persistent fetal circulation syndrome

In newborn, particularly full-term, infants the normal postnatal dilatation of the pulmonary arteries sometimes fails to occur. These infants, therefore, develop pulmonary hypertension, usually within hours after birth. This situation may be fatal in a high percentage of the cases. The cause of this syndrome varies but may be meconium aspiration. However, there is a group of infants in whom there is no explanation for the persistent neonatal pulmonary hypertension.[26]

Morphology

In fatal cases the pulmonary arteries show an increased muscularity. Not only is their media thicker than normal, but intra-acinar branches, which in the normal newborn are non-muscular, now have a distinct muscular coat bounded by elastic laminae. The pulmonary veins are normal in these instances.[27]

Some rare vascular diseases

Misalignment of lung vessels

This rare syndrome consists of persistent neonatal pulmonary hypertension, based on a congenital vascular anomaly by which the pulmonary veins lie in close association with the pulmonary arteries, that is next to the bronchi and bronchioli.[28] Even intra-acinar arterioles and venules lie in this juxtaposition (Fig. 7.6). In addition there is an abnormal maturation of the lung tissue with irregular air spaces and interstitial fibrosis of alveolar walls. The changes of this syndrome may affect both lungs; in these instances the condition is fatal within a few weeks. Sometimes there is only one lung or one lobe involved with survival till the age of one year and possibly longer.

Media defects

This is also a very rare congenital anomaly of the lung vessels, consisting of defects in the media of both muscular pulmonary arteries (Fig. 7.7) and pulmonary veins. There is always a sharp demarc-

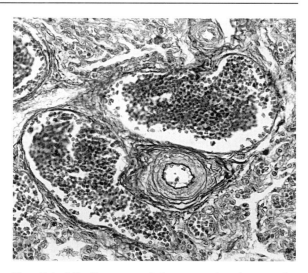

Fig. 7.6 Misalignment of lung vessels. A muscular pulmonary artery with prominent medial hypertrophy lies close to two wide, thin-walled pulmonary veins.

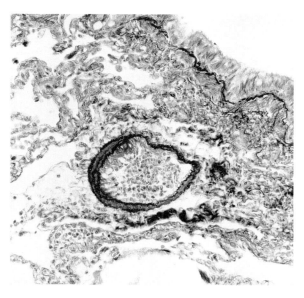

Fig. 7.7 Medial defects of lung vessels. In a muscular pulmonary artery there is an interruption of the muscular coat so that on one side the wall consists of elastic laminae only. There is some intimal fibrosis but only on the thick part of the wall.

ation of the defect, in which the media is absent or very thin, to the rest of the muscular coat which is often thick and contracted. Corresponding with the thick parts of the media, a peculiar form of intimal fibrosis, mostly with cells arranged perpendiculary to the wall, may develop. This intimal fibrosis may cause severe narrowing of the arterial lumen with subsequent pulmonary hypertension.[29]

Capillary haemangiomatosis

Pulmonary capillary haemangiomatosis is a disease of the lung vessels which has been reported so far in patients varying in age from 12 to 71 years. Its nature is not entirely clear. It may well be hamartomatous in origin but gradually it behaves as a neoplasm of low-grade malignancy.

The hallmark of the condition is a proliferation of capillaries or capillary-like channels (Fig. 7.8), filling alveolar spaces and gradually invading alveolar walls, pleura, interlobular septa, bronchi and vessels, particularly pulmonary veins.[30] In this way pulmonary hypertension may be the result. Death from respiratory failure or from right ventricular insufficiency is the usual outcome. Recurrent haemoptyses may be the first sign of the disease.[31]

Pulmonary haemosiderosis is regularly observed in these cases. Sometimes there are clusters of wide, thin-walled vessels with fibrous septa in between.[32] Also groups of enlarged and wide bronchial arteries, as well as lymphangiectases, possibly as expression of a hamartomatous nature, have been described.[31]

Vasculopathy in lung fibrosis

Diseases of the lung tissue may affect the lung vessels in various ways. Pulmonary emphysema and chronic bronchitis often result in hypoxic arteriopathy (p.154) and in adventitial fibrosis.[33] Sarcoidosis and tuberculosis may be accompanied by a granulomatous vasculitis, and pulmonary vasculitis is a characteristic feature of Wegener's granulomatosis. Particularly, fibrosis of lung tissue, by whatever cause and whether focal or diffuse interstitial, is almost always associated with alterations in pulmonary arteries and veins.

These lesions are often very striking. In muscular pulmonary arteries, they consist of a prominent increase in medial thickness and in eccentric, non-laminar intimal fibrosis, which is often severe. Similarly, in pulmonary veins, medial thickening occurs sometimes in combination with arterialization, while also here intimal fibrosis may be pronounced. The combination of these arterial and venous changes, when observed in a lung biopsy specimen, may strongly suggest congestive vasculopathy. This suggestion can be reinforced when the alveolar walls in the specimen are thickened and fibrotic, since interstitial lung fibrosis occurs often in cases of chronic congestion. In fact, it is sometimes impossible to differentiate between congestive vasculopathy and the vascular path-

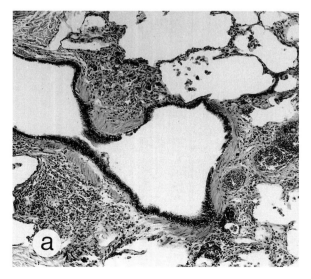 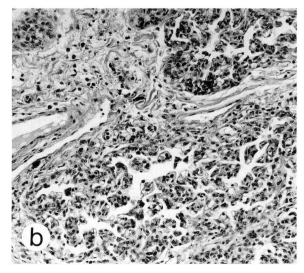

Fig. 7.8 Capillary haemangiomatosis. (a) Proliferating capillary-like channels encroach upon a bronchus and associated artery; (b) the proliferating mass at higher magnification.

ology associated with lung fibrosis. However, the vascular lesions in lung fibrosis are found only within, or immediately adjacent to areas of fibrotic lung tissue; in normal areas the vessels are also normal.[1] In contrast, in congestive vasculopathy the vessels are involved irrespective of the condition of the tissue in which they are lying. Usually, an adequate biopsy specimen from fibrotic lung tissue contains some non-fibrotic tissue as well, so that the nature of the vascular alterations can be established.

Another common mistake is that the pathologist, impressed by the severity of the changes in pulmonary arteries and veins, is convinced that the patient must have a prominent elevation of pulmonary arterial pressure. Pulmonary hypertension in patients with lung fibrosis does occur but is not common. It certainly cannot be deduced from a lung biopsy, since it depends on the degree to which the lungs have been affected by the fibrotic process.

Non-thrombotic embolism

While thromboembolism is the most common and the most important form of pulmonary embolism, there are several other forms less often observed but not necessarily rare or inconsequential.[3] Any particle or mass entering the systemic veins or the right side of the heart, may end up lodging in the pulmonary arterial tree. The chance that the pathologist finds such emboli in a biopsy specimen or in a few blocks taken from lobectomy or autopsy material, is limited unless these emboli are fairly numerous. Even then in some instances it may be difficult to demonstrate them as in air or fat embolism.

Air or fat embolism

Air embolism is best demonstrated at autopsy by opening the right ventricle of the heart under water so that the air escapes as bubbles. A frothy appearance of the blood in the lung vessels is often a striking feature. Microscopically the air may leave globular empty spaces within the blood of pulmonary arterial branches but such a finding may also be produced by fat embolism, when slides from routinely embedded tissue are used. If fat embolism is suspected, frozen section should be stained by specific fat stains. Suspicion may arise when the cut surface of the lung tissue shows an

oily shine. Cells collected from bronchoalveolar lavage fluid may contain intracellular fat droplets, allowing a diagnosis of fat embolism.[35]

Amniotic fluid embolism

Massive amniotic fluid embolism is rare but has serious implications including respiratory distress, shock and death. Moreover, if the patient survives, afibrinogenaemia and disseminated intravascular coagulation form another risk. Particulate components of amniotic fluid can be found lodged in alveolar capillaries or in small pulmonary arterial branches, and also in cases in which this form of embolism is clinically insignificant. These particles include epithelial squames, lanugo hairs, meconium and mucus. Also trophoblast and, rarely, decidua may reach the pulmonary arteries as emboli.

Tissue embolism

Tissue emboli may be derived from a wide variety of sources. Only bone marrow emboli[36] and tumour emboli are regularly found. Tumour emboli are occasionally so numerous that they cause pulmonary hypertension and right cardiac failure.[37] Emboli of brain, liver, skin and bone usually result from trauma. They are all rare and generally without clinical significance.

Foreign body embolism

Foreign body emboli vary greatly in nature and size, from bullets and pieces of catheters to talc or starch granules. Usually the clinical significance of foreign body embolism is limited since only occasional and usually minor pulmonary arterial branches are blocked in this way. Cotton wool or gauze fibres are often recognized in lung tissue from patients who underwent multiple intravenous injections or cardiac catheterization (Fig. 7.9).[38] One of the few forms of foreign body embolism that may have serious clinical consequences is sometimes observed in drug addicts. Filler material or other additives to drugs, used for intravenous injection may contain fibres, starch or talc particles which cause a granulomatous reaction in small pulmonary arterial branches with obstruction of their lumen.[39] This may result in severe pulmonary hypertension and death.[40]

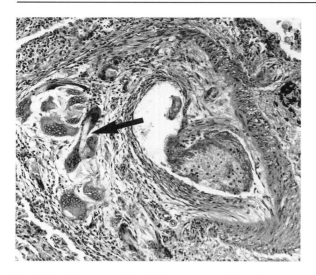

Fig. 7.9 Cotton wool embolism to muscular pulmonary artery. A foreign body granuloma with giant cells has destroyed most of the arterial wall, the remnant of which is to the right. Cross-section of cotton wool fibre indicated by arrow.

Primary pulmonary hypertension

There are rare instances in which severe and even fatal pulmonary hypertension develops in the absence of a recognizable cause. Neither heart or lungs nor embolism, toxins or other agents can be held responsible for the pronounced elevation of pulmonary arterial pressure. Despite numerous studies, our knowledge about this condition is limited. Moreover, the terminology and definition have led to some confusion. Most patients are young adults, although children and older age groups are not exempt. Women are affected two to four times more often than men.

Primary pulmonary hypertension is not a single disease but a syndrome, or a group of diseases sharing the elevated pulmonary arterial pressure and the clinical findings, but not a morphological substrate, pathogenesis or aetiology. Therefore, the pathologist who is reporting on the vascular changes in autopsy or biopsy specimens should avoid the term primary pulmonary hypertension. Instead an indication should be given as to the nature of the underlying pathological picture, which may vary considerably.[41,42]

Vasculopathy in diminished pulse pressure

A pulmonary stenosis, which is usually congenital and which may be valvular, infundibular or within the pulmonary arterial tree, is associated with a diminished pulse pressure whenever the stenosis is severe enough. The systolic peaks have largely disappeared from the pressure curve, which may be almost flat. This has a pronounced effect on the arterial smooth muscle cells. The media of the muscular pulmonary arteries becomes thinner than normal (Fig. 7.10) and may even disappear completely.[43] Mild changes may be difficult to recognize since the pulmonary arterial media is normally thin, but medial atrophy is often very striking.

The pulmonary arteries are wider than normal in this condition, so that they may be distinctly greater in calibre than the bronchi and bronchioli they accompany. Alveolar capillaries and pulmonary venules and veins are also wide. This is even the case in tetralogy of Fallot and related conditions, in which the pulmonary arterial flow is diminished. Moreover, in these instances it has been shown that in the kidneys the glomerular capillaries are dilated, suggesting that an increased blood volume may play a part in the general vascular dilation.

Another feature, observed particularly in tetralogy of Fallot and in the presence of a high haematocrit, is thrombotic lesions in the muscular pulmonary arteries. Such lesions, which vary from recent thrombi and organized intimal patches to delicate intravascular fibrous septa (Fig. 7.10), may be found in infants but are far more common beyond the age of 15 years.[44]

Pathology of bronchial vasculature

In patients with systemic hypertension, the media of bronchial arteries can be distinctly increased in thickness. Enlargement and tortuosity of bronchial arteries is observed in patients with bronchiectasis. Often there is also fibrosis of both media and intima in these instances. In various forms of cardiac or pulmonary disease, the bronchial arteries show intimal bundles or layers of longitudinal smooth muscle cells, which may even occlude the lumen completely. The significance of this intimal smooth muscle is unknown, but a thrombotic origin is a distinct possibility since such smooth muscle cells develop often in patches of post-thrombotic intimal fibrosis. When there is an

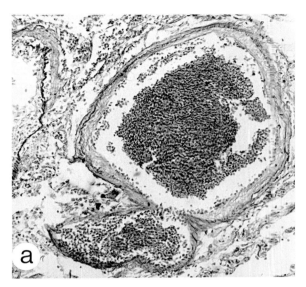 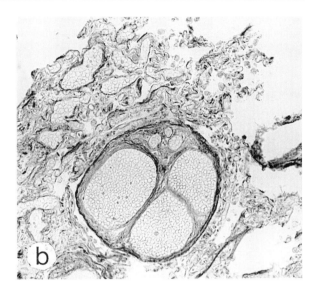

Fig. 7.10 Lesions in diminished pulse pressure (tetralogy of Fallot). (a) Wide and very thin-walled muscular pulmonary artery which in some areas has lost its muscular coat; b) thin-walled muscular pulmonary artery with delicate intravascular fibrous septa due to recanalization of organized thrombus.

impediment to the pulmonary arterial flow, as in tetralogy of Fallot, numerous anastomoses between bronchial and pulmonary arteries and between arteries and veins may develop.

References

1. Wagenvoort CA, Mooi WJ. *Biopsy Pathology of the Pulmonary Vasculature.* London: Chapman & Hall, 1989.
2. Hatano S, Strasser T. *Primary Pulmonary Hypertension.* Geneva: World Health Organization, 1975.
3. Pääkkö P, Sutinen S, Remes M, Paavilainen T, Wagenvoort CA. A case of pulmonary vascular occlusive disease: comparison of post-mortem radiography and histology. *Histopathology* 1985; **9**, 253–262.
4. Fishman AP. Unexplained pulmonary hypertension. *Circulation* 1982; **65**, 651-652.
5. Wagenvoort CA. The terminology of primary pulmonary hypertension. In: *Pulmonary Circulation. Advances and Controversies*, Wagenvoort CA, Denolin H (eds). Amsterdam: Elsevier, 1989, pp. 191–197.
6. Rich S. Primary pulmonary hypertension. *Progress in Cardiovascular Diseases* 1988; **31**, 208-238.
7. Gurtner HP, Gertsch M, Salzmann C, Stucki P, Wyss F. Häufen sich die primär vaskulären Formen des chronischen Cor pulmonale? *Schweiz Med Wschr* 1968; **98**, 1579-1589.
8. Lopez-Sendon J, Gomez-Sanchez MA, Mestre de Juan MJ, Coma-Canella I. Pulmonary hypertension in the toxic oil syndrome. In: *The Pulmonary Circulation: Normal and Abnormal*, Fishman AP (ed.). Philadelphia: University of Pennsylvania Press, 1990, pp. 385-395.
9. Wagenvoort CA, Wagenvoort N, Draulans-Noe Y. Reversibility of plexogenic pulmonary arteriopathy following banding of the pulmonary artery. *J Thoracic Cardiovasc Surg* 1984; **87**, 876-886.
10. Wagenvoort CA. Open lung biopsies in congenital heart disease for evaluation of pulmonary vascular disease. Predictive value with regard to corrective operability. *Histopathology* 1985; **9**, 417-436.
11. Heath D, Edwards JE. The pathology of hypertensive pulmonary vascular disease: a description of six grades of structural changes in the pulmonary arteries with special reference to congenital cardiac septal defects. *Circulation 1958;* **18**, 538-547.
12. Wagenvoort CA, Mulder PGH. Thrombotic lesions in primary plexogenic arteriopathy. Similar pathogenesis or complication? *Chest* 1993; **103**, 844-849.
13. Diggs LW. Sickle cell crises. *Am J Clin Pathol* 1965; **44**, 1-19.
14. Arciniegas E, Coates EO. Massive pulmonary arterial thrombosis following pneumonectomy. *J Thoracic Cardiovasc Surg* 1971; **61**, 487-489.
15. Moser KM, Auger WR, Fedullo PF, Jamieson SW. Chronic thromboembolic pulmonary hypertension: clinical picture and surgical treatment. *Eur Respir J* 1992; **5**, 334-342.

16. Wagenvoort CA. Lung biopsies in the differential diagnosis of thromboembolic versus primary pulmonary hypertension. *Progr Respir Res* 1980; **13**, 16–21.

17. Wagenvoort CA, Mooi WJ. Controversies and potential errors in the histological evaluation of pulmonary vascular disease. In: *Pulmonary Circulation. Advances and Controversies*, Wagenvoort CA, Denolin H (eds). Amsterdam: Elsevier, 1989, pp. 7–26.

18. Heath D, Williams DR. *Man at High Altitude.* Edinburgh: Churchill Livingstone, 2nd edn. 1981.

19. Hasleton PS, Heath D, Brewer DB. Hypertensive pulmonary vascular disease in states of chronic hypoxia. *J Pathol Bacteriol* 1968; **95**, 431–440.

20. Heath D, Longitudinal muscle in pulmonary arteries. *J Pathol Bacteriol* 1964; **85**, 407–412.

21. Wagenvoort CA. Pathology of congestive pulmonary hypertension. *Progr Respir Res* 1975; **9**, 195–202.

22. Kay JM, Edwards FR. Ultrastructure of the alveolar-capillary wall in mitral stenosis. *J Pathol* 1973; **111**, 239–245.

23. Ramirez A, Grimes ET, Abelmann WH. Regression of pulmonary vascular changes following mitral valvuloplasty. An anatomic and physiologic study. *Am J Med* 1968; **45**, 975–982.

24. Wagenvoort CA, Wagenvoort N, Takahashi T. Pulmonary venoocclusive disease. Involvement of pulmonary arteries and review of literature. *Hum Pathol* 1985; **16**, 1033–1041.

25. Capewell SJ Wright AJ, Ellis DA. Pulmonary venoocclusive disease in association with Hodgkin's disease. *Thorax* 1984; **39**, 554–555.

26. Heymann MA, Soifer SJ. Persistent pulmonary hypertension of the newborn. In: *The Pulmonary Circulation. Normal and Abnormal*, Fishman AP (ed). Philadelphia: University of Pennsylvania Press, 1990, pp. 371–83.

27. Murphy JD, Rabinovitch M, Goldstein JD, Reid LM. The structural basis of persistent pulmonary hypertension of the newborn infant. *J Pediatr* 1981; **98**, 962–967.

28. Wagenvoort CA. Misalignment of lung vessels. A syndrome causing persistent neonatal pulmonary hypertension. *Hum Pathol* 1986; **17**, 727–730.

29. Wagenvoort CA. Medial defects of lung vessels. A new cause of pulmonary hypertension. *Hum Pathol* 1986; **17**, 722–726.

30. Wagenvoort CA, Beetstra A, Spijker J. Capillary haemangiomatosis of the lungs. *Histopathology* 1978; **2**, 401–406.

31. Wagenaar SS, Mulder JJS, Wagenvoort CA, Van den Bosch JJM. Pulmonary capillary haemangiomatosis diagnosed during life. *Histopathology* 1989; **14**, 212–214.

32. Heath D, Reid R. Invasive pulmonary haemangiomatosis. *Br J Dis Chest* 1985; **79**, 284–294.

33. Andoh Y, Shimura S, Aikawa T, Sasaki H, Takishima T. Perivascular fibrosis of muscular pulmonary arteries in chronic obstructive pulmonary disease. *Chest* 1992; **102**, 1645–1650.

34. Wagenvoort CA, Mooi WJ. Vascular diseases. In: *Pulmonary Pathology* 2nd edn, Dail DH, Hammer SP, (eds). New York: Springer-Verlag, 1993, 985–1025.

35. Chastre J, Fagon JY, Soler P et al. Bronchoalveolar lavage for rapid diagnosis of the fat embolism syndrome in traumatic patients. *Ann Intern Med* 1990; **113**, 583–588.

36. Havig O, Gruner OPN. Pulmonary bone marrow embolism. A histological study of a non-selected autopsy material. *Acta Pathol Microbiol Scand* 1973; **81**, 143–148.

37. Shriner RW, Ryu RH, Edwards WD. Microscopic pulmonary tumor embolism causing subacute cor pulmonale: a difficult antemortem diagnosis. *Mayo Clinic Proc* 1991; **66**, 143–148.

38. Dimmick JE, Bove KE, McAdams AJ, Bensing G. Fiber embolization: a hazard of cardiac surgery and catheterization. *New Engl J Med* 1975; **292**, 685–687.

39. Lamb D, Roberts G. Starch and talc emboli in drug addicts' lungs. *J Clin Pathol* 1972; **25**, 876–881.

40. Tomashefski JF, Hirsch CS. The pulmonary vascular lesions of intravenous drug abuse. *Hum Pathol* 1980; **11**, 133–145.

41. Wagenvoort CA, Wagenvoort N. Primary pulmonary hypertension: a pathologic study of the lung vessels in 156 clinically diagnosed cases. *Circulation* 1970; **42**, 1163–1184.

42. Palevsky HI, Schloo BL, Pietra GG et al. Primary pulmonary hypertension. Vascular structure, morphometry, and responsiveness to vasodilator agents. *Circulation* 1989; **80**, 1207–1221.

43. Wagenvoort CA, Nauta J, Van der Schaar PJ, Weeda HWH, Wagenvoort N. Vascular changes in pulmonic stenosis and tetralogy of Fallot studied in lung biopsies. *Circulation* 1967; **36**, 924–932.

44. Rich AR. A hitherto unrecognized tendency to the development of widespread pulmonary vascular obstruction in patients with congenital pulmonary stenosis (tetralogy of Fallot). *Bull Johns Hopkins Hosp* 1948; **82**, 389–401.

8

Paediatric lung disease

M Malone ⎯⎯⎯⎯⎯⎯⎯⎯⎯⎯⎯⎯⎯⎯⎯⎯⎯⎯⎯⎯⎯⎯⎯⎯⎯⎯⎯⎯⎯⎯⎯

Introduction

Lung pathology in the paediatric age group is seen in four clinical settings: perinatal lung pathology, which is usually postmortem material; lobectomy specimens for cystic disease, bronchiectasis and tumours; open lung biopsy for 'medical' conditions; and transbronchial biopsies, most commonly used in heart–lung transplant patients. This chapter will deal with the subject under these headings, making only brief mention of those conditions described by other contributors.

Perinatal lung pathology

General considerations

In the interpretation of perinatal lung pathology, two considerations are important.

Because of interpretative limitations of the histopathology, a clear understanding of the clinical history is mandatory. The history should include the antenatal history with accurate gestational dates, and specific reference to such events as premature rupture of membranes and intercurrent maternal infections; a detailed history of labour and delivery including reference to meconium staining of liquor and intrapartum fetal distress; and postpartum history, referring to establishment of spontaneous respiration, and annotation of all events in the special care baby unit, including details of ventilation, pneumothoraces and septicaemic episodes.

Such is the overlap between the morphological changes seen in different clinico-pathological syndromes and iatrogenic effects that the histopathologist should be wary of being too dogmatic. It is frequently difficult and even impossible to say which changes are primary and which are due to the effects of treatment. However, giving due consideration to the history may permit a reasonable assessment of the possible pathogenic mechanisms which have pertained in the individual case.

Development of the immature lung

Since any examination of the perinatal lung must include an assessment of the appropriateness of maturation for gestational age, it is necessary to consider briefly the development of the immature lung, particularly in the third trimester. There are five phases of normal human lung growth.

1. In the embryonic phase (4–6 weeks' gestation) the proximal airways develop.
2. In the pseudoglandular phase (7–16 weeks' gestation) the conducting airways develop.
3. In the canalicular phase (17–27 weeks' gestation) the basic structure of the acinus appears, and small blood vessels canalize the acinar mesenchyme (Fig. 8.1a). In this phase, the channels are lined first by simple cuboidal epithelium, later by type 2 alveolar epithelial cells which are associated with surfactant synthesis. Type 1 cells develop from type 2 cells and are flat epithelial cells. Late in this period, there are sufficient numbers of these acini present for independent extrauterine existence to become a possibility.
4. The saccular phase (28–35 weeks' gestation) is marked by an abrupt change in the appearance

of the lung (Fig. 8.1b). The mesenchymal tissue present between the developing air spaces becomes much less prominent. The saccular walls have a double capillary configuration, that is, capillaries are aligned to each epithelial surface and form two capillary networks which are separated by a band of interstitial tissue. Cuboidal epithelial cells are only found in distal air spaces.

5. The alveolar phase (from 32 weeks' gestation) is characterized by the development of polygonal terminal air space structures lined by flattened epithelial cells with a single capillary network in their thin walls. Alveoli are plentiful in the normal-term lung. Alveolar acquisition continues through early childhood, until the age of about 2 years.

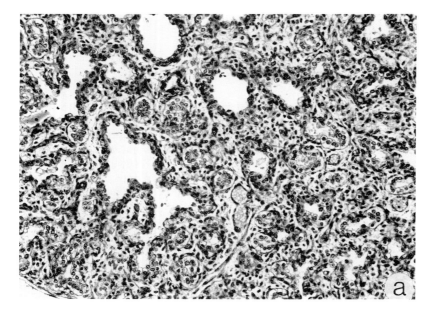

Fig. 8.1a Fetal lung 22 weeks' gestation – canalicular phase. Channels lined by cuboidal epithelium separated by prominent mesenchymal tissue. Scattered developing air spaces lined by type 2 epithelial cells are present in the mesenchyme.

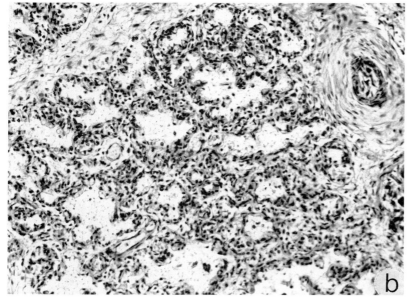

Fig. 8.1b Fetal lung 29 weeks' gestation – saccular phase. Mesenchymal tissue is much less prominent, with air spaces lined by flat type 1 cells. Note muscular pulmonary artery. Same magnification as Fig. 1a.

The pulmonary vasculature becomes progressively more complex with acinar development. The pulmonary trunk and main pulmonary arteries of the fetus are of the elastic type. Below a diameter of about 100 μm, the pulmonary arteries become muscular arteries with thick media and narrow lumen and an internal and external elastic lamina. As the muscular arteries decrease in diameter there is a relative increase in the muscle. These small muscular arteries lie adjacent to terminal bronchioles and represent the major resistance sites within the pulmonary vasculature.

After birth, there is an abrupt fall in pulmonary vascular resistance when respiration begins and the lungs expand. The small muscular pulmonary arteries open up. Within the first 2 or 3 weeks after birth there is a rapid thinning of the media followed by a more gradual thinning until, after approximately 1 year, the relative medial thickness is in the range of 5% of the external vascular diameter, which is the same value as for the adult. Numerous new vessels develop during the first 2 years of life, contributing to a further fall in pulmonary vascular resistance.

Assessment of lung development

There are various methods for the assessment of adequacy of lung development. Lung weight : body weight ratio is a crude indicator, with ratios less than 0.015 below 28 weeks' gestation and less than 0.012 at 28 weeks or more said to define pulmonary hypoplasia.[1] The simplest of the methods used to quantify the degree of acinar development is the radial count. This is the number of alveoli transected by a perpendicular line drawn from the centre of a terminal respiratory bronchiole to the nearest connective tissue septum.[2,3] Mean values range from 2.2 at 24–27 weeks' gestation, 4.4 at birth, 5.5 at 1 week–4½ months, 6.6 at 5–9 months and 7.2 at 2 years. These values are for uninflated specimens.

Perinatal lung pathology can be considered under three broad headings:

1. the pathology of prematurity;
2. meconium aspiration syndrome;
3. the pathology of the mature lung in which spontaneous respiration could not be established, that is, the mature baby who required ventilation from birth, and who could never be weaned off the ventilator.

Prematurity

Hyaline membrane disease

This is a disorder related primarily to deficiency of pulmonary surfactant. It occurs predominantly in premature babies. Other predisposing factors include maternal diabetes, twins, and birth by Caesarean section without a trial of labour. Female and non-white infants appear to be relatively protected, as are infants with a variety of prenatal stresses including amniotic fluid infection and growth retardation.

Surfactant deficiency may not be the sole cause of hyaline membrane formation. An early morphological feature of hyaline membrane disease is necrosis of the bronchiolar epithelium.[4] This has been attributed to pulmonary circulatory problems and may explain the development of hyaline membranes in some full-term infants with abruptio placentae, placenta praevia and fetal malposition, as well as their occurrence in some stillborn infants.

Clinically the affected infant develops respiratory distress at or soon after birth. Chest X-ray shows ground-glass opacification of the lung fields. Grossly, the lungs have a solid or liver-like consistency. Microscopically the most important feature is uniform alveolar collapse.[5] The dilated air spaces seen are alveolar ducts and small bronchioles, and these are lined by the characteristic eosinophilic membranes (Fig. 8.2). The membranes are composed of necrotic alveolar lining cells, amniotic fluid constituents and fibrin. They form as early as 2–3 hours after the onset of the respiratory distress syndrome. They are well formed by 8–12 hours, and in the absence of high oxygen tensions and mechanical ventilatory pressures, will begin resolution by 24–48 hours. Yellow pigmentation of hyaline membranes results from incorporation of unconjugated bilirubin and may be seen in infants with kernicterus, intraventricular haemorrhage, intrahepatic bile stasis, and disseminated intravascular coagulation. Hyaline membranes may also be altered by the presence of bacteria, rendering them fragmented, granular and faintly basophilic.

Since the 1980s, there have been numerous clinical trials of exogenous pulmonary surfactants for the treatment of respiratory distress syndrome in preterm neonates. Analysis of the results yields conflicting evidence. While clinical studies suggest that surfactant treatment is efficacious,[6] some pathological data demonstrate no significant

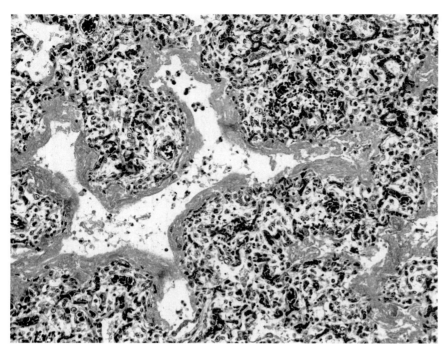

Fig. 8.2   Hyaline membrane disease. There is alveolar collapse. Bronchioles and alveolar ducts are lined by hyaline eosinophilic membranes.

difference in the involvement of lung tissue obtained at autopsy by hyaline membrane disease,[7] while other pathological studies suggest that the severity of the hyaline membrane disease at autopsy is less severe in surfactant-treated babies.[8] Treatment with exogenous surfactant results in accumulation of yellow birefringent material in air spaces, not to be confused with meconium aspiration syndrome (Fig. 8.3a,b).

Bronchopulmonary dysplasia (BPD)
This is a complex, multistage structural and physiological alteration of the lungs which occurs as a prolonged response to acute injury in the neonatal period. Factors in its pathogenesis are high-pressure ventilation,[9] oxygen toxicity,[10] pulmonary infections,[11] pulmonary oedema (secondary to a patent ductus arteriosus, which frequently occurs during the clinical course of BPD) and nutritional deficiencies (fat, glycogen, vitamins A and E, minerals, and trace elements such as iron, copper, zinc and selenium).

The typical infant with BPD is premature, has developed hyaline membrane disease, and has required prolonged ventilatory support. The inci-

dence increases with decreasing birth weight, ranging from 10% in infants over 2500 g to 40% or even higher as birth weight declines towards 1000 g. The clinical course is frequently complicated by repeated infections, pneumothorax, aspiration and patent ductus arteriosus, making interpretation of histological changes difficult.

A striking feature of BPD is the segmental nature of the disease. There is usually a notable difference in the degree of injury from one lobe to another and also within a single lobe. The major pathological changes are seen at the level of the distal bronchioles and alveolar ducts. The changes observed at different phases of the disease (acute, reparative and long-standing healed BPD) form a spectrum.[12,13]

The histological changes seen in infants dying early in the course of BPD (1–2 weeks) are those of hyaline membrane disease together with necrotizing bronchiolitis and bronchiolitis obliterans. There may be acute pulmonary interstitial emphysema with large cystic air-filled spaces extending along interlobular septa, which become lined by multinucleate giant cells (Fig. 8.4a,b).

During the reparative stage (2–4 weeks) there is

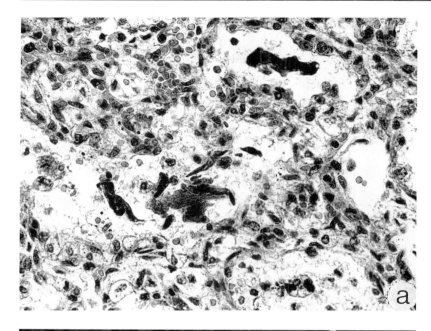

Fig. 8.3a Lung from a premature baby dying after treatment with exogenous surfactant for respiratory distress syndrome. Air spaces are filled with yellow material (black in the reproduction).

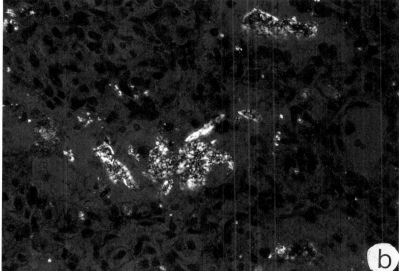

Fig. 8.3b The same field after polarization showing birefringence of the material.

interstitial fibrosis. Alveolar spaces are lined by type 2 pneumocytes. Small round cells in the interstitium are myofibroblasts.

In chronic (long-standing) healed BPD in infants dying many months after the onset of the disease, the lungs macroscopically show a characteristic 'cobblestoning' of the pleura. Microscopically, there is prominent interstitial fibrosis and smooth muscle hyperplasia in the septal wall. Alternating areas of overexpanded and collapsed air spaces are seen. There may be squamous metaplasia of the bronchial epithelium. Morphological evidence of pulmonary hypertension (medial hyperplasia and intimal thickening of small arteries) may be present.

Long-term complications of BPD include tracheal stenosis, apnoeic spells and sudden unexpected death, respiratory failure during viral infection, bronchiolitis obliterans and interstitial fibrosis.

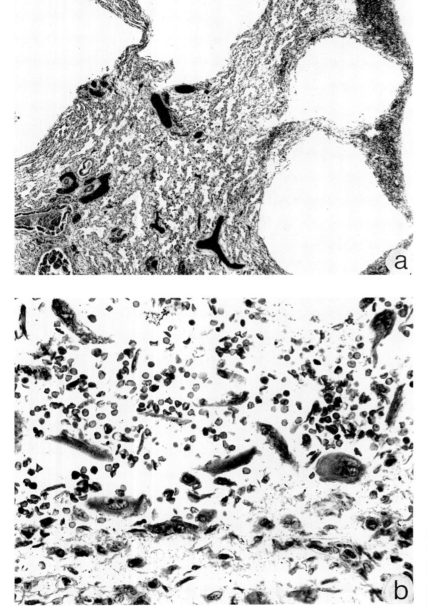

Fig. 8.4a,b Pulmonary interstitial emphysema. (a) Low-power view showing large cystic spaces within the interstitium. (b) High-power view showing multinucleated giant cells lining the cystic spaces.

Massive pulmonary haemorrhage
Focal alveolar or interstitial haemorrhage is a frequent histological finding in the lungs of infants with hyaline membrane disease or BPD. Massive bleeding into airways is a distinct clinico-pathological entity.[14] The affected infants are predominantly male, premature and small for gestational age, and have suffered some perinatal stress. Clinically, the affected infant collapses suddenly, usually in the first week of life, with bloodstained fluid pouring from the trachea and nose. Microscopically, there are confluent areas of haemorrhage with blood filling alveolar spaces and dissecting into the interstitium.

Meconium aspiration

Passage of meconium into the amniotic fluid occurs as a result of a wide variety of perinatal stresses. Symptomatic aspiration of meconium mainly occurs in mature or postmature infants, and the mortality is still high. The infant and placenta are coated in meconium. Microscopically, airways are obstructed by masses of fetal squames and aggregated mucus (Fig. 8.5).

Perinatal infection

Perinatal pneumonia may be congenital or acquired and is usually seen in the context of a disseminated infection.

Congenital pneumonia may occur as a result of transplacental infection or may be acquired during birth as a result of infection by organisms from the birth canal. Congenital pneumonia as a result of transplacental infection may be viral (cytomegalovirus, parvovirus,[16] varicella[17]) or bacterial (*Listeria monocytogenes*).[18] Mulberry-like aggregates of red blood cells with a bluish hue on haematoxylin and eosin staining in the blood vessels of stillborn babies are an artefact of maceration and should not be mistaken for organisms (Fig. 8.6). Although their appearance would suggest that they are lightly calcified, in fact they fail to stain with the von Kossa and Alizirin red stains.

Infections acquired during birth may be bacterial (group B streptococcus),[19] viral (herpes type 2)[20] or fungal (*Candida albicans*)[21] and are frequently associated with prolonged rupture of membranes and chorioamnionitis.

Late onset perinatal pneumonia (usually in the context of a sick premature infant in the Special Care Baby Unit) includes infection caused by *Staphylococci*,[22] *Escherichia coli* and pseudomonas, as well as by *Candida albicans* and cytomegalo-virus.

Gestationally mature baby in whom spontaneous respiration could not be established

Pulmonary hypoplasia
This is defined as an abnormal reduction in the weight and/or volume of the lung without the absence of any of its lobes. Bilateral pulmonary hypoplasia is usually associated with oligohydramnios (renal agenesis, polycystic and dysplastic

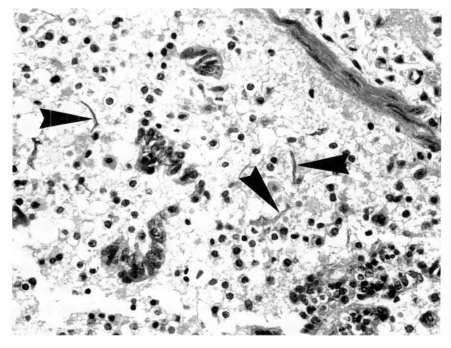

Fig. 8.5 Meconium aspiration syndrome. Airway containing plate-like fetal squames (arrowheads) and mucus.

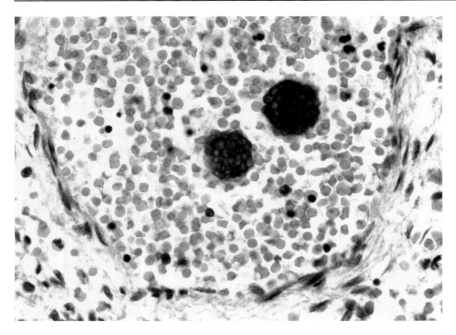

Fig. 8.6 Haematoxyphil 'mulberry-like' aggregates of fetal red blood cells in a lung vessel from a macerated stillbirth of 29 weeks' gestation. These were present in other organs and are an artefact of maceration, not to be mistaken for organisms. (Reproduced by kind permission of Dr Beth Peters, St Andrew's Hospital, Billericay, Essex.)

kidneys, urinary outlet obstruction and prolonged rupture of membranes), thoracic wall abnormalities, lung compression by abdominal contents, abnormalities of the central nervous system, and conditions such as exomphalos and hydrops fetalis. Unilateral pulmonary hypoplasia is usually due to diaphragmatic hernia. The lung weight : body weight ratio is reduced.

Histologically, appearances vary depending on the aetiology. Lungs from all conditions associated with oligohydramnios dating from earlier than 20 weeks' gestational age show failure of growth and maturation of the peripheral parts of the acinus with delay in development of the blood–air barrier and delay in epithelial maturation.[23] Most alveoli are lined by cuboidal epithelium, and capillaries are embedded within the interstitial tissue with a cell structure seen between the blood in the capillaries and the alveolar lumen instead of a thin blood–air barrier. In contrast, the hypoplastic lungs of infants with normal or increased amniotic fluid usually have a histological structure which is appropriate for the gestational age.[24]

Persistent fetal circulation (persistent pulmonary hypertension of the newborn)

This disorder represents failure of the pulmonary circulation to adapt normally to extrauterine life. It is usually a secondary condition, occurring in congenital diaphragmatic hernia, and in many term infants with meconium aspiration. Persistent fetal circulation is also seen in many cardiac anomalies causing pulmonary hypertension, but by definition these are not included in the syndrome.[25] When it occurs in patients with no recognized antecedents, it is considered to be primary or idiopathic. The infants present with cyanosis, respiratory distress and evidence of right-to-left shunt with elevated pulmonary artery pressure. There is associated dilatation of the foramen ovale and a widely patent ductus arteriosus.

Histologically, the pulmonary vasodilatation and decrease in vascular smooth muscle that normally occur rapidly after birth are not seen. Instead, these vessels remain closed and thick-walled, and additional smooth muscle is added to both small pulmonary arteries and along normally nonmuscularized small intra-acinar blood vessels (Fig. 8.7). These histological changes may also be seen in infants with sepsis, pneumonia or respiratory disease associated with prematurity. Intravascular thrombi are frequently associated with idiopathic persistent fetal circulation.

Alveolar capillary dysplasia (malalignment of lung vessels)
This is a rare developmental abnormality of the pulmonary vasculature resulting in intractable pulmonary hypertension from birth. Histologically, lobules composed of small, rounded air spaces, generally lined by cuboidal epithelium, are separated by prominent walls which are deficient in capillaries. Mature alveoli are not seen. Pulmonary veins are malpositioned, being rare in interlobular septa, and are found accompanying small pulmonary arteries adjacent to airways.[26,27]

Abnormality of surfactant production and/or type
This subgroup includes congenital pulmonary alveolar proteinosis and other rare disorders of surfactant production and/or type. Several familial cases of pulmonary alveolar proteinosis have been described, particularly in children, and the condition may mimic neonatal respiratory distress syndrome.[28,29] Microscopically, the alveoli are filled with dense eosinophilic granular material that is PAS-positive and diastase-resistant, and stains positively with fat stains. Within the granular material are larger lamellated eosinophilic bodies. Cholesterol clefts, macrophages and degenerating cells are also present. Electron microscopy shows widespread cellular degeneration and an electron-dense granular background with numerous lamellar bodies up to 5 μm in diameter. The lamellar bodies resemble those found in type 2 pneumocytes. Cells free within the alveolar material include degenerating type 2 pneumocytes and macrophages. Analysis of the intra-alveolar material in pulmonary alveolar proteinosis shows it to be similar to surfactant.

We have seen postmortem material from three patients in whom we have made the diagnosis of an abnormality of surfactant production and/or type, and without the features of pulmonary alveolar proteinosis. In one, the alveolar spaces were filled with foamy material which histochemical staining demonstrated to contain lecithin and sphingomyelin. Electron microscopy showed a profusion

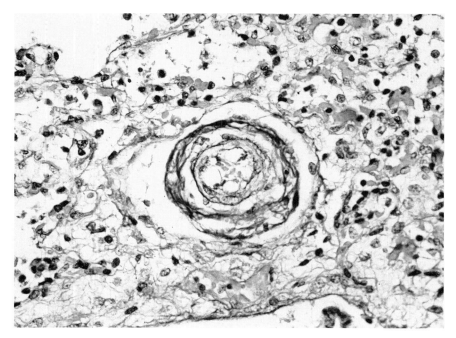

Fig. 8.7 Persistent fetal circulation. Small pulmonary artery from the lung of a baby dying at 3 weeks of age. The lumen is barely patent, the vessel is thick-walled and there is smooth muscle hypertrophy. Postmortem examination revealed patency of the foramen ovale and ductus arteriosus. Reticulin stain.

of abnormal lamellar bodies, both within type 2 alveolar epithelial cells and air spaces (Fig. 8.8). The other two patients were brothers who in life had been demonstrated to produce biochemically abnormal surfactant. Postmortem histology showed the effects of long-term ventilation, and no specific features. No abnormality of lamellar bodies was seen ultrastructurally.

Maturational arrest

Having ruled out the above causes of failure to establish spontaneous respiration in a gestationally mature baby, the diagnosis of maturational arrest should be considered. This may be difficult to establish if ventilation has been prolonged, but may be suggested if the radial alveolar count is reduced (see above) and air spaces are lined by primitive cuboidal cells.

Extracorporeal membrane oxygenation (ECMO)

Chau *et al.*[30] have described the autopsy findings in 17 infants receiving ECMO therapy for meconium aspiration, diaphragmatic hernia and other causes of pulmonary hypoplasia, and congenital heart disease. They noted the presence of interstitial and intra-alveolar haemorrhage along with hyaline membrane formation during the first few days of therapy. They described the hyperplasia of type 2 alveolar cells and bronchial epithelial cells after 2 days of ECMO therapy in some patients, and by 7 days in all patients. Squamous metaplasia of bronchial epithelium was also seen in the majority of cases, and one case developed mucinous metaplasia as well. Clusters of calcified material were present in the alveoli of 5 of 18 cases. Interstitial fibrosis was also a consistent finding after 7 days of ECMO therapy.

Paediatric surgical pulmonary pathology

Cystic lung lesions in infants and children

General considerations
Causes of lung cysts in infants and children differ

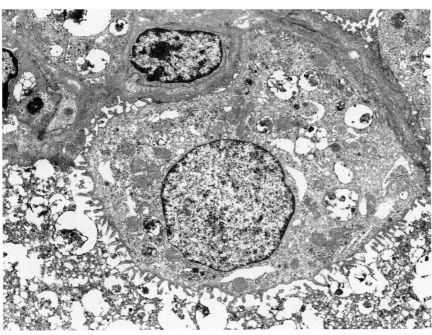

Fig. 8.8 Electron micrograph showing abnormal surfactant aggregates in a type 2 pneumocyte. Note the absence of the normal whorled lamellar pattern. Surfactant material of the same density and morphology is seen in the alveolar lumen, and histochemical staining showed this material to contain sphingomyelin and lecithin. From the lung of a gestationally mature baby in whom spontaneous respiration was never established.

from those in adults.[31] Pulmonary cysts in infants and children are either congenital or acquired, with congenital cystic adenomatoid malformation and interstitial pulmonary emphysema accounting for the vast majority of cases in infants, and post-infectious cysts (poststaphylococcus, intralobar sequestration) seen most frequently in older children.

Points of interest in the differential diagnosis of pulmonary cysts in children relate to the age at presentation, location within the lung, the nature of the epithelial lining, the character of the subepithelial stroma, the nature of the airway connection, if any, and the blood supply (Table 8.1).

Pulmonary sequestrations

Sequestration of a pulmonary lobe or segment implies absence of a normal connection to the airways system of the surrounding lung and a blood supply via systemic vessels. Classically, two types are recognized, intralobar and extralobar.

Intralobar sequestration is more common than extralobar, and is probably inflammatory rather than congenital in origin.[32] It is confined to the lower lobes in 98% of cases. It is thought to be a lesion acquired through the process of chronic inflammation of the lower lobe with development of a pleuritis and formation of a richly vascular granulation tissue deriving its blood supply from hypertrophied pulmonary ligament arteries. With progression and resolution of the pneumonia one or more dominant systemic arteries assumes a substantial supply to the chronically infected segment of lung. The clinical presentation is of chronic cough, fever, and/or recurrent pneumonia in older children and adults. Chest X-ray shows a lower lobe mass. The systemic blood supply is identified at surgery. Macroscopically, the sequestered segment of lung usually has a thickened pleura. Cut section reveals a dense fibrous parenchyma with, frequently, multiple cysts filled with viscid fluid or occasionally gelatinous material. Microscopically, the features are of chronic pneumonitis and bronchiectasis.

Extralobar sequestration is a congenital malformation, presenting most commonly in the first 6 months of life with dyspnoea, cyanosis and feeding difficulties. It is frequently associated with other anomalies, notably diaphragmatic hernia, congenital lymphangiectasia, lobar emphysema and congenital cystic adenomatoid malformation. It is found in both thorax and abdomen. It can occur anywhere in the thorax. The vast majority receive their arterial blood supply directly from the aorta, and venous drainage is via the systemic system to the right atrium. Macroscopically, it consists of a mass in a range of 3 to 6 cm diameter. On sectioning, it may resemble normal lung parenchyma. Well formed bronchi are occasionally seen. In a few cases, the lesion is grossly cystic. Microscopically, it consists of dilated bronchioles, alveolar ducts and alveoli (Fig. 8.9). A well-formed bronchus is found in about 50% of cases.

Bronchogenic cyst

As with extralobar sequestration, bronchogenic cysts are thought to develop as remnants of an aberrant budding of the foregut during early embryonic life. They present between the ages of 7 and 10 years as an anterior mediastinal cyst. Cases described as occurring within the lung substance (central bronchogenic cyst) probably represent type 1 congenital adenomatoid malformation. The wall is muscular or fibrous with a lining of bronchial epithelium. Identification of cartilage is necessary to make the diagnosis, but it may be difficult to find without multiple sections. There may be a peripheral rim of poorly developed lung tissue.

Congenital cystic adenomatoid malformation

Congenital cystic adenomatoid malformation is a rare hamartomatous lesion which usually presents as respiratory distress within the first month of life. It can now be diagnosed on antenatal ultrasound scan[33] and surgery can be performed early in neonatal life. The lesion is usually restricted to a single lobe. Cases in which involvement is extensive or bilateral can result in fetal hydrops with polyhydramnios and stillbirth. There is an association with all forms of cystic renal disease. The macroscopic appearance varies from a completely solid lobe to a multiloculated cyst within a lobe. Stocker *et al.*[34] divided congenital cystic adenomatoid malformation into three different categories.

The type 1 lesion consists macroscopically of multilocular large cysts (Fig. 8.10a), and microscopically of large cysts lined by ciliated columnar to pseudostratified tall columnar epithelium overlying a fibromuscular layer. Clusters of mucus-secreting cells are seen within the cyst walls (Fig. 8.10b).

The type 2 lesion varies in size from less than 1 cm in diameter to a lesion occupying a whole lobe, or even a whole lung. It consists of multiple

Table 8.1   Cystic lung lesions in infants and children

Entity	Age at presentation	Location	Nature of cysts	Cyst wall stroma	Airway connection	Blood supply	Comments
Intralobar sequestration	Older children and adults	Lower lobes.	Dilated muscle-filled bronchioles.	Destroyed lung parenchyma.	None.	Systemic.	Probably inflammatory rather than congenital in origin.
Extralobar sequestration	Neonate	Anywhere in thorax. Rarely abdomen.	Dilated bronchioles, alveolar ducts and alveoli. May be inflamed. May be immature and disorderly.	NA	None.	Arterial supply from aorta. Drain to right atrium.	Associated with diaphragmatic hernia.
Bronchogenic cyst	7–10 years	Anterior mediastinal cyst.	Lined by ciliated pseudostratified columnar epithelium. May be inflamed.	Smooth muscle, fibrous tissue, cartilage.	May or may not communicate with tracheobronchial tree.	NA	Identification of cartilage mandatory for diagnosis.
CCAM, type 1	Neonate	Usually restricted to a single lobe.	Large cysts, lined by cilinated columnar and pseudostratified epithelium. Mucus-secreting cells.	Fibromuscular. No cartilage.	Cysts communicate with bronchial tree.	NA	Considerable overlap between 3 types of CCAM. Areas of normal lung between cysts. May give rise to pulmonary sarcomas.
CCAM, type 2	Neonate	Usually restricted to a single lobe.	Cysts >1cm in diameter. Lined by ciliated cuboidal to columnar epithelium.	Fibromuscular. May see striated muscle.	Cysts communicate with bronchial tree.	NA	As above.
CCAM, type 3	Neonate	Usually restricted to a single lobe.	Cysts <0.5cm in diameter. Cuboidal epithelium. Resembles fetal lung.	NA	No communication.	NA	As above.
Congenital lobar emphysema	Neonate	Upper or middle lobe.	Over-distended alveoli.	NA	NA	NA	Aetiology frequently not identified.
Pulmonary interstitial emphysema	Young children	Any lobe.	Loose connective tissue and compressed parenchyma. Multinucleated giant cells.	NA	None.	NA	History of ventilation in the neonatal period.
Post-infarction peripheral cysts.	Up to 2 years	Subpleural.	Attenuated type 1 and type 2 alveolar cells.	NA	None.	NA	Association with trisomy 21 and complex congenital heart disease.
Pulmonary mesenchymal cystic hamartoma	Up to 4 years	Peripheral.	Lined by low cuboidal and ciliated columnar epithelium.	Layer of poorly differentiated mesenchymal cells immediately beneath epithelium.	In continuity with walls of small airways and alveoli.	NA	Association with blastomas and malignant mesenchymal tumours.

Abbreviations: NA, not applicable; CCAM, congenital cystic adenomatoid malformation.

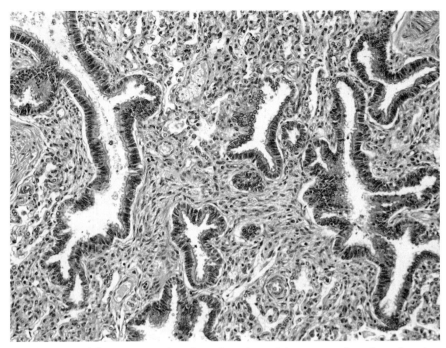

Fig. 8.9 Extralobar sequestration. Intrathoracic nodule separate from lung found at repair of diaphragmatic hernia. Photomicrograph showing dilated bronchioles, alveolar ducts, alveoli and immature mesenchyme.

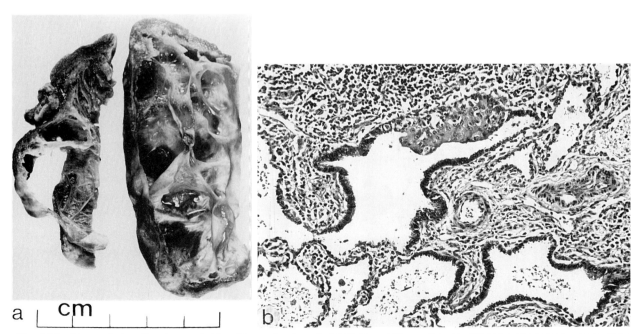

Fig. 8.10 Congenital cystic adenomatoid malformation type 1. (a) Gross specimen, showing dilated cystic spaces, greater than 1cm in diameter. (b) Photomicrograph showing cysts lined by ciliated cuboidal to columnar epithelium overlying a fibromuscular layer. Note mucus-secreting cells in one of the cyst walls.

evenly spaced cysts that rarely exceed 1 cm in diameter. Microscopically, the cysts are lined by ciliated cuboidal to columnar epithelium overlying a fibromuscular layer. Occasional striated muscle fibres may be seen. Mucus-secreting cells are not found. In types 1 and 2, the cysts communicate with the bronchial tree and there may be areas of intervening normal lung between the cysts.

The type 3 lesion is a bulky firm mass, consisting of numerous cysts less than 0.5 cm in diameter. Microscopically, the lesion resembles fetal lung at the early canalicular stage of intrauterine development (16–20 weeks' gestation) with the cysts being lined by cuboidal epithelium.

In practice, there is considerable overlap between the three types, and an individual case may show features of all three in different areas. Moerman et al.[35] performed postmortem bronchography in two cases, and serial sections of the bronchial tree proximal to the malformation in four cases of congenital cystic adenomatoid malformation; segmental bronchial absence or atresia was shown in all of them. Moerman et al. suggest that bronchial atresia is the primary defect leading to the development of congenital cystic adenomatoid malformation.

Congenital lobar emphysema (congenital pulmonary overinflation)

The term 'emphysema' is a misnomer in this condition, since there is no destruction of lung parenchyma (Fig. 8.11). Congenital lobar emphysema usually presents as respiratory distress in neonates. Initial chest X-ray in the neonatal period may show a 'white-out' of the affected lobe because of retained alveolar fluid. Later chest X-rays show overinflation. An upper or the middle lobe are most commonly affected. The basic defect is partial obstruction of the lobar bronchus with resultant air-trapping and collapse. There may be external compression by vessels, or defective or absent cartilage causing expiratory collapse of the lobar bronchus.

In practice, the cause is frequently not identified. Surgical removal of the affected lobe is the usual form of therapy, although in a few cases the disorder has apparently resolved spontaneously.[36]

Rarely, a true structural anomaly of the lobar parenchyma, the so-called polyalveolar lobe, can be diagnosed by use of radial alveolar counts (see above) and is a rare cause of congenital overinflation of the lung.[37] The number of alveoli per acinus is increased.

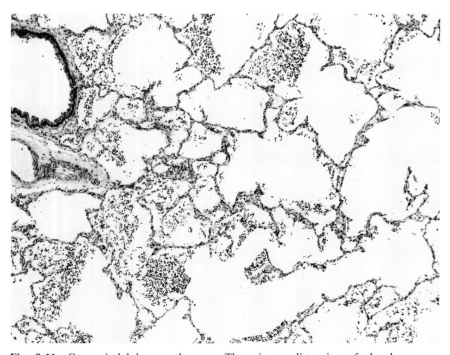

Fig. 8.11 Congenital lobar emphysema. There is overdistension of alveolar spaces without destruction of lung parenchyma.

Pulmonary interstitial emphysema

This has been mentioned above in the context of bronchopulmonary dysplasia. The most common form is that of diffuse multilobar involvement, but localized unilobar involvement may also occur, resulting in mediastinal shift and necessitating lobectomy. Histologically, air-filled spaces follow interlobular septa and airways. Subpleural blebs may be seen. The spaces may be so large as to suggest a cystic anomaly of the lung, but the clinical history of respiratory distress and high-pressure ventilation in the neonatal period should point to the correct diagnosis.

Post-infarction peripheral cyst of the lung

Six cases of post-infarction peripheral cysts of the lung have been described.[38] Particularly striking is the presence of trisomy 21 in three of these patients, all of whom had significant cardiovascular malformations. Age at diagnosis ranged from 15 days to 2 years. In all cases, the cysts were present in more than one lobe, and were subpleural in location. Chest X-ray showed patchy cystic change. Macroscopically, air-filled cysts up to 0.5 cm in diameter lie just beneath the pleura and are limited to the outer 1 cm of the lung. Microscopically, the cyst walls are formed by the intact pleura and the interlobular septa, and are lined by attenuated type 1 and type 2 alveolar lining cells. The pathogenesis is thought to be anoxia and infarction in the peripheral lung secondary to a severe reduction in pulmonary artery blood flow, either as a result of congenital heart disease, or pulmonary artery thrombosis and recanalization.

Pulmonary mesenchymal cystic hamartoma

This entity was first described by Mark in 1986.[39] The lesion, which is multifocal and bilateral, consists of complex multilocular peripheral pulmonary cysts in continuity with the walls of small terminal airways and alveoli. The cystic spaces are lined by low cuboidal and ciliated columnar epithelium with an underlying stromal layer of small, poorly differentiated mesenchymal cells immediately beneath the epithelium. The significance of the lesion is in its relationship to pulmonary mesenchymal tumours, including sarcomas and a number of the 'blastomas' described in children, which have now been documented as arising concurrently with or in pre-existing lung cysts, the majority probably in mesenchymal cystic hamartomas (see below).

Bronchiectasis

General considerations

An adequate clinical history is essential for the pathological interpretation of a lobectomy specimen for bronchiectasis. In particular, it is important to know the age at onset and the history of antecedent infections or of foreign body inhalation since the pathology is for the most part non-specific.

Cystic fibrosis

Cystic fibrosis is the most common cause of bronchiectasis in children in the Western world. The protein encoded by the cystic fibrosis gene functions as a chloride channel regulated by cyclic AMP (cAMP).[40] In cystic fibrosis cells, cAMP-regulated chloride secretion is absent, and there is a three-fold increase in sodium reabsorption, rendering secretions viscid. Lung specimens from children with cystic fibrosis present to pathologists either as native specimens from patients undergoing heart–lung transplantation, transbronchial biopsies of transplanted lung as part of the routine follow-up of such patients (see below), or at postmortem examination. Mucoid impaction, areas of atelectasis and overinflation, bronchiectasis and necrotizing pneumonia are features of lung involvement in cystic fibrosis. Recurrent staphylococcal and pseudomonas infections are a prominent feature. Allergic bronchopulmonary aspergillosis may also occur.

Post-infectious bronchiectasis

This is the most common cause of lobectomy for bronchiectasis in children. Whooping cough and measles were common antecedent infections, but with the development of vaccines, these two causes have become less frequent. Adenovirus infection is now an important cause of bronchiectasis, while staphylococcal infection is a frequent bacterial cause.

Bronchiectasis due to bronchial obstruction

Foreign body inhalation is an important cause of bronchiectasis or atelectasis leading to lobectomy in children.

Congenital bronchiectasis

This forms part of a spectrum with congenital lobar emphysema (see above). It is considered to be due to a deficiency in bronchial cartilage. The compliant bronchial wall collapses during expiration with distal air-trapping and eventual bronchiectasis. Clinical presentation is in early infancy. As with congenital lobar emphysema, the cause is not always identifiable in a lobectomy specimen.

Kartagener's syndrome (immotile cilia syndrome)
Kartagener's triad consists of sinusitis, bronchiectasis and situs inversus. This diagnosis is made on electron microscopic examination of cilia in a nasal mucosal biopsy specimen. Abnormalities include absence of both inner and outer dynein arms, absence of spoke heads and absence of one or both central microtubules or the central sheath.[41]

Tumours

General considerations
Metastatic tumours account for the majority of lung tumours in children. Formerly, these were typically documented at postmortem examination. However, the current practice of aggressive chemotherapy and excision of metastatic nodules yields many wedge resections for pathological examination with particular reference to tumour viability and response to chemotherapy. Childhood tumours with a predilection to metastasize to the lung are Wilms' tumour and hepatoblastoma. Osteogenic sarcoma, rhabdomyosarcoma and angiosarcoma also metastasize to the lung. Neuroblastoma hardly ever involves the lungs.

Primary lung tumours are uncommon in the paediatric age group. They include endobronchial and parenchymal tumours, as well as tumours originating outside the lung (chest wall or thoracic cavity) which may extensively involve the lung, necessitating lobectomy. They also include nonneoplastic conditions or pseudotumours presenting with a discrete mass on chest X-ray. In fact, these conditions are more common than true primary lung tumours in children. Type 3 congenital cystic adenomatoid malformations, intralobar and extralobar sequestration, an abscess or a fungal infection may all simulate a neoplasm.

Endobronchial tumours

Epithelial tumours

  1. *Bronchial carcinoid*
Although the bronchial carcinoid is the most common primary malignant epithelial tumour of the lung in children and adolescents, it is estimated that only 2–4% of all bronchial carcinoids are diagnosed in the paediatric population.[42] The tumour is identical to the classic carcinoid described in adults, and the prognosis is similar, that is, a 5 year survival greater than 90%. The atypical carcinoid virtually does not occur in children.

  2. *Mucoepidermoid carcinoma*
Virtually all mucoepidermoid carcinomas described in children are histologically low-grade tumours of good prognosis. Dyskeratosis, cytological atypia and abundant mitoses are not seen.[43]

  3. *Adenoid cystic carcinoma*
These account for less than 5% of malignant epithelial endobronchial tumours in children. The tumour resembles that seen in adults. Treatment is by resection, and while metastases may occur, prognosis is good.[44]

Mesenchymal tumours

  1. *Chondroma*
While occasionally seen as an isolated endobronchial lesion, chondromas are most often associated in young girls with gastric epithelioid leiomyosarcoma and functioning extra-adrenal paraganglioma (Carney's triad).[45] Pulmonary chondromas are soft to firm, bosselated masses composed of a mixture of mature cartilage, bone, vascular adipose tissue and stellate cells in a myxoid stroma.

  2. *Granular cell tumour*
Approximately 10% of granular cell tumours are located in the upper or lower respiratory tract. One of the youngest examples was in a 5-year-old girl.[46] Large granular polyhedral cells are arranged in sheets or as single cells throughout the bronchial submucosa. Immunohistochemically the cells reveal positive staining for S-100 protein, and ultrastructurally the intracellular granules consist of membrane-bound vacuoles containing cellular debris.

  3. *Lipoma*
Bronchial lipoma is a rare tumour, and is probably hamartomatous in nature.[47] It consists predominantly of mature fat together with clefts lined by ciliated or nonciliated epithelium and islands of normal cartilage.

  4. *Inflammatory pseudotumour (see below)*
This may also uncommonly present as an endobronchial tumour.

Parenchymal tumours

Epithelial tumours

1. *Squamous cell carcinoma*
Squamous cell carcinoma is exceedingly rare in childhood.[48]

2. *Adenocarcinoma*

a. Bronchioloalveolar adenocarcinoma
Bronchioloalveolar adenocarcinoma is the most common form of bronchogenic carcinoma in children. It presents with cough, pneumonia and chest pain, but, if peripheral, it may be asymptomatic. It can occur as a single peripheral nodule, multiple nodules or a diffuse pneumonic-like infiltrate.[49] Microscopically, well-differentiated mucin-containing columnar cells line the walls of terminal air spaces without invading the stroma. Inflammatory cells, fat-laden macrophages and areas of interstitial fibrosis are frequently associated with this tumour. Psammoma bodies may be present. The differential diagnosis includes conventional adenocarcinoma and metastatic well-differentiated adenocarcinoma. In childhood, the most likely metastatic tumour to show this appearance is papillary carcinoma of thyroid, which can be confirmed by immunostaining for thyroglobulin. Bronchioloalveolar adenocarcinoma is thought to be of bronchiolar epithelial (including Clara cell) origin. Treatment is by resection, and prognosis is favourable if resection is complete and no metastatic disease is present at the time of diagnosis.

b. Well-differentiated adenocarcinoma
Well-differentiated adenocarcinoma resembling fetal lung is a unique type of adenocarcinoma of young adults. The youngest patient described with this tumour was 12 years old.[50] Microscopically, the tumour resembles the canalicular stage (17–27 weeks' gestation) of normal fetal lung development, with more complex glands and cytological atypia. There is a similarity to the pulmonary blastoma of adulthood but it lacks the blastomatous stroma. The other differential diagnosis is congenital cystic adenomatoid malformation, but this lacks atypia. It has a good prognosis.

Mesenchymal tumours

1. *Spindle cell tumours*

a. Inflammatory pseudotumour
Inflammatory pseudotumour is the most common spindle cell tumour of the lung in childhood.[51] Malaise, pyrexia, weight loss, hypergammaglobulinaemia and raised sedimentation rate are present in approximately 60% of cases. These symptoms characteristically resolve dramatically with excision of the lesion. The lesions are otherwise detected on the basis of an incidental finding on chest X-ray which shows a solitary intraparenchymal mass with calcification (25–30% of cases). Microscopically, a spindle cell proliferation with an admixture of plasma cells, lymphocytes and foamy histiocytes is seen (Fig. 8.12). The prognosis is favourable in the overwhelming majority of cases provided surgical resection is complete. The lesion may recur locally after incomplete resection, and locally aggressive behaviour has been documented.[52,53]

b. Fibromatosis
Fibromatosis of the lung is a very rare tumour, and is only seen in the context of generalized fibromatosis.[54] The lesions may be endobronchial or parenchymal. Microscopically, the tumours consist of interlacing bundles of spindle cells arranged in a herringbone pattern. Tumour cells are positive for vimentin, but negative for muscle-specific actin, desmin, *Ulex europaeus*, neuron-specific enolase, S-100 protein, Leu-7, α-1-antichymotrypsin and factor VIII-related antigen.

c. Leiomyoma and Leiomyosarcoma
Leiomyoma and leiomyosarcoma[55] of the lung have both been described in childhood. Overall, leiomyosarcoma has been reported more often than leiomyoma in children, including congenital and neonatal examples. Mitotic activity is the most accurate and reproducible predictor of behaviour. Tumours having less than 3 mitoses per 10 high-power fields behave in a benign fashion.[56] In childhood, leiomyosarcomas do not demonstrate the aggressive clinical behaviour seen in adults.

2. *Vascular tumours*

a. Haemangiomatosis
Haemangiomatosis is a localized or diffuse vascular proliferation,[57] characterized by the presence of capillary-sized vascular spaces with a lining of plump endothelial cells. It may present with symptoms suggestive of idiopathic pulmonary haemosiderosis, that is, cough, lethargy, dyspnoea and haemoptysis.

b. Sclerosing haemangioma
Sclerosing haemangioma, though probably an

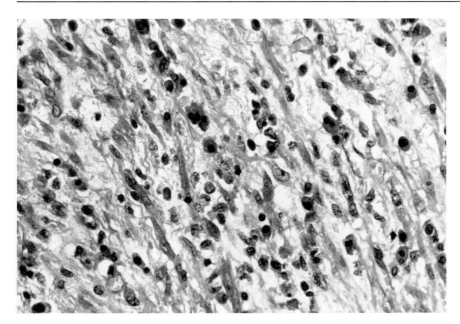

Fig. 8.12   Inflammatory pseudotumour. There is proliferation of cytologically bland fibroblasts with an associated plasma cell infiltrate.

epithelial tumour, is conventionally considered a vascular neoplasm which typically presents as an incidental finding on chest X-ray, frequently in children. The plump lining cells have the ultra-structural features of type 2 pneumocytes,[58] and show positive immunoreactivity for cytokeratin and epithelial membrane antigen.

c. Epitheliod haemangioendothelioma
Epithelioid haemangioendothelioma is the only potentially malignant vascular neoplasm that occurs with any frequency in the paediatric age group. It occurs primarily in females and tends to be multifocal. Histologically, pale-staining myxohyaline nodules fill and replace alveoli. Involvement of bronchioles, lymphatics and blood vessels is common. Centrally these nodules may be necrotic, while peripherally they are composed of plump histiocytoid cells with granular to vacuolated cytoplasm. Typically, the cells are uniform and lack cytological atypia or mitotic activity. The tumour cells form small intracellular lumina which are seen as clear spaces or vacuoles. Immunochemically, factor VIII-related antigen can be demonstrated within the cytoplasm, with accentuation of the staining around the cytoplasmic mini-lumen. The cells also bind *Ulex europaeus*. Immunostaining for cytokeratins is negative. Elec-

tron microscopically the cells have well-developed basal lamina, pinocytotic vesicles and occasional Weibel–Palade bodies. Mortality has been reported as 35%.[59]

Tumours arising in congenital lung cysts

Pulmonary blastoma, primary pulmonary rhabdomyosarcoma and malignant mesenchymoma of lung have a propensity to occur in association with congenital lung cysts.

1. *Pulmonary blastoma*
Spencer[60] proposed the designation pulmonary blastoma for a neoplasm in which the histological pattern was reminiscent of first trimester lung: solid blastema showing epithelial differentiation in the form of bronchiolar structures in a background of embryonic mesenchyme. Only 20% of patients with a pulmonary blastoma are less than 20 years old. It is now recognized that pulmonary blastoma in children differs in many ways from adult-type pulmonary blastoma.[61-63] In children, the tumour may be intrapulmonary or extrapulmonary, whereas the adult pulmonary blastoma frequently presents as a polypoid endobronchial tumour and never occurs outside the lung. The childhood form is commonly associated with a cystic malformation of the lung, which is not the case in adults. Con-

genital cystic adenomatoid malformation and mesenchymal cystic hamartoma are the most common associations. In the adult form, the immature blastema, stroma and epithelium all show overt histological features of malignancy. In childhood, the tumour consists of blastema and mesenchymal components only with no evidence of epithelial differentiation (Fig. 8.13a). Tubular epithelial structures represent entrapped epithelium of pre-existing bronchiolar origin. Blastemal and epi-

thelial cells in the adult form show positive immunochemical staining for epithelial markers, while staining for epithelial markers is negative in childhood pulmonary blastoma. Most pulmonary blastomas in childhood are predominantly mesenchymal tumours capable of differentiation along several pathways, notably cartilage and skeletal muscle. Metastases are invariably purely sarcomatous. The prognosis is uniformly poor in both childhood and adult forms.

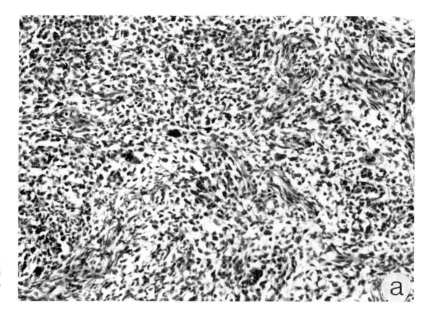

Fig. 8.13a Pulmonary blastoma. Low-power view showing featureless primitive small round cell tumour. No malignant epithelial elements seen.

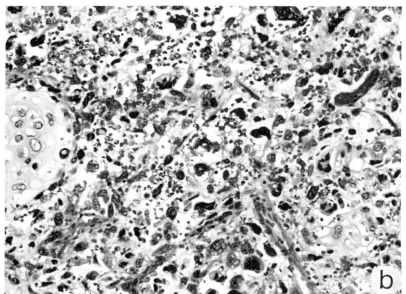

Fig. 8.13b Malignant mesenchymal tumour which arose in a pre-existing cystic lesion. Note skeletal muscle and chondroid differentiation.

2. *Primary pulmonary rhabdomyosarcoma and malignant mesenchymoma of lung*

There are several reports of primary pulmonary rhabdomyosarcoma[64,65] and malignant mesenchymoma of lung[66] (Fig. 8.13b) arising in association with pre-existing or simultaneously recognized cystic disease of the lung. Primary pulmonary rhabdomyosarcoma has been described in patients with an age range of 18 months to 13 years. Histologically, the majority are embryonal rhabdomyosarcomas, containing well-differentiated rhabdomyoblasts and showing cross-striations in a number of tumour cells. Less commonly, they are predominantly composed of poorly differentiated cells with foci of bizarre giant cells. Diagnosis is made in these cases by immunohistochemical demonstration of desmin in tumour cells, and electron microscopical demonstration of myofilaments. Malignant mesenchymoma of lung contains chondrosarcomatous and occasionally fibrosarcomatous elements in addition to rhabdomyosarcoma (Fig. 8.13b).

The underlying pathogenesis of the relationship between congenital lung cysts and pulmonary blastoma, primary pulmonary rhabdomyosarcoma and malignant mesenchymoma has yet to be elucidated. One theory is the malignant transformation of a multipotent stromal mesenchymal cell within the cyst wall. Another is the malignant transformation of pre-existing foci of heterotopic skeletal muscle. Ueda *et al.*[64] suggested that certain pulmonary malformations are intrinsically susceptible to carcinogens, perhaps by serving as a vehicle for retention and persistence of unstable mesenchymal cells.

Extrapulmonary tumours

1. *Primitive neuroectodermal tumour (PNET); Askin tumour*

This tumour typically occurs in the thoracopulmonary region, frequently paravertebrally, and in the lung with or without rib involvement. It is a tumour of children and adolescents, and is now considered to represent the same tumour type as Ewing's sarcoma. Morphologically it is a malignant small round cell tumour made up of monotonous cells with inconspicuous cytoplasm or a small amount of glycogen-rich cytoplasm (Fig. 8.14a). Small dark pyknotic cells are scattered throughout. Mitoses are inconspicuous. Large haemorrhagic areas and foci of coagulative necrosis are other frequent features. Immunohisto-

chemically, staining with a panel of neural antibodies will show positivity for at least one in up to 50% of cases.[67] Vimentin expression is variable. Focal keratin expression has been reported. The recently described antibody to the MIC-2 gene product is extremely useful in these tumours since tumour cells show a characteristic membrane staining (Fig. 8.14b). Other small round cell tumours of childhood do not react with the antibody.[68] Cytogenetically, PNET shares a specific chromosomal translocation with Ewing's sarcoma, that is, a reciprocal translocation between the long arms of chromosomes 11 and 22.[69]

2. *Extrapulmonary blastoma*

As discussed above, pulmonary blastoma of childhood may occur in an extrapulmonary site with extensive involvement of the underlying lung.

Open-lung biopsy

Reasons for performance of open-lung biopsy in the paediatric age group are listed in Table 8.2. Most of the topics are covered in other chapters, and the account here is confined to a few comments.

Immunocompromised host

In childhood, this group includes congenital immunodeficiency syndromes, the acquired immunodeficiency syndrome (AIDS) and iatrogenic immunodeficiency as a result of chemotherapy for haematological malignancy.

In addition to opportunistic infections, pulmonary alveolar proteinosis is a rare cause of respiratory symptoms in these patients, and probably represents a nonspecific reaction pattern to a variety of different injuries. The histological and electron microscopical appearances are identical to those described above for the congenital form. Its appearance in the lung biopsy from an immunocompromised host should prompt a vigorous search for infection, and serial sections may be necessary to locate a small focus of organisms that is overshadowed by an extensive pulmonary alveolar proteinosis reaction.

Bronchiolitis obliterans

Clinically, patients with bronchiolitis obliterans have features of obstructive airways disease.

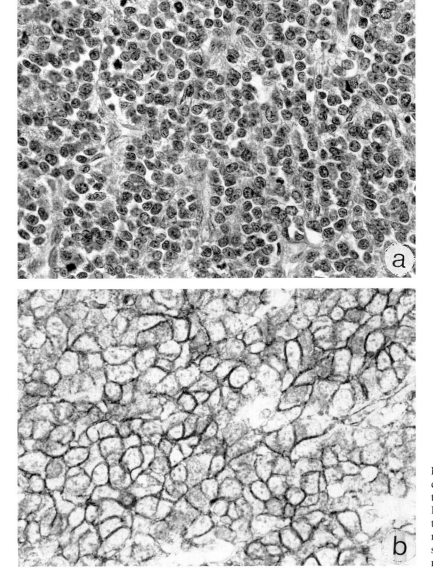

Fig. 8.14 Primitive neuroecto-dermal tumour of chest wall (Askin tumour) which invaded lung. (a) Haematoxylin and eosin stained section showing undifferentiated small round cell tumour. (b) Immuno-staining for MIC-2 shows strong membrane staining.

Aetiological factors in the paediatric age group include bronchopulmonary dysplasia (see above), infections (especially adenovirus, measles, myco-plasma and respiratory syncytial virus), bron-chiectasis (see above), and graft-versus-host disease following bone marrow transplantation and heart-lung transplantation. Histologically, the features are of granulation tissue plugs in small bronchi and bronchioles which may extend to alveolar ducts, and in advanced disease, complete destruction of bronchi or bronchioles, leaving only a scar.

Table 8.2 Reasons for open-lung biopsy in children

Pulmonary hypertension in congenital heart disease
The immunocompromised host
Bronchiolitis obliterans
Interstitial fibrosis
Inhalational lipid pneumonia
Idiopathic pulmonary haemosiderosis
Congenital pulmonary lymphangiectasia
Lymphangiomatosis

Interstitial fibrosis

Clinically, patients with interstitial fibrosis have features of restrictive lung disease. Causes of interstitial fibrosis in childhood include broncho-pulmonary dysplasia (see above), Langerhans' cell histiocytosis, and desquamative pneumonia/fibrosing alveolitis.

Langerhans' cell histiocytosis
The term Langerhans' cell histiocytosis (LCH) embraces several previous eponyms including histiocytosis X, Hand–Schuller–Christian disease, eosinophilic granuloma of bone and Letterer–Siwe disease.[70] Many tissues and organs can be affected. Clinical features include a skin rash, lytic bone lesions, proptosis, lymphadenopathy, hepatosplenomegaly and diabetes insipidus. Pulmonary involvement leads to breathlessness and a honeycomb appearance on chest radiograph.

In children LCH of the lung usually occurs in the context of multisystem disease, and lung biopsy is rarely undertaken as the diagnosis can be made on a skin biopsy. Histologically, the pulmonary nodules are discrete and roughly symmetrical with a fibrotic centre, and with cellular peripheral interstitial extensions. The characteristic LCH cells are more easily found at the periphery (Fig. 8.15a,b).

Lesions examined early in the course of the disease are usually cellular and locally destructive and feature aggregates of LCH cells. These pathological Langerhans' cells have homogeneous pink cytoplasm and lobulated nuclei, many showing a longitudinal groove resulting in a 'coffee bean' appearance. Mitoses are not seen, and phagocytosis by LCH cells is unusual. The numbers of eosinophils, macrophages and lymphocytes are variable. Multinucleated giant cells are frequently encountered. Later in the course of the disease, lesions become less cellular and contain fewer LCH cells and more macrophages, leading on to fibrosis. In long-standing disease, LCH cells may no longer be demonstrable.

Immunochemically, approximately 60% of LCH cells show cytoplasmic staining for S-100 protein. A more sensitive marker is peanut agglutinin, which shows a characteristic membrane staining. If frozen tissue is available, LCH cells display positive staining with anti-CDI antibody. They are weakly HLA-DR positive, and negative for the macrophage markers CD68, Mac 387, lysozyme and α-1-antitrypsin and anti-1-antichymotrypsin. Monocytes, macrophages and multinucleated giant cells within the infiltrate show positivity for these macrophage markers. Electron microscopically, LCH cells contain Birbeck granules which occur only in the cytoplasm of normal or pathological Langerhans' cells.

The differential diagnosis of LCH in the lung is with a granulomatous response to infection, and desquamative interstitial pneumonia. Opportunistic infection should always be borne in mind whenever histiocytosis is found, and should be excluded by appropriate stains. Positive identification of LCH cells is the mainstay of the diagnosis. Alveolar macrophages are positive for macrophage markers, and may also be S-100 positive. This antigen is therefore not reliable for the differentiation of LCH and desquamative interstitial pneumonia. The characteristic pattern of staining of LCH cells with peanut agglutinin is not shown by any other cell in the lung, and electron microscopy shows Birbeck granules. In the late stages of LCH, the lung may be so fibrotic and lacking in LCH cells that the distinction from fibrosing alveolitis cannot be made.

Desquamative pneumonia/fibrosing alveolitis
This is a rare, generally sporadic, disorder in children. An autosomal recessive disease that has a worse prognosis than the sporadic form is increasingly recognized.[71] In the early stage of the disease, large numbers of macrophages are seen within the alveolar spaces. The alveoli are lined by prominent type 2 pneumocytes, and the name desquamative interstitial pneumonia came from the early idea that these cells were shed into the alveolar spaces. In fact, it is a misnomer. These appearances are associated with an interstitial, predominantly lymphocytic, infiltrate. In the later stages, interstitial fibrosis predominates. Patients with the desquamative pattern show a significantly better response to treatment with steroids than those with the fibrosing pattern.

Inhalational lipid pneumonia

This occurs following milk inhalation and is seen with any condition in which there is persistent vomiting or oesophageal reflux. Cough, wheezing and dyspnoea are the presenting clinical features, and in patients with asymptomatic oesophageal reflux, the clinical differential diagnosis includes extrinsic allergic alveolitis. However, the latter is extremely rare in childhood. Open-lung biopsy shows numerous intra-alveolar macrophages and

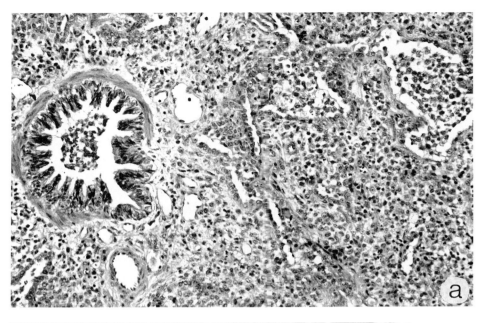

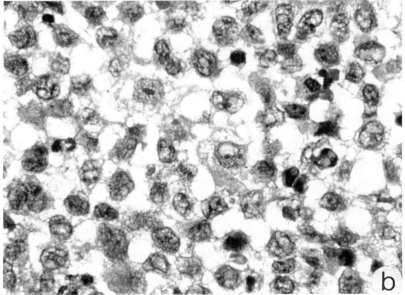

Fig. 8.15 Langerhans' cell histiocytosis. (a) Low-power view showing interstitial and intra-alveolar infiltrate. (b) High-power view showing characteristic morphology of Langerhans' cells with bland grooved 'coffee bean' nuclei, and ample eosinophilic cytoplasm. Immunostaining for S-100 protein and peanut agglutinin was positive, and electron microscopy showed Birbeck granules.

foreign body giant cells containing and surrounding cholesterol clefts, with marked type 2 pneumocyte hyperplasia. In the later stages, there is associated interstitial fibrosis (Fig. 8.16).

Idiopathic pulmonary haemosiderosis

This occurs primarily in children between 1 and 6 years of age, but can be seen in children as young as 4 months. Sex incidence is equal, and 15–20% of cases occur in adolescents and young adults. Symptoms include cough, lethargy, dyspnoea and haemoptysis, and there may also be fever, lymphadenopathy and hepatosplenomegaly. Radiologically, patchy or diffuse pulmonary infiltrates are seen, which in the late stages may progress to perihilar shadowing or a pattern of diffuse interstitial disease. Hypochromic microcytic

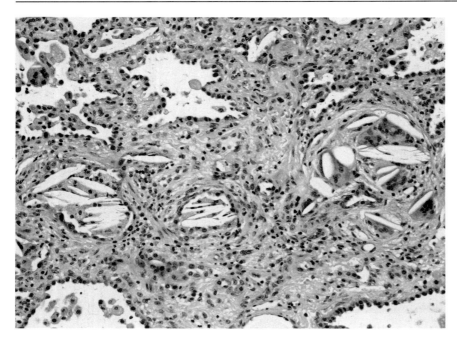

Fig. 8.16 Inhalation lipid pneumonia. Open-lung biopsy from a 6-year-old child with dyspnoea and wheezing, reticular shadowing on chest X-ray, and abnormal pulmonary function tests. Photomicrograph showing macrophages and foreign body giant cells containing cholesterol clefts. There is type 2 pneumocyte hyperplasia and interstitial fibrosis. Subsequent investigation showed gastro-oesophageal reflux, and a long history of loud snoring when asleep was elicited.

anaemia is seen in virtually all cases, and 12–15% show eosinophilia. While some children may die of massive haemorrhage shortly after presentation, most have a history of progressive respiratory insufficiency leading to death 2–5 years after diagnosis.

Bronchoalveolar lavage demonstrates large numbers of haemosiderin-laden macrophages. Lung biopsy shows focal consolidation due to massive accumulations of haemosiderin-laden macrophages which obliterate alveolar spaces and are associated with interstitial fibrosis. Stainable iron is present in alveolar macrophages, free in connective tissues and encrusting elastic fibres of small blood vessels and alveolar septa. There is mild to moderate hyperplasia of type 2 pneumocytes, peribronchial lymphoid hyperplasia, and a mastocytosis which is concentrated in close proximity to small pulmonary vessels. Electron microscopy shows abnormalities of the alveolar capillary membrane which may be primary or secondary. The presumed defect is thought to be an abnormality of the alveolar air–blood barrier involving either the basement membrane or endothelial cells, but the nature of the defect, and whether it is primary or acquired, remain unclear.[72]

Congenital pulmonary lymphangiectasia

Pulmonary lymphangiectasia, or dilatation of pre-existing lymphatics in the pleura and interlobular septa is usually associated with congenital heart disease involving obstruction of pulmonary venous drainage (Fig. 8.17). Primary pulmonary lymphangiectasia is extremely rare. It is always congenital, affecting one or more lobes or even both lungs. The condition often leads to stillbirth or early neonatal death.

Lymphangiomatosis

This is a rare disorder characterized by proliferating lymphatic channels in osseous or extra-osseous tissue in a diffuse fashion.[73] It represents a generalized abnormality of lymphatics and presents in childhood or early adulthood. The

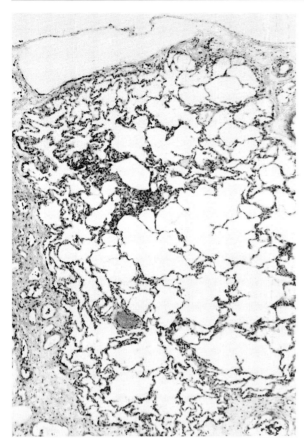

Fig. 8.17 Secondary lymphangiectasia in a case of congenital heart disease. Note the dilated subpleural lymphatic (top).

outcome is related to the extent of dissemination; it is favourable when disease is limited to the bones (except cervical vertebrae) but generally poor when vital organs are affected. Patients characteristically present with recurrent chylous pleural effusions. Physical examination may reveal skin lesions, hepatosplenomegaly and lymphadenopathy. Skeletal survey may show multiple lesions in bone. Lung biopsy shows proliferating lymphatics in parietal and visceral pleura and in interlobular septa. As the lymphatic channels are frequently collapsed and lined by scanty endothelial cells, immunochemical staining for endothelial markers is useful.[73]

Transbronchial biopsy

This technique has been increasingly used in paediatric practice in recent years, mainly in moni-

toring pulmonary graft rejection and infection in recipients of heart–lung transplants, but also in immunocompromised patients at risk of opportunistic infections and in patients with fibrosing alveolitis.[74]

References

1. Wigglesworth JS, Desai R. Use of DNA estimation for growth assessment in normal and hypoplastic fetal lungs. *Arch Dis Child* 1981; **56**, 601–605.
2. Emery JL, Mithal A. The number of alveoli in the terminal respiratory unit of man during late intrauterine life and childhood. *Arch Dis Child* 1960; **35**, 544–547.
3. Cooney TP, Thurlbeck WM. The radial alveolar count method of Emery and Mithal – a reappraisal. I. Post natal lung growth. *Thorax* 1982; **37**, 572–579.
4. de la Monte SM, Hutchins GM, Moore GW. Respiratory epithelial cell necrosis is the earliest lesion of hyaline membrane disease of the newborn. *Am J Pathol* 1986; **123**, 155–160.
5. Lauweryns JM. 'Hyaline membrane disease' in newborn infants. Macroscopic, radiographic and light and electron microscopic studies. *Hum Pathol* 1970; **1**, 175–204.
6. Seo IS, Gillim SE, Mirkin LD. Hyaline membranes in postmature infants. *Pediatr Pathol* 1990; **10**, 539–548.
7. Gitlin JD, Soll RF, Parad AB *et al*. Randomized controlled trial of exogenous surfactant for the treatment of hyaline membrane disease. *Pediatrics* 1987; **79**, 31–37.
8. Hagstrom N, Waters BL. Morphometric analysis of exogenous surfactant's affect on neonatal lung: a retrospective case-controlled postmortem study. *Pediatr Pathol* 1993; **13**, 112 (abstract).
9. Pinar H, Makarova N, Ruben L, Singer DB. Pathology of the lung in surfactant-treated neonates. *Pediatr Pathol* 1993; **13**, 105–106 (abstract).
10. Taghizadeh A, Reynolds EOR. Pathogenesis of bronchopulmonary dysplasia following hyaline membrane disease. *Am J Pathol* 1976; **82**, 241–264.
11. Stocks J, Godfrey S, Reynolds EOR. Airway resistance in infants after various treatments for hyaline membrane disease; special emphasis on prolonged high levels of inspired oxygen. *Pediatrics* 1978; **61**, 178–183.
12. Stenmark KR, Veelkel NF. Potential role of inflammation and lipid mediators in the pathophysiology of bronchopulmonary dysplasia. In: *Bronchopulmonary Dysplasia*, Bancalari E, Stocker JT (eds). Washington DC: Hemisphere, 1988, pp. 58–77.
13. Anderson WR, Engel RR. Cardiopulmonary sequelae of reparative stages of bronchopulmonary dysplasia. *Arch Path Lab Med* 1983; **107**, 603–608.

14. Stocker JT. The pathology of long-standing 'healed' bronchopulmonary dysplasia. A study of 28 infants 3–40 months of age. *Hum Pathol* 1986; **17**, 943–961.
15. Trompeter R, Yu VYH, Aynsley-Green A, Roberton NRC. Massive pulmonary haemorrhage in newborn infants. *Arch Dis Child* 1975; **50**, 123–127.
16. Arnand A, Gray ES, Brown T, Clewley JP, Cohen BJ. Human parvovirus infection in pregnancy and hydrops fetalis. *New Engl J Med* 1987; **316**, 183–186.
17. Greenspoon JS, Masaki DI. Fetal varicella syndrome. *J Pediatr* 1988; **112**, 505–506.
18. Vawter GF. Perinatal listeriosis. *Perspect Pediatr Pathol* 1981; **6**, 153–156.
19. Craig JM. Group B beta hemolytic streptococcal sepsis in the newborn. *Perspect Pediatr Pathol* 1981; **6**, 139–151.
20. Singer DB. Pathology of neonatal herpes simplex virus infection. *Perspect Pediatr Pathol* 1981; **6**, 243–278.
21. Whyte RK, Hussain Z, deSa DJ. Antenatal infections with *Candida*. *Arch Dis Child* 1982; **57**, 528–535.
22. Noel GJ, Edelson PJ. *Staphylococcus epidermidis* bacteraemia in neonates: further observations and the occurrence of focal infection. *Pediatrics* 1984; **74**, 832–837.
23. Wigglesworth JS, Desai R, Guerrini P. Fetal lung hypoplasia: biochemical and structural variations and their possible significance. *Arch Dis Child* 1981; **56**, 606–615.
24. Silver MM, Vilos GA. Pulmonary hypoplasia in neonatal hypophosphatasia. *Pediatr Pathol* 1988; **8**, 483–493.
25. Haworth SG. Primary and secondary pulmonary hypertension in childhood: a clinicopathological reappraisal. *Curr Top Pathol* 1983; **73**, 92–152.
26. Wagenvoort CA. Misalignment of lung vessels. A syndrome causing persistent neonatal pulmonary hypertension. *Hum Pathol* 1986; **17**, 727–730.
27. Langstom C. Misalignment of pulmonary veins and alveolar capillary dysplasia. *Pediatr Pathol* 1991; **11**, 163–170.
28. Teja K, Cooper PH, Squires JE, Schnatterly PT. Medical intelligence. Pulmonary alveolar proteinosis in four siblings. *New Engl J Med* 1981; **305**, 1390–1392.
29. Coleman M, Dehner LP, Sibley RK *et al.* Case reports. Pulmonary alveolar proteinosis: an uncommon cause of chronic neonatal respiratory distress. *Am Rev Respir Dis* 1980; **121**, 583–586.
30. Chau P, Shen-Schwartz S *et al.* Pulmonary epithelial changes with extracorporeal membrane oxygenation (ECMO) therapy: analyses of 17 autopsy cases. *Mod Pathol* 1991; **4**, 2P.
31. Soosay GN, Baudouin SV, Hanson PJV *et al.* Symptomatic cysts in otherwise normal lungs of children and adults. *Histopathol* 1992; **20**, 517–522.
32. Stocker JT. Sequestrations of the lung. *Semin Diagnost Pathol* 1986; **3**, 106–121.
33. Adzick MS, Harrison MR, Glick PL. Fetal cystic adenomatoid malformation: prenatal diagnosis and natural history. *J Pediatr Surg* 1985; **20**, 483–488.
34. Stocker JT, Drake RM, Madewell JE. Cystic and congenital lung disease in the newborn. *Perspect Pediatr Pathol* 1978; **4**, 93–101.
35. Moerman P, Fryns J-P, Vandenberghe K *et al.* Pathogenesis of congenital cystic adenomatoid malformation of the lung. *Histopathol* 1992; **21**, 315–322.
36. Morgan WJ, Leman RJ, Rojas R. Acute worsening of congenital lobar emphysema and subsequent spontaneous improvement. *Pediatrics* 1983; **71**, 844–848.
37. Tapper D, Schuster S, McBride J *et al.* Polyalveolar lobe: anatomic and physiologic parameters and their relationship to congenital lobar emphysema. *J Pediatr Surg* 1980; **15**, 931–937.
38. Stocker JT. Postinfarction peripheral cysts of the lung. *Pediatr Pathol* 1987; **7**, 111–117.
39. Mark EJ. Mesenchymal cystic hamartoma of the lung. *New Engl J Med* 1986; **315**, 1255–1259.
40. Davis PB. Cystic fibrosis from bench to bedside. *New Engl J Med* 1991; **325**, 575–576.
41. Afzelius BA. The immotile cilia syndrome and other ciliary diseases. *Int Rev Exp Pathol* 1979; **19**, 1–43.
42. Andrassy RJ, Feldtman RW, Stanford W. Bronchial carcinoid tumours in children and adolescents. *J Pediatr Surg* 1977; **12**, 513–517.
43. Mullins JD, Barnes RP. Childhood bronchial mucoepidermoid tumours. A case report and review of the literature. *Cancer* 1979; **44**, 315–322.
44. Hartmann GE, Schochat SJ. Primary pulmonary neoplasms of childhood: a review. *Ann Thorac Surg* 1983; **36**, 108–120.
45. Carney JA. The triad of gastric epithelioid leiomyosarcoma, pulmonary chondroma and functioning extra-adrenal paraganglioma: five-year review. *Medicine (Baltimore)* 1983; **62**, 159–169.
46. Sawada K, Fukuma S, Karasawa K, Suchi T. Granular cell myoblastoma of the bronchus in a child: a case report. *Jpn J Surg* 1981; **11**, 111–114.
47. Tomashefski JR Jr. Benign endobronchial mesenchymal tumours. Their relationship to parenchymal pulmonary hamartomas. *Am J Surg Pathol* 1982; **6**, 531–540.
48. McKneally MF. Lung cancer in young patients. *Ann Thorac Surg* 1983; **36**, 505–507.
49. Donaldson JC, Kaminsky DB, Elliott RC. Bronchiolar carcinoma. Report of 11 cases and review of the literature. *Cancer* 1978; **41**, 250–258.
50. Manning JT Jr, Ordonez NG, Rosenberg HS, Walden WE. Pulmonary endodermal tumour resembling fetal lung – report of a case with immunohistochemical studies. *Arch Pathol Lab Med* 1985; **109**, 48–50.

51. Monzon CM, Gilchrist GS, Burgert EO Jr *et al.* Plasma cell granuloma of the lung in children. *Pediatrics* 1982; **70**, 268–274.

52. Muraoka S, Sato T, Takahashi T, Ando M, Shimoda A. Plasma cell granuloma of the lung with extrapulmonal extension. Immunohistochemical and electron microscopic studies. *Acta Pathol Jpn* 1985; **35**, 933–944.

53. Hutchins GM, Eggleston JC. Unusual presentation of pulmonary inflammatory pseudotumour (plasma cell granuloma) as esophageal obstruction. *Am J Gastroenterol* 1979; **17**, 501–508.

54. Roggli VL, Kim H-S, Hawkins E. Congenital generalised fibromatosis with visceral involvement: a case report. *Cancer* 1980; **45**, 954–960.

55. Jimenez JF, Uthman EQ, Townsend JW, Gloster ES, Seibert JJ. Primary bronchopulmonary leiomyosarcoma in childhood. *Arch Pathol Lab Med* 1986; **110**, 348–351.

56. Guccion JG, Rosen SH. Bronchopulmonary leiomyosarcoma and fibrosarcoma. A study of 32 cases and review of the literature. *Cancer* 1972; **30**, 386–847.

57. Whittaker JS, Pickering CAC, Heath D, Smith P. Pulmonary capillary haemangiomatosis. *Diagn Histopathol* 1983; **6**, 77–80.

58. Heard BE. Benign sclerosing 'haemangioma' of the lung. *Histopathol* 1986; **10**, 541–542.

59. Dail DH, Leibow AA, Gmelich JJ *et al.* Intravascular, bronchiolar and alveolar tumour of the lung (IVBAT). An analysis of twenty cases of a peculiar sclerosing endothelial tumour. *Cancer* 1983; **51**, 452–464.

60. Spencer H. Pulmonary blastoma. *J Pathol Bacteriol* 1961; **82**, 161–165.

61. Manivel JC, Priest JR, Watterson J *et al.* Pleuropulmonary blastoma. The so-called pulmonary blastoma of childhood. *Cancer* 1988; **62**, 1516–1526.

62. Cohen M, Emms M, Kaschula RQC. Childhood pulmonary blastoma: a pleuropulmonary variant of adult-type pulmonary blastoma. *Pediatr Pathol* 1991; **11**, 737–749.

63. Hachitanda Y, Aoyama C, Sato JK, Shimada H. Pleuropulmonary blastoma in childhood. A tumour of divergent differentiation. *Am J Surg Pathol* 1993; **17**, 382–391.

64. Ueda K, Gruppo R, Unger F, Martin L, Bove K. Rhabdomyosarcoma of the lung arising in congenital cystic adenomatoid malformation. *Cancer* 1977; **40**, 383–388.

65. Allan BT, Day DL, Dehner LP. Primary pulmonary rhabdomyosarcoma of the lung in children: report of two cases presenting with spontaneous pneumothorax. *Cancer* 1987; **59**, 1005–1011.

66. Domizio P, Leisner RJ, Dicks-Mireaux C, Risdon RA. Malignant mesenchymoma associated with a congenital lung cyst in a child. Case report and review of the literature. *Pediatr Pathol* 1990; **10**, 785–797.

67. Malone M. Soft tissue tumours in childhood. *Histopathol* 1993; **23**, 203–216.

68. Ambros IM, Ambros PF, Strehl S *et al.* MIC 2 is a specific marker for Ewing's sarcoma and peripheral primitive neuroectodermal tumors. *Cancer* 1991; **67**; 1886–1893.

69. Stephenson CF, Bridge JA, Sandberg ES. Cytogenetic and pathologic aspects of Ewing's sarcoma and neuroectodermal tumours. *Hum Pathol* 1992; **23**, 1270–1277.

70. Malone M. The histiocytoses of childhood. *Histopathol* 1991; **19**, 105–119.

71. Buchino JJ, Keenan WJ, Algren JT, Bove KE. Familial desquamative interstitial pneumonia occurring in infants. *Am J Med Genet* (suppl) 1987; **3**, 285–291.

72. Cutz E. Idiopathic pulmonary hemosiderosis and related disorders in infancy and childhood. *Perspect Pediatr Pathol* 1987; **11**, 47–81.

73. Ramani P, Shah A. Lymphangiomatosis. Histologic and immunohistochemical analysis of four cases. *Am J Surg Pathol* 1993; **17**, 329–335.

74. Whitehead B, Scott JP, Helms P, Malone M *et al.* Technique and use of transbronchial biopsy in children and adolescents. *Pediatr Pulmonol* 1992; **12**, 240–246.

9

Handling of resection and autopsy lung specimens

A Kennedy

Introduction

The examination of large specimens of lung, whether from the postmortem room or the operating theatre, requires considerable care and a good deal of planning. First, it must always be remembered that all such specimens are potentially infected and that tuberculosis is a well-recognized hazard as a cause of laboratory acquired infection. The pathologist has a duty to protect both himself and other laboratory staff but, at the same time, it must be remembered that the specimen may be the sole source of material for successful bacteriological investigation. Even when it is known that the lung contains a tumour, there may be coexistent tuberculosis. Tuberculous infection commonly complicates some forms of industrial lung disease so that all fresh specimens must be regarded as being infected and must be handled accordingly. Anatomically the lung is a very complex organ and its structure is far from uniform; the specimen must therefore be handled in such a way that any disease process can be studied with regard to its relationships with different components of the lung and with regard to the gross areas of the lung which may be involved. While it is useful to have a standard routine, it is essential that each case should be approached in a flexible way so that the method can be adapted to the problems that are specific to the individual case.

General considerations

The more information which is available before the examination begins, the more easily the procedure can be planned. In the case of resection specimens there may already have been a biopsy so at least the nature or localization of the main pathological process will be known, but the clinical, radiological and laboratory findings should be available at the stage when the initial handling of the specimen is being considered.

A knowledge of the clinical history and radiological appearances is particularly valuable in the postmortem room as the whole of the respiratory tract is available for examination. If the sites and distribution of the lesions is known at the outset it may be possible to handle the two lungs differently accordingly to the circumstances and the amount of time which is available. In cases where death has occurred suddenly clinical information may be very incomplete and, occasionally, it may be misleading. The pathologist must glean what information he can from the external appearances of the body.

The height and weight should be recorded and the presence of dependent oedema may be related to cardiac failure, cor pulmonale or deep venous thrombosis. In some clinical contexts the possibility of deep venous thrombosis and pulmonary embolism may suggest itself on external examination alone, particularly if oedema affects one leg only. In superior vena caval obstruction the face and arms are swollen and plethoric. Careful inspection of the extremities may show clubbing of the fingers or toes or hypertrophic osteoarthropathy. While inspecting the hands it is valuable to record evidence of rheumatoid arthritis or scleroderma which, like lupus erythematosus, may also produce characteristic lesions on the face. In coal miners the skin of the hands will usually be marked by coal dust tattoos, which are also found over the knees, forearms and back (Fig. 9.1).

Pleural surfaces

Before examining the lung itself the condition of the pleural cavities and surfaces must always be recorded, even if they are completely normal; this fact on its own may be of considerable medico-legal significance especially in cases of alleged asbestos exposure. The state of inflation of the lungs must be recorded; in status asthmaticus they will be overinflated and they fill the pleural cavities, while in pneumothorax the lung will be small and collapsed.

If pneumothorax is suspected the thorax should be opened under water to test for the presence of air in the pleural cavities. This is easy in the case of a small child as the whole body can be immersed in a large sink after reflection of the skin and muscle; each pleural cavity is then opened in turn with a scalpel. In adults this technique is impossible but if an assistant holds up the reflected skin and muscle

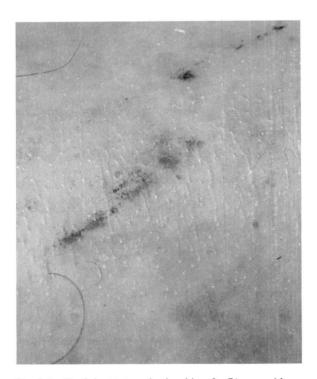

Fig. 9.1 Coal dust tatoos in the skin of a 71-year-old ex-miner. These were found over the backs of the hands, the forearms, the knees and the shins.

on each side of the thorax the resulting space can be filled with water; the pleural cavity is then opened through an intercostal space while watching for the escape of air bubbles.

The size and nature of any pleural effusion must be recorded; this is most easily done if the autopsy room is equipped with a length of plastic tubing, a peristaltic pump and a large trap which will catch the fluid and retain it for measurement. If the fluid appears to be purulent some should be retained, with sterile precautions, for bacteriological examination. Bloodstained effusions, except in the immediate post-thoracotomy situation, often signify malignancy, and chylous effusions are usually associated either with lymphatic obstruction or with rupture of the thoracic duct.

Similarly the site, extent and nature of any pleural adhesions must be recorded along with an inspection of the parietal pleura and ribs for the presence of tumour. It is particularly important to record the presence or absence of any pleural plaques on the parietal pleura or on the domes of the diaphragm. Fibrous plaques on the pleura may be of post-inflammatory origin but these are usually just areas of pleural thickening which are usually found on the visceral pleura over some old lung lesion such as an infarct. In contrast the plaques associated with asbestos exposure are often very characteristic in that they are on the parietal pleura, they are very white, they are hyaline in consistency and they sometimes have raised irregular borders.

The general anatomy of the thoracic cavity should be assessed with particular regard to the situs of the heart, great vessels and azygos vein. Interstitial emphysema of the chest wall or the mediastinum may be found in postoperative cases, following trauma, or after positive pressure ventilation at high pressure when it may be detected by the crepitant feel of the tissues beneath the fingers and by the presence of small bubbles of air in the soft areolar tissues.

When all such observations have been made the heart and lungs can be removed *en bloc* in continuity with the organs of the throat, either by themselves or along with the abdominal organs by the Rokitansky technique. The choice of method is a matter of individual preference which may, and should, be varied accordingly to the circumstances of the case. Except in cases in which there is disease of the aorta or oesophagus, there are no special advantages to the Rokitansky technique as far as the pulmonary pathologist is concerned and it must not be used in cases of known or suspected tuberculosis (see below).

After removal of the thoracic organs, the remainder of the pleural cavities can be inspected for the presence of tumour or fracture of the ribs, vertebrae or sternum.

Separation of the heart and lungs

At some stage of the investigation it will be necessary to separate the heart and lungs so that each may be examined in detail but, before taking this important and irreversible step, it is essential to examine the arrangement of the heart and great vessels as precipitate separation of the two systems may cause vital evidence to be lost. The first step is to open the pericardium and to examine its contents; some major anomalies of the heart will be visible at this stage. The specimen may then be turned over and the oesophagus can be opened from behind; it is then carefully dissected off the back of the pericardium and trachea so as to leave the trachea intact. It should then be possible to examine the vascular connections of the lungs.

This is not the place for a description of congenital heart disease but it must be remembered that many minor congenital anomalies of the cardiovascular system, which may have an important bearing in the pulmonary circulation, can persist into adult life undetected. Apart from septal defects it is important to be aware of the possibility of partial anomalous pulmonary venous drainage which has similar circulatory effects to an atrial septal defect. Therefore it is essential to check that all of the pulmonary veins are normally situated and that they all drain into the left atrium; in case of doubt the atria and veins must be opened and their connections checked using a soft probe.

Anomalies of the pulmonary arterial supply must also be identified. In cases of sequestration the anomalous segment is usually supplied by a systemic artery arising from the aorta (Fig. 9.2) and, rarely, a pulmonary artery may be missing so that the affected lung is supplied only by large bronchial arteries (Fig. 9.3). If all such vascular anomalies are not identified before the heart and lungs are separated, their subsequent identification may be impossible.

Fig. 9.2 Resection specimen showing systemic vessel supplying lung from pleural surface. (Courtesy of M.N. Sheppard, Royal Brompton Hospital.)

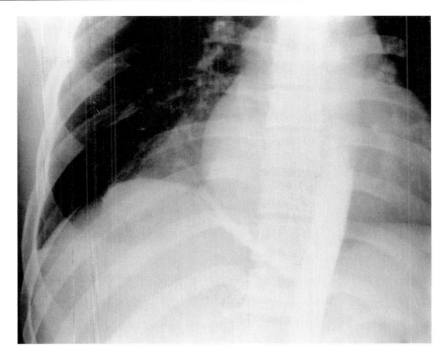

Fig. 9.3 Congenital absence of the right pulmonary artery. This patient underwent pneumonectomy because of impaired exercise tolerance which was due to a gross discrepancy between perfusion and diffusion. The right lung was supplied only by bronchial arteries. Aortogram showing bronchial artery supplying right lung.

The heart may now be removed for separate examination but, in so doing, it is important that no fragments of embolus are lost when the pulmonary trunk is divided. In any case of diffuse parenchymal lung disease which may be associated with pulmonary hypertension, it will be necessary to obtain the differential weights of the cardiac ventricles. The condition of the mediastinal and subcarinal lymph nodes must be recorded and a decision can then be made as to whether the lungs should remain attached to the trachea or separated; this depends on a large number of considerations such as the presence of a tumour at the carina and the degree of stenosis of the main bronchi due to enlargement of the hilar lymph nodes. The remainder of the great vessels, thymus, mediastinal tissues, the larynx and the thyroid may now be removed. Alternatively, each lung may be cut off the mediastinal structures using a large knife or scissors, but this must be done in such a way that a sufficient length of bronchus remains for the lung to be cannulated and fixed by inflation.

Once the lungs have been separated, the visceral pleura can be examined and the sites of any scars, tears or adhesions may be noted. Many lesions, including tumours, areas of consolidation and pneumoconiotic nodules can often be identified by palpation at this stage. Fibrosis is sometimes very obvious from the texture and a 'morocco leather' appearance of the pleura over the lower lobes is seen in cases of fibrosing alveolitis (Fig. 9.4).

Radiology

The radiological investigation of specimens is usually only required when it is necessary to make direct comparisons with *in vivo* radiographs. It may also be useful in explaining any unusual radiological findings or for demonstrating the relationship of a previously identified lesion to the bronchial tree or to the vasculature. These procedures need to be considered before fixation as flooding the alveoli with fixative renders the whole specimen more radio-opaque. If examination of

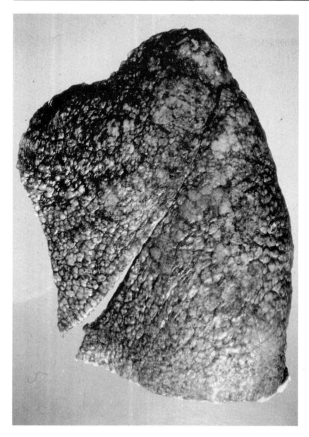

Fig. 9.4 Surface of left lung showing the 'cobblestone' or 'morocco leather' appearance due to underlying fibrosing alveolitis. (Courtesy of M.N. Sheppard, Royal Brompton Hospital.)

the whole specimen is required, then it will be necessary to distend it with air or formalin vapour before the plates are exposed. Fixation by formalin vapour can be achieved with special equipment[1] but this is a relatively troublesome technique as leakage can present serious problems and the escape of formalin vapour represents a serious hazard to laboratory staff.

Postmortem bronchography is a relatively simple matter which is best carried out after fixation. The bronchi are filled either with a conventional barium sulphate contrast medium or the barium sulphate can be suspended in warm gelatin which has the advantage that the contrast medium will not run out again once the gelatin has cooled and set. While excellent results may be obtained with a clinical radiological machine, it is usually much more convenient to use a laboratory X-ray

cabinet such as the Faxitron. Injection of the vessels is usually more troublesome due to the presence of clots but it may be useful in tracing and identifying the vascular supply of a lesion particularly in the case of arteriovenous malformations.

Fixation

The fixation of any pathological specimen is a matter of enormous importance as errors made at this stage can seldom be rectified and premature or inappropriate fixation may make some lines of investigation, e.g. bacteriological culture, impossible. The choice of fixative and the method used is dependent on what information it is hoped to obtain from the specimen, but it also depends on the size and integrity of the specimen itself. In general, the preferred method in most cases is fixation by perfusing the bronchial tree with buffered formol saline.[2]

If it is planned to study the microanatomy of the lung lobules, small airways or small blood vessels, then it is absolutely essential that at least part of the specimen is fixed in a state of inflation. If this is not done the terminal bronchioles, alveolar ducts and alveoli will collapse and become unrecognizable so that only the most gross features of emphysema can be identified and it will be impossible to differentiate arterioles from venules.[2,3]

In order to achieve such preservation the bronchial tree must be filled with 4% formaldehyde in buffered saline at a sufficient pressure to expand all the air spaces; the pressure usually recommended is 25 cm of water. The lung should then be left to fix in a large container of formol saline for 24 hours but 48 hours is better and 72 hours is necessary if it is planned to prepare whole lung sections by embedding in gelatin using the Gough–Wentworth techniques.[4,5]

This seldom presents a problem in the postmortem room as one lung may be cut and examined immediately and the other retained for proper fixation and subsequent slicing. In nearly all cases it is much easier to inflate the left lung as the left main bronchus is larger and therefore more easily connected to the inflation device. If it is hoped to make quantitative measurements of alveolar volume or surface area, then inflation must be carried out at the standard pressure of 25 cm of water. Some form of automatic inflation device is required to do this and such a machine is worth building if lungs are to be inflated on a regular basis. The usual method is to use a small pump in order to maintain a constant head of pressure in the excised lung, which floats in a bath of fixative while its main bronchus is connected to an upper reservoir via a cannula. A variety of such machines have been described and most incorporate a switch which stops the pump for as long as the level in the reservoir is maintained.

When cutting the lung from the trachea the main bronchus must be left as long as possible and divided close to the trachea in order to provide a sufficient length to secure the cannula in place. Before cannulation of the main bronchus it helps to remove any soft tissue or lymph nodes from the hilum so that the cannula can be tied into place. The combination of a hard cannula and a ligature will crush the bronchial wall and the cannula itself will abrade the bronchial mucosa. It is therefore essential to take any histological samples from the main bronchus before cannulation. It is worthwhile having a wide choice of conical cannulae so that it is possible to accommodate main bronchi of any size. Following inflation the lung can be left attached to the machine for as long as is convenient.

Often the connection of a lung to an inflation machine is impossible, especially if it is a surgical specimen of the right lung in which case the main bronchus may be very short and in some specimens none of it is present. It will then be necessary to inflate each segment by hand by inserting a nozzle, connected to a large syringe, into each lobar or segmental bronchus in turn. If it is necessary to inflate a specimen by hand it is not possible to judge the pressure accurately but the aim should be to expand the lung until the pleural surface is smooth. If a machine is not available and the specimen is inflated by hand then it is a good idea to leave the cannula in place and occlude it so that there is no tendency for the elastic recoil of the parenchyma to drive some of the fixative out again.

Apart from difficulties in cannulation there are other problems associated with fixation by inflation. The two most frequent are leaks and blockage of a major bronchus. In all cases in which good fixation of the parenchyma may be required it is essential to avoid damaging the pleural surfaces or main bronchus when the specimen is removed from the chest. In necropsy cases this is the sole responsibility of the pathologist but surgeons have other more important priorities. If there are only one or two small pleural tears caused by the division of adhesions, the process of inflation should not be seriously compromised. However, if a leak becomes large once the specimen is under pressure, then it is sometimes possible to seal it with a carefully placed suture.

In cases where a bronchus is blocked by a tumour it may be possible to proceed as described above but the fixative may only reach those segments which are not occluded and fixation of the centre of a large solid lesion may be incomplete in its centre. In the worst cases, in which a tumour completely occludes the main bronchus, it may be necessary to abandon the attempt at inflation and simply cut the specimen into parasaggital slices, which may then be fixed by simple immersion; this is seldom necessary. More widespread bronchial obstruction by mucus as in asthma or chronic obstructive airways disease seldom causes serious difficulties, but artefacts can be produced by plugs of mucus or detached epithelium being driven down into the peripheral parts of the lung. If there is widespread occlusion of small bronchi the specimen may be fixed by perfusing the pulmonary artery with fixative.

A problem which is sometimes met in the postmortem room is the lung which is already fully expanded but which is nevertheless airless due to either pulmonary oedema or widespread consolidation. Fixing such specimens by inflation can still be attempted but it is much slower and less effective. In some cases relatively little may be lost by slicing such specimens in the unfixed state as the exudate in the air spaces holds them open in the inflated position so that the lobular architecture can still be identified when the sections have been cut and stained.

Sampling before fixation

As the range of available laboratory techniques expands there is a growing number of instances in which it is necessary to obtain unfixed material for microbiological culture, electron microscopy, immunohistochemistry or tissue culture. In many cases such sampling of the fresh specimen need not interfere with the proper fixation of the rest of the specimen, provided that any such samples are taken by the pathologist in person or under his supervision. This will ensure that no important diagnostic material is lost and that the subsequent fixation and handling of the specimen are not compromised.

If some research protocol requires the provision

of pieces of apparently normal lung, then these can be obtained from some peripheral part of the specimen such as the lingula, right middle lobe or the acute margins of a larger lobe. The defect can then be closed with a suture and the remainder of the specimen inflated in the usual way.

If it is required to obtain material from some discrete solid lesion such as a tumour or a granuloma, this may be done by incising the lesion in the area where it most closely approaches the pleura. It is usually possible to inflate the rest of the specimen after this. If the lesion is deeply situated, such an incision may produce too large a leak to enable the whole specimen to be inflated. The untouched segments can still be filled by hand using a syringe with a large nozzle which is inserted into each lobar or segmental bronchus in turn.

Bronchi, blood vessels and parenchyma

Before dissecting the lung or the lobe, the pleural surfaces should be carefully inspected for the presence of scars, fibrosis or the 'morocco leather' appearance seen in diffuse fibrosis with honeycomb change. Solid lesions can be felt even if they are not visible at this stage.

The parenchyma is most easily assessed in a series of closely placed parasaggital slices of the whole fixed lung. The lung is placed on the dissecting surface with the hilar surface downwards and with the dorsal aspect to the right while the left hand holds the anterior border (Fig. 9.5). A series of parallel longitudinal slices are then cut using a long brain knife; the knife must be very sharp and the slices should be 1–2 cm thick. This is easy in a lung which is consolidated or which is well fixed but it may be much more difficult in the unfixed lung or in cases in which there is much emphysema. The minimum of force should be used and a sawing motion is best avoided.

The cut surfaces may now be inspected with the naked eye or with a hand lens; the extent and type of any emphysema should be noted along with any areas of fibrosis, consolidation, tumours, scars or cavities. Samples must be taken for histological examination and, if necessary, the counting of asbestos fibres, but these should be removed in such a way that at least one slice is retained intact for photography, whole lung sectioning or museum mounting.

While such an approach is the best for examining

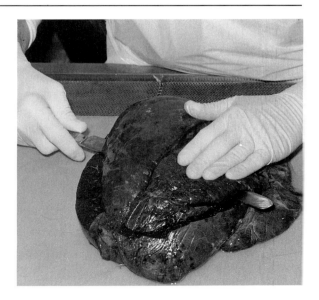

Fig. 9.5 Obtaining parasaggital slices. The anterior border is steadied with the left hand and a long knife is used with the lung resting on its mediastinal surface. This photograph is of the same lung as Figs 9.6 and 9.7 and it shows how the lung can be sliced even after the bronchi and arteries have been explored.

diffuse parenchymal disorders, the dissection of the bronchial and arterial trees requires a different approach. If these are the main areas of interest it may be necessary to open them longitudinally, but this does not prevent the cutting of slices later if the task is done in the right way.

For the dissection of the bronchi the specimen is placed with the hilum uppermost. Each lobar, segmental and subsegmental bronchus may now be opened longitudinally using a pair of round-pointed scissors (Fig. 9.6). The site and nature of any tumour or obstruction is noted as well as the nature of the bronchial contents and the condition of the mucosa. The dissection is carried as far peripherally as possible. When a solid lesion is encountered an attempt must be made to establish whether it has a bronchial connection and which bronchi are involved. This task may be made easier by passing a blunt-ended malleable probe down the bronchi and guiding it into the lesion before the dissection is begun. Alternatively the probe may be used to guide a well-placed slice with a knife.

If it is necessary to open the arteries longitudinally in the search for emboli, this may be done even if the bronchi have already been opened as described above. The lung is placed with the hilar

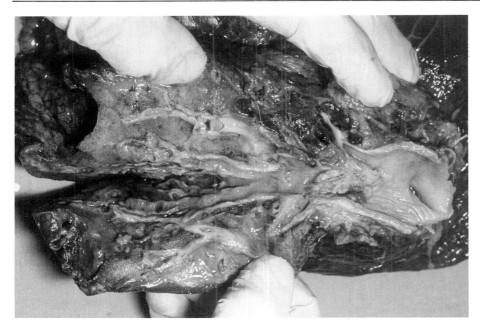

Fig. 9.6 Male aged 65. Pneumonectomy specimen. The bronchi have been opened longitudinally to show a sessile squamous cell carcinoma in the right lower lobe bronchus.

surface downwards and the arteries are approached through the oblique fissure. Gentle dissection in the depths of the fissure will reveal the pulmonary artery after which it and its branches may be opened longitudinally with scissors in the same way as the bronchi (Fig. 9.7).

By approaching the bronchi from the hilum and the arteries from the fissure the two systems may be examined without the cuts crossing each other and causing the specimen to fall apart; even after such extensive dissection it should still be possible to make at least some parasagittal slices through the whole lung.

Emphysema

The extent, severity and type of emphysema is most readily assessed in parasaggital slices of lung, but these must have been fixed in inflation and without such material the study of emphysema is impossible. If any form of quantitation is to be attempted, whether by point counting, measuring mean linear intercepts, comparison with standards or by using computerized image analysis, the lung must have been fixed at a standard pressure. For a detailed account of the use of such methods the

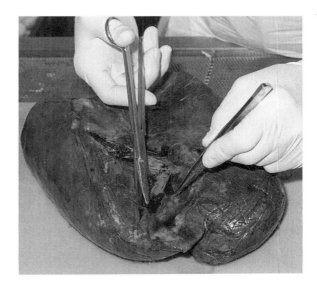

Fig. 9.7 After opening the bronchi the specimen is turned over and the main pulmonary artery is exposed in the oblique fissure; the arteries are then traced peripherally in the same way as the bronchi.

reader is referred to a number of authors.[3,6-11]

For the pathologist who has neither the time nor the facilities for the more elaborate methods, the method described by Heard provides a simple and reproducible method which requires little special equipment.[7] Complete, well-fixed parasaggital slices are required; each of these is divided into six areas and the degree of emphysema in each is assessed usually on a scale of 0–3; the result is expressed as a fraction of the maximum total score of 18 or as a percentage (Fig 9.8).[7]

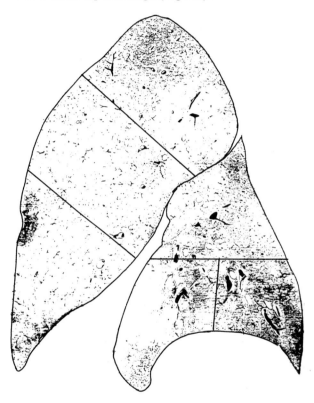

Fig. 9.8 The estimation of emphysema by Heard's method.[7] The emphysema in the six areas shown is graded on a scale of 0–3 and the total score expressed as a fraction of 18. The Edinburgh group have used a series of longitudinal lines to divide the lung into approximately equal areas.[11]

While emphysema can be assessed by the naked eye examination of the cut surface, the accuracy of the observation is improved if the fixed slice is floated in a large bath of clean water and the lesions examined with a hand lens. The water supports the walls of any cavities which may otherwise collapse if the slice is viewed in air. The appearances are further improved by impregnating the slice with barium sulphate which renders the surviving parenchyma pale grey while the cavities

appear as black holes (Fig. 9.9). Air spaces larger than 1 mm in diameter are considered emphysematous.

Gillooly *et al.*[11] have emphasized that these methods which rely on the low-power examination of lung slices are insensitive for the detection of early emphysema. They have developed a rapid automated technique for measuring air space wall surface area per unit volume of lung tissue. Using this automated method they have found that, by the time the lesion becomes visible in lung slices, as much as 75% of the alveolar wall surface has already been lost.

An attempt should be made to assess the types of emphysema present and their distribution within the lung, although this may be very difficult in severe cases as very extensive emphysema of any sort may appear to be panacinar (Fig. 9.10). The classification of the emphysema depends upon the observer's ability to identify the lung lobules so that the lesions may be classified as panacinar, paraseptal or centrilobular. Individual lung lobules are best identified in the areas just beneath the pleural surfaces, where they can be identified with a hand lens. They are between 1 and 2 cm in diameter and they have sharp boundaries formed by fibrous septa. In the centre of the lobule there is a central bronchiole and its accompanying branch of the pulmonary artery. From this central point the terminal bronchioles spread outwards to ventilate about five acini; the vessels seen at the edges of the lobule are venous but as the pulmonary vasculature is so thin-walled it is usually only possible to differentiate between arterioles and venules from their positions within the lobule. This is why it is so important that at least some of the lung tissue is fixed in inflation if it is necessary to study the vasculature. In lung fixed without distension the lobules collapse and without knowledge of their positions, the identities of individual vessels may become unrecognizable.

The distribution of emphysema also is useful and should be recorded. Centrilobular emphysema tends to affect the upper lobes more than the lower while the presence of panacinar emphysema should prompt a search for evidence of α-1-antitrypsin deficiency. Finally, the presence of irregular emphysema in relation to old scars, granulomata or fibrosis should be recorded.

Adequate histological samples may be taken from the main and lobar bronchi as these will be needed for assessment of the pathology of the bronchial wall and for quantitation of the degree

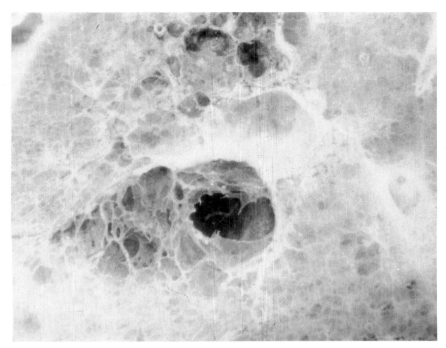

Fig. 9.9 Lung parenchyma fixed in inflation, impregnated with barium sulphate and viewed under water. The cavities of centrilobular emphysema are held open and they are seen as dark holes.

of mucous gland hyperplasia. There are a variety of methods for such quantitation, employing simple measurement of the gland depth, radial intercepts or point counting.[12–14]

Tumours and other discrete lesions

The pathologist has a central role in staging tumours pathologically. Any solid lesion in the specimen must be identified, its size recorded, its relation to the pleura described and its position in relation to the bronchial and lobar anatomy must be recorded. Involvement of the pleural or mediastinal surfaces of the lung and any extension to the resected margin of the bronchus are of considerable importance in dealing with surgical resections. However, it is also essential to examine the parts of the lung which appear not to be involved by the main lesion as it is common for more than one process to be present. A lung or lobe resected for malignant disease may also have evidence of chronic obstructive airways disease as well as some industrial lung disease such as evidence of silicosis or asbestos exposure. Such information must be

recorded as it may have considerable medico-legal significance at a later date should the patient or the executors wish to sue a former employer for a possible role in the causation of the tumour.

Having identified the site of the main lesion by gentle palpation, its relationship to any pleural scars or pieces of adherent parietal pleura should be assessed. Next the cut end of the bronchus should be examined for any evidence of tumour in the lumen or infiltration of the bronchial wall. If the resected margin of the bronchus has not already been sampled, a circumferential slice should be removed for histological examination. Such sections may show whether the tumour has permeated beyond the limits of the operation but they may also be used for quantitative assessment of the size of bronchial mucus glands.

In all cases it is important to examine, identify and sample any attached lymph nodes (Fig. 9.11).[15] This is easy in autopsy specimens but relatively few nodes may be attached to surgical specimens, particularly if the operation has been restricted to a lobectomy. The identification of small metastases in thoracic nodes requires microscopy. This is because in adults the nodes are

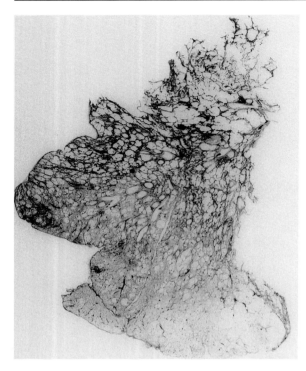

Fig. 9.10 Gough–Wentworth section. This patient was a man aged 62 who had received radiotherapy for pulmonary emphysema as part of an MRC trial. The destruction of the parenchyma is so severe, particularly in the upper lobe, that it is no longer possible to classify it as centrilobular.

usually anthracotic and fibrous due to the large amount of inhaled debris which they contain, so that the naked eye appearances may be deceptive.

Having completed the external examination of the specimen the size, shape, position and anatomical relationships of the tumour must be recorded. If the tumour is not visible in the cut end of the bronchus the bronchus may be opened longitudinally with scissors until the tumour is reached, with or without the use of a probe as described above. An attempt must be made to find out whether the lesion is connected with a bronchus and, if so, which bronchi are involved. The distance between the edge of the tumour and the bronchial margin should be recorded. The bronchial mucosa should also be carefully inspected for roughness or irregularity and blocks should be taken which may show the presence of any *in situ* change.[16]

The lesion should be measured in three dimensions and it is important to be able to state whether it reaches the pleura or crosses a fissure and involves another lobe. The naked eye appearance and shape of the tumour should be noted; some form a large spherical mass, some are small and stellate, while others form a sheath around the bronchus.

Often it is necessary to revise one's estimate of the size of the lesion at the stage of microscopy as it often happens that what appears to be a large mass may consist largely of obstructive pneumonitis. The underlying tumour may be quite small, but by causing bronchial obstruction a large inflammatory mass has been produced. Finally the uninvolved parts of the lung must be examined and sampled in order to search for any coexistent diffuse parenchymal disease or any evidence of intrapulmonary spread of the primary lesion.

Selection of blocks and special stains

In the course of this chapter there are a number of references to the materials which should be retained for histology and it seems appropriate to summarize these.

1. If possible a complete parasaggital slice of lung which has been fixed in inflation should be preserved sealed in a plastic bag, mounted as a museum specimen or a Gough–Wentworth section may be prepared.
2. In cases of parenchymal disease at least one block should be taken from each of the main bronchopulmonary segments; these should be taken from the subpleural areas (Fig. 9.12). If this is impractical then at least one block should be taken from each lobe – preferably two, one from the apex and the other from the base of each lobe.
3. Transverse section of the main bronchus.
4. Transverse blocks should be taken across the area of the main segmental and subsegmental bronchi in each lobe. These blocks may include the bronchopulmonary lymph nodes.
5. The hilar and any mediastinal or subcarinal nodes should be sampled (Fig. 9.11). In the case of specimens which contain solid tumour the following additional material is necessary.
6. Lesion with overlying pleura.
7. Lesion with involved bronchi. A minimum of four blocks should be taken; one should be perpendicular to the bronchial wall which in the

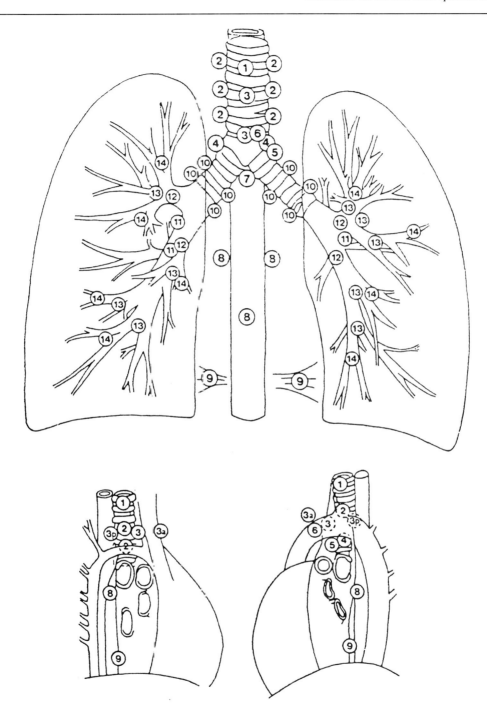

Fig. 9.11 The location and nomenclature of the thoracic lymph nodes. Key: 1, superior mediastinal or highest mediastinal; 2, paratracheal; 3, pretracheal (3a, anterior mediastinal; 3p, retrotracheal or posterior mediastinal); 4, tracheobronchial; 5, subaortic (Botallo's); 6, para-aortic (ascending aorta); 7, subcarinal; 8, para-oesophageal (below carina); 9, pulmonary ligament; 10, hilar; 11, interlobar; 12, lobar (upper, middle, lower); 13, segmental; 14, subsegmental. Nakute lymph node map.

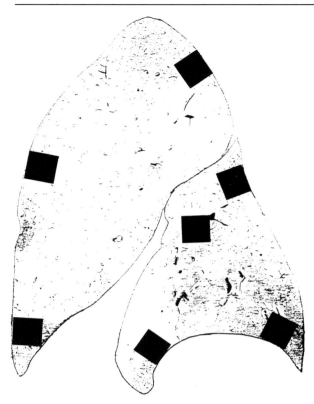

Fig. 9.12 Large thin blocks approximately 30 × 25 mm should be cut from the subpleural areas of the apical anterior and lingular segments of the upper lobe, as well as the apical and basal segments of the lower lobes. A more centrally placed block is taken to sample the segmented airways and blood vessels. The right lung is sampled in the same way with the middle lobe being treated in the same way as the lingula. This should be regarded as the minimum adequate sampling for parenchymal disease and many cases require rather more blocks than this.

case of a central tumour should show the depth of invasion of the tumour beyond the bronchus.
8. Proximal resected margin of the bronchus.
9. As in the case of parenchymal disease, at least one block should be taken from each lobe.

Handling of cases of suspected occupational lung disorder

These cases are often the subject of civil litigation and it is common for months or even years to pass before the legal proceedings start. Although the pathologist may not be directly involved in the proceedings in person, he has a duty to the patient,

or more commonly, the patients' executors, to see that sufficient properly preserved pathological material is kept to provide the necessary evidence.

Well-fixed lung slices or Gough–Wentworth sections should be retained. As paraffin blocks are easily stored and transported, it is worth taking a sufficient number to support the conclusions which have been reached. This is particularly important in cases of suspected asbestos exposure which are dealt with at the end of the chapter.

Exposure to agents such as silica causes secondary dust lesions which are round or stellate and are palpable, often feeling as though the pleura contains small beads or ball bearings. Some estimate of their sizes and numbers should be recorded.[17] Dust lesions which are greater than 1 cm in diameter are regarded as progressive massive fibrosis (PMF). The number, sites, size and colour of these lesions should be recorded as well as any evidence of cavitation. As in the case of emphysema, the identification of fibrosis is often made more obvious by impregnation with barium sulphate and the process can be graded histologically. Dust lesion can be recorded by their size on a 0–3 scale and the percentage of lobules affected in each lobe also recorded on a similar scale (see Table 9.1).

Special stains

The choice of special stains is a study in itself and most pathologists have their own preferences. However, certain techniques are essential additions to routine haematoxylin and eosin when pulmonary diseases and tumours are being investigated.

Elastic tissue
No case of pulmonary vascular or parenchymal disease has been properly investigated until the elastic tissue has been stained. The choice of stain is a personal matter but a well-performed Verhoeffs' elastic stain with van Gieson as a counterstain is difficult to surpass. An excellent alternative is Miller's elastic stain, which may be combined with both van Gieson and Alcian blue. Elastic stains are essential for the investigation of vasculitides, pulmonary hypertension and embolic disease.

Mucin stains
A reliable mucin stain is essential in the investigation of almost all lung tumours. In the author's

Table 9.1 Naked eye assessment of fibrotic lung disease

Type of lesion	Size of lesion	Grade
Primary dust foci	1–3 mm	1
	4–5 mm	2
	>5 mm	3
Proportion of lobules affected	<1/3	1
	1/3–2/3	2
	>2/3	3
Secondary (palpable) dust foci	<0.5 cm	
	0.5–2 cm	
	>2 cm	

(The number and type of foci should be recorded for each size)

Emphysema

Type: centrilobular, pan acinar, etc.

Severity	3 mm	1
	4–5 mm	2
	>5 mm	3
Extent		0–3

(Each lobe may be graded separately)

Interstitial fibrosis

Severity (graded histologically)

Absent	0
Slight reticulum or collagen around respiratory bronchioles	1
Fibrosis extends into adjacent ducts and alveoli	2
Fibrosis extends into adjacent respiratory bronchioles	3
Widespread fibrosis and/or honeycombing	4

After Gibbs and Seal.[17]

experience the most useful is the periodic acid Schiff (PAS)–Alcian blue combination which stains both acid and neutral mucins. This is more reliable than the combined mucin and keratin stain described in the World Health Organization (WHO) classification of lung tumours.[18,19] More elaborate mucin stains are sometimes required in some cases, such as mesothelioma, but most of the diagnostic problems may be more easily resolved by immunohistochemistry or electron microscopy.

Keratin
The stain suggested by the WHO classifications[18] may be used to demonstrate keratin but most laboratories have a cytological department which uses Papanicolaou's stain as a routine; this is just as useful when applied to sections as it is in cytological preparations.

Perl's reaction
The detection of ferrous iron is essential in all vascular disorders, including pulmonary hypertension, parenchymal disorders and in industrial lung disease in which it helps in the identification of ferruginous bodies.

Unstained sections
If sections are cut at a thickness of 20–30 µm, cleared, mounted and examined unstained, any ferruginous bodies remain intact and can be seen as

golden-brown beaded rods. This method is much more sensitive than the examination of conventional 4 μm sections stained with haematoxylin and eosin or by Perl's method.

Polarized light
In all cases in which industrial disease is a possibility, the material should be examined using polarized light, which demonstrates foreign material such as silica or talc. It is important to use a high-intensity light source. This is also valuable in the examination of cases of suspected drug addiction.

Microorganisms
A full range of stains for microorganisms must be available. In many cases immunological methods are available; these are particularly useful in cases of suspected cytomegalovirus (CMV) or *Pneumocystis carinii* infection. All pulmonary granulomata must be considered as being tuberculous until proved otherwise and all such lesions should as a routine be stained both by the Ziehl–Neelsen technique and using PAS or Grocott stains to demonstrate fungi.

Neuroendocrine tumours
The use of the silver impregnation techniques in the examination of lung tumours is now almost obsolete. Pulmonary carcinoids, being of foregut origin, are seldom argentaffin and argyrophilia is very inconstant. For the identification of carcinoids electron microscopy or immunostaining for neuroendocrine markers is more reliable.

Immunohistochemistry
It is only possible to give a few generalizations about this enormous range of methods as the pulmonary pathologist may meet almost any systemic disease which may involve the lung secondarily. The main problems which influenced the choice of antisera used may be summarized as follows:

1. Antisera for the identification of specific organisms (e.g. CMV, *P. carinii*).
2. A panel of antisera which distinguishes epithelial from mesodermal and lymphoid neoplasms (e.g. cytokeratin, epithelial membrane antibody, carcino-embryonic antigen, vimentin, desmin, leucocyte common antigen, pan B cell, pan T cell).
3. For the detection of tumour specific antigens (e.g. α-fetoprotein, human chorionic gonado-

trophin (HCG), thyroglobulin, prostate-specific antigen).
4. The differentiation of mesothelioma from carcinoma (Au A1, BER.EP4, CEA).

Handling of tuberculous cases

Pulmonary tuberculosis continues to pose a serious threat to health and, despite the success of recent decades, the prevalence of the disease is increasing again in some parts of the world and in some sectors of society.[20,21] Laboratory staff, whether technical or medical, need to be aware of this risk, especially when they are involved in handling unfixed material from the lung. All specimens of lung must be regarded as potential sources of infection as, even if there is an established diagnosis of malignant disease, it is not rare for tuberculosis to be coexistent especially in material from elderly patients.

If surgical specimens have been fixed by inflation the risks of infection are almost entirely eliminated and such cases can be handled in the usual way. However, if the specimen requires dissection before fixation, then this should be done in a bacteriological cabinet with exhaust ventilation. The pieces selected for histological examination can then be fixed in formalin and the remains of the main specimen immersed in a liberal quantity of formol saline for storage. All equipment used should either be disposable or it must be cleaned and sterilized preferably by autoclaving; the cabinet itself must then be cleaned and disinfected.

Any samples required for bacteriological examination must be appropriately labelled and dated without contaminating the outsides of the containers. The containers must then be enclosed in plastic bags for transfer to the bacteriology department.

If frozen sections are required by the clinical staff, or by the prosector in autopsy cases, then the samples selected must be immersed in formol saline and boiled for 5 minutes before freezing. The quality of the frozen section suffers but this precaution avoids the contamination of the cryostat and its cabinet which would otherwise become highly dangerous to subsequent users.

Precautions in the postmortem room

Some pathologists seem reluctant to carry out autopsies in cases of suspected tuberculosis but, if care is taken, the work is done methodically and the working conditions are good, the risks can be minimized. The establishment of a diagnosis of tuberculosis is a fact of enormous practical importance in terms of public health as it is an indication for contact tracing amongst the family and associates of the deceased. However, in all cases it is necessary to balance the risks against the possible benefits.[22]

Tuberculosis may be encountered in almost any autopsy and the pathologist must be prepared to modify his technique in the middle of the procedure as circumstances dictate. The guiding principles are:

1. to reduce the number of staff exposed to a minimum;
2. to reduce the area and amount of equipment contaminated;
3. to minimize the time for which exposure occurs; and
4. to fix the lungs promptly so that they may be dissected later.

In cases in which pulmonary tuberculosis is suspected the Rokitansky technique should be avoided but the body may be opened in the usual way provided that the rib cage is left intact. The abdominal organs are then inspected, removed and dissected after which attention can be paid to the skull and brain. Only after completion of the rest of the examination is the chest opened. If there is no empyema and the pericardium appears normal the heart may be removed and dissected, after which, last of all, the lungs and trachea are removed from the chest. After taking any bacteriological specimens needed the lungs are inflated with formol saline via the trachea and they are then left to fix in a large container filled with fixative. After 2 or 3 days' fixation the lungs and trachea may be examined in the usual way.[23,24]

Cases of suspected asbestos exposure

Industrial diseases are covered fully in Chapter 10 but the following practical points should be remembered. The importance of recording the condition of the pleural cavities has already been mentioned but, in addition to the usual histological samples, it is necessary to retain material for the estimation of the asbestos fibre burden. The Pneumoconiosis Committee of the American College of Pathologists have recommended that 15 tissue

blocks should be taken from suspected asbestos cases. If this is too much a minimum of 1 block from each lobe should be taken, paying particular attention to the subpleural areas of the basal segments.

A quick method of demonstration is to examine fresh samples of lung fluid using the light microscope. The fluid may be obtained by scraping the cut surface of the lung with a knife or by simply squeezing a piece of lung and transferring the fluid to a slide; if sufficient fluid can be obtained it may be centrifuged to concentrate the deposit. Red blood cells may be troublesome but these may be lysed by mixing the fluid with distilled water. The deposit is sealed under a coverslip with wax and then examined using the highpower objective of the microscope. A more permanent preparation may be made by drying some of the deposit onto a slide using heat after, which it may be stained by Perl's method.

If a more quantitative technique is required some form of digestion process is required. Blocks of lung are selected from the parenchyma of both lungs, weighed and then dried to constant weight. Another accurately weighed piece of lung is then digested with either hypochlorite solution[25,26] or with potassium hydroxide.[27] The resulting deposit is then resuspended and the number of fibres are counted in a Fuchs–Rosenthal counting chamber. Alternatively the suspension may be filtered through a millipore filter and the chamber fibres found on the filter can be counted using phase contrast illumination or electron microscopy. These quantitative methods are much more sensitive than the examination of histological sections as both coated and uncoated fibres are counted.[28]

Acknowledgements

I am indebted to the Department of Medical Illustration, Northern General Hospital NHS Trust, Sheffield for assistance with the illustrations and to Mrs L.A. Norcliffe who word processed the manuscript.

References

1. Wright BM, Slavin G, Kreel L, Callan K, Sandin B. Postmortem inflation and fixation of human lungs. A technique for pathological and radiological correlations. *Thorax* 1974; **29**, 189–194.
2. Quantitative assessment of chronic non-specific lung disease at necropsy. Report by Panel on

Pathology of the Medical Research Council Committee on Research in Chronic Bronchitis, April 1972. *Thorax* 1975; **30**, 241–251.

3. Weibel ER. Principles and methods for the morphometric study of the lung and other organs. *Lab Invest* 1963; **12**, 131–155.

4. Gough J. The pathogenesis of emphysema. In: *The Lung*, Liebow and Smith DE (eds). Baltimore: Williams and Wilkins, 1968.

5. Gough J, Wentworth ID. Thin sections of whole organs mounted on paper. In: *Recent Advances in Pathology*, Harrison CV (ed.) London: Churchill, 1960.

6. Dunnill MS. Evaluation of a simple method of sampling the lung for quantitative histological analysis. *Thorax* 1964; **19**, 443–448.

7. Heard BE, Izikawa T. Pulmonary emphysema in fifty consecutive male necropsies in London. *J Pathol Bacteriol* 1964; **88**, 423–431.

8. Thurlbeck WM, Dunnill MS, Hartung W, Heard BE, Heppleston AG, Ryder RC. A comparison of three methods of measuring emphysema. *Hum Pathol* 1970; **1**, 215–226.

9. Turner P, Whimster WF. Volume of emphysema. *Thorax* 1981; **36**, 932–937.

10. Nagai A, Yamawaki I, Thurlbeck WM, Takizawa T. Assessment of lung parenchymal destruction by using routine histologic tissue sections. *Am Rev Respir Dis* 1989; **139**, 313–319.

11. Gillooly M, Lamb D, Farrow AS. New automated technique for assessing emphysema on histological sections. *J Clin Pathol* 1991; **44**, 1007–1011.

12. Alli AF. The radial intercepts method for measuring bronchial mucous gland volume. *Thorax* 1975; **30**, 687–692.

13. Dunnill MS, Massarella GR, Anderson JA. A comparison of the quantitative anatomy of the bronchi in normal subjects, in status asthmaticus, in chronic bronchitis, and in emphysema. *Thorax* 1969; **24**, 176.

14. Takizawa T, Thurlbeck WM. A comparative study of four methods of assessing the morphologic changes in chronic bronchitis. *Am Rev Respir Dis* 1971; **103**, 774–783.

15. Naruke T, Suemasu K, Ishikawa S. Lymph node mapping and curability at various levels of metastasis in resected lung cancer. *J Thorac Cardiovasc Surg* 1978; **76**, 832–839.

16. Carter D. Pathological examination of major pulmonary specimens resected for neoplastic disease. *Pathol Ann* 1983; 315–332.

17. Gibbs AR, Seal RME. Examination of lung specimens. *ACP Broadsheet* no. 123, January, 1990.

18. WHO. *Histological Typing of Lung Tumours*, 2nd edn. Geneva: WHO, 1981.

19. Kennedy A, Burgin PD. A comparison of different methods of detecting mucin in adenocarcinomas of the lung. *Br J Dis Chest* 1975; **69**, 137–143.

20. Styblo K. The impact of HIV infection on the global epidemiology of tuberculosis. *Bull Int Union Against Tuberculosis Lung Dis* 1991; **66**, 27–32.

21. Murray IF. An emerging global programme against tuberculosis: agenda for research, including the impact of HIV infection. *Bull Union Against Tuberculosis Lung Dis* 1991; **66**, 207–209.

22. Reichert CM. New safety considerations for the acquired immunodeficiency syndrome autopsy. *Arch Pathol Lab Med* 1992; **116**, 1109.

23. *A Handbook of Mortuary Practice and Safety*. London: The Royal Institute of Public Health and Hygiene, 1992.

24. Health Services Advisory Committee. *Safe Working and the Prevention of Infection in the Mortuary and Postmortem Room*. London: HMSO, 1991.

25. Morgan A, Holmes A. Concentrations and dimensions of coated and uncoated asbestos fibres in the human lung. *Br J Industr Med* 1980; **37**, 25–32.

26. Morgan A, Holmes A. Distribution and characteristics of amphibole asbestos fibres, measured with the light microscope, in the left lung of an insulation worker. *Br J Industr Med* 1983; **40**, 45–50.

27. Gold C. The quantitation of asbestos in tissue. *J Clin Path* 1968; **21**, 537.

28. Ashcroft T, Heppleston AG. The optical and electron microscopic determination of pulmonary asbestos fibre concentration and its relation to the human pathological reaction. *J Clin Pathol* 1973; **26**, 224–234.

10

Pneumoconiosis

A R Gibbs

Introduction

There are many definitions of the term pneumoconiosis; in this chapter I use it broadly to mean any pattern of pulmonary reaction caused by exposure to minerals. Inhalation of mineral particles in humans can cause a wide range of pathological responses which depends on cumulative dose, size, shape, surface properties and durability of the mineral concerned. Practically all human exposures are to mixtures of minerals and there may be complex interactions between minerals which may alter the response to a specific mineral. Generally the disease produced is named after the mineral which is considered to have mainly caused the pulmonary reaction, for example silicosis in granite workers and kaolinosis in china clay workers. Sometimes an atypical reaction to a mineral(s) may be due to individual susceptibility, for example Caplan lesions occurring in coal miners with rheumatoid disease. As well as cumulative exposure, the nature of exposure will also influence the response, relatively acute exposures will often produce different reactions than chronic lower exposures. A well-recognized example is the lipoproteinosis pattern produced by high exposures to silica, which has been described in sandblasters,[1] and the nodular fibrotic lesions (chronic silicosis) observed in slate workers with long term exposures.[2]

Other than when a suspected exposure to minerals causes a neoplasm it is uncommon for biopsies to be undertaken for pneumoconioses, although sometimes they may be performed when an exposure is not suspected. However, a pneumoconiosis can sometimes be present in conjunction with other unrelated pathology – for example in my practice it is not uncommon to see mild simple coal workers' pneumoconiosis in the background lung tissue of a lobectomy or pneumonectomy specimen containing a lung cancer.

Endoscopic bronchial and transbronchial biopsies are unsuitable for accurate assessment of pneumoconioses because they are frequently unrepresentative, pneumoconiotic lesions are often larger than the biopsy itself and no assessment of severity of disease can be made. Chronic berylliosis is one exception to this because it produces a granulomatous pattern. If open-lung biopsies are taken a reasonable appreciation of the type, distribution and severity of lesions can be made, which are essential for diagnostic and medico-legal purposes.

Whenever the diagnosis of a pneumoconiosis is considered on a biopsy it is important to obtain as much information as possible about the work history. This should be comprehensive from the first day the individual commenced work, since the relevant exposure may have been decades before clinical presentation. Also it should be remembered that some pneumoconioses can be caused by indirect (bystander) exposure – examples include mesothelioma and chronic beryllium disease.[3,4] Therefore on occasions it may be necessary to enquire about the occupations of the relatives also. Radiological correlation is also necessary – if a clinically significant fibrotic disease is being considered it is important to know if the lesions are diffuse or localized; this cannot always be established by the biopsy alone.

In this chapter I shall concentrate mainly on the patterns of reaction that may occur and give guidelines for their ascription to mineral(s) exposure.

The general approach to interpretation of the biopsy is no different from that of other pulmonary disease; a low-power examination first to decide the distribution of the lesions followed by higher-power examination to determine the cellular and connective tissue components of the lesions. A useful simple technique which should be employed for screening every biopsy is polarized light microcopy, since the finding of birefringent particles may be the first feature to alert the pathologist that he is looking at some form of pneumoconiosis. However, although the relative brightness and shape of the particles may give some indication of the identity of the mineral particles present, it is not sufficient in itself. There are several hundred types of mineral particles that are birefringent and modern analytical techniques are necessary to identify them precisely. Also the mere presence of various types of minerals in the lung tissues may not indicate that the lesions have been caused by them – quantitation may be necessary to elucidate their role.

Generally mineral analysis is applicable to large biopsies only and is unsuitable for endoscopic biopsies. The most accurate methods utilize electron microscopy and energy dispersive X-ray spectometry and diffraction techniques on digested tissue samples. It is a very skilled procedure and should only be performed by laboratories who have experience in this field. Up until recently the majority of analyses have been performed on tissue specimens where asbestos has been implicated in the aetiology. However, mineral analysis can also be performed for non-fibrous minerals and for fibres other than asbestos.

The simplest method of analysis in asbestos-related cases is examination of tissue sections for asbestos bodies by light microscopy. This can be aided by the use of a Perl's stain. Providing strict criteria are used for their identification, an assessment of their number can be very useful in cases of asbestosis and mineral analysis may not be required. If asbestos bodies are sparse or other types of ferruginous bodies are present, then mineral analysis will be necessary if a diagnosis of asbestosis is being considered. Malignant mesotheliomas can arise after exposure to relatively low doses of asbestos and in some cases no asbestos bodies will be found on conventional sections by light microscopy, whereas often they will be demonstrated by the sophisticated electron microscopic techniques. If there is any question about the accuracy of the occupational history of asbestos exposure, an electron microscopic mineral analysis is recommended.

There are light microscopic methods for quantification of asbestos bodies and fibres on tissue digests which can be useful in assessing the severity of exposure. However, they are limited in so far as they cannot provide a breakdown of fibre type and they are relatively insensitive. These should also be performed by laboratories which have experience and have reference ranges for normals and the different pathologies.

If the clinico-pathological features are insufficient for the diagnosis of a particular pneumoconioi or there is a discrepancy between them and the occupational history, then mineral analysis should be considered. The choice of method should be discussed with a laboratory experienced in these types of problem.

Patterns of response

The pathological responses of the lung to mineral exposures can include a variety of cellular accumulations, fibrosis and neoplasms and in turn this will determine clinically whether the disease is asymptomatic, associated with a mild or severe chronic disability or life threatening.[5,6] I shall not consider lung cancers or mesotheliomas here (see Chapters 11 and 12). It is useful to consider the lesions according to the predominant anatomical compartment in which they are situated and this is the way I shall group them. However, it should be realized that some pneumoconioses can show more than one pattern which often can be seen concomitantly; for example it is not uncommon to observe a mild degree of interstitial fibrosis in addition to the typical centrilobular macular lesions in coal workers' pneumoconiosis. Since many of the pathological reactions of the lung to minerals are no different to those from other causes, several of the detailed pathological descriptions will be found in other chapters of this book. However, where there are particular differences or difficulties in pathological diagnosis the diseases will be described in detail.

Intra-alveolar disorders

Pulmonary diseases with a predominantly intra-alveolar distribution are an uncommon response to inhaled particles. They include alveolar proteinosis, desquamative interstial pneumonia (DIP) and giant cell interstitial pneumonia (GIP). Why these types of unusual response occur is not understood but usually they have been associated with relatively high exposures and fine mineral particle size. These pathological reactions are not caused exclusively by mineral exposures but if encountered an occupational cause should be considered and ideally a mineralogical analysis of the lung tissue performed. There may be clues in the biopsy that can suggest the cause; for example in alveolar proteinosis due to silica there may be poorly formed silicotic nodules present. Table 10.1 shows the claimed associations between the different types of intraalveolar disorder and mineral exposures.[7-15] There appears to be a strong association between hard metal exposure and GIP (Fig. 10.1).[16]

Table 10.1 Claimed associations between certain intraalveolar reactions and mineralogical exposures

Alveolar proteinosis	Silica, metal fumes, cement, talc, aluminium
DIP	Asbestos, silica, talc, tungsten carbide, aluminium silicates
GIP	Tungsten carbide (hard metal)

DIP = desquamative interstitial pneumonia;
GIP = giant cell interstitial pneumonia.

Pulmonary alveolar proteinosis (pulmonary alveolar lipoproteinosis, pulmonary alveolar phospholipoproteinosis)
The disease is rare and is characterized by an intraalveolar accumulation of granular eosinophilic acellular material which stains positively with PAS

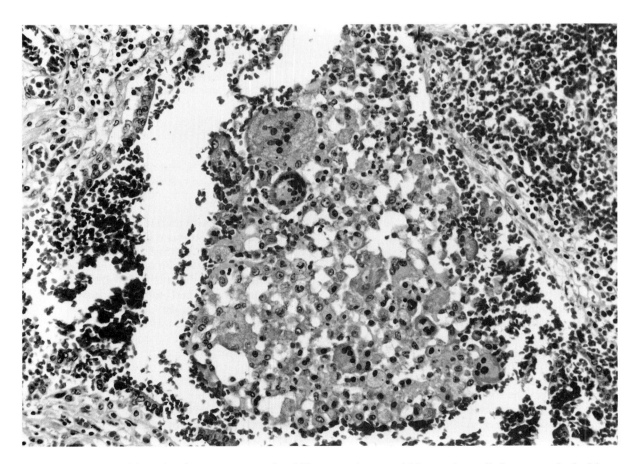

Fig. 10.1 An area of lung showing numerous eosinophilic macrophages and bizarre giant cells in a man who had been exposed to hard metal for many years. In other areas of the lung there was severe interstitial fibrosis.

even after diastase treatment. The architecture of the lung remains intact although sometimes the interstitium may be widened by fibrosis. Acicular (needle-shaped) spaces and cell ghosts may be present within the intra-alveolar material. The alveolar lining cells may be hyperplastic and cuboidal.

The disease has been associated with a variety of mineral exposures, immunological deficiencies, malignancies, infectious agents or may be idiopathic.[11-13] An exposure to an industrial agent should always be excluded. When caused by occupational exposure there may be clues in the lung tissues; for example silicotic nodules which may be poorly formed may also be present when the disease is caused by exposure to silica.

Mixed intra-alveolar and interstitial disorders

The typical disease characterized by this pattern is diffuse alveolar damage and the appearances do not differ from diffuse alveolar damage caused by agents other than minerals (see p. 130). It is an acute response and it is usually consequent to high exposures and develops rapidly following exposure. It has been associated with exposures to fumes of nitrogen, cadmium, beryllium, mercury, nickel, uranium and ammonia.[14,15] The outcome of this reaction depends on the severity of the exposure and varies from complete resolution to death of the individual.

Interstitial disorders

Diffuse interstitial inflammatory processes can be a consequence of inhalation of fibrous and/or non-fibrous particles. Significant fibrosis which causes clinical disability is usually a result of severe and prolonged exposure and such fibrosis takes many years to develop. Table 10.2 lists minerals which have been associated with this reaction.[16-23] A light microscopic grading system had been developed for asbestosis based on the severity of fibrosis (Table 10.3).[22,23] Studies have shown that the degree of fibrosis correlates with the mineral fibre burden within the lung.[24] This grading system can also be applied to interstitial disease caused by other minerals. The fibrosis itself is not characteristic for any one mineral and is similar to that unrelated to mineral exposure (cryptogenic fibrosing alveolitis), but there are useful markers which may be present which indicate the likely cause. These include ferruginous bodies, the presence of strongly birefringent particles and

Table 10.2 Minerals associated with diffuse interstitial fibrosis

Mineral	Other pathological features
Asbestos	Asbestos bodies
Talc	Numerous doubly refractile particles; ferruginous bodies; foreign body granulomas
Kaolin	Numerous doubly refractile particles; occasional ferruginous bodies
Mica	Numerous doubly refractile particles; ferruginous bodies; foreign body granulomas
Fuller's earth	Numerous doubly refractile particles
Aluminium	Black dust
Coal	Macules, nodules and mass lesions
Silica	Whorled nodules, mass lesions

Table 10.3 Histological grading system for interstitial fibrosis

Grade 1	Fibrosis confined to the walls of the bronchioles
Grade 2	Fibrosis extends from the walls of the bronchioles into the walls of the alveolar ducts, atria and adjacent alveoli but adjacent acini are not joined up
Grade 3	There is further increase of the interstitial fibrosis with linking up of adjacent bronchioles
Grade 4	Severe, widespread fibrosis with severe distortion of lung architecture

foreign body granulomas. If strongly birefringent particles are conspicuous within the lung tissue it is likely that predominantly non-fibrous silicates such as talc, mica or kaolin may be the cause. Some investigators claim that pleural plaques are useful in determining whether pulmonary fibrosis is due to asbestos exposure; this is inaccurate since (1) pleural plaques can occur after very light exposures to asbestos, (2) there are other causes, and (3) there is no correlation between the degree of interstitial fibrosis and the extent of pleural plaques in asbestos-exposed subjects.[25]

Ferruginous bodies are golden brown beaded or dumbbell, straight or curved shaped structures which form on fibres or plates. One of the most important types is the so-called asbestos body which is the hallmark of asbestos exposure, but ferruginous bodies can also be formed on non-asbestos fibrous or platy dusts. The latter include talc, mica, kaolin, carbon, metal oxides, fibrous zeolites, glass fibres, rutile and mullite. Asbestos bodies are characterized by straight thin transparent fibrous cores (Fig. 10.2) whereas the other types can have black, yellow, round, platey, fibrous, solid, lacey or irregular cores.[26,27]

Silicatoses
Usually the interstitial lesions caused by inhalation of talc, mica or kaolin are very cellular (Fig. 10.3) and are associated with high mineral burdens within the lung tissues.[18,19] Therefore there is usually a good occupational history of heavy dust exposure. If the disease has been caused by a primary exposure such as mining or processing of the material, there is usually little doubt as to which mineral or minerals are responsible. However, if exposure has occurred in a secondary industry, e.g. the rubber industry, then there may be doubt as to the causative mineral – this is particularly true for talc, where the commercial term is used loosely and may refer to a product that may not even contain the mineral talc.[28] The only way to determine the cause may be by mineralogical analysis of the dust within the lung tissue since the pathological appearances are not distinctive. These diseases can be grouped under the umbrella term silicatoses. The presence of ferruginous bodies and or foreign

body granulomas in this context point to talc or mica as the aetiology; these features are absent or inconspicuous in cases of kaolinosis in my experience.[29]

Asbestosis
Exposure to asbestos can cause pleural effusions, pleural plaques, diffuse pleural fibrosis, malignant mesothelioma, rounded atelectasis, pulmonary fibrosis (asbestosis) and lung cancer. The latter two conditions require much heavier exposures than the other sequelae. The term asbestosis should only be applied to the presence of lung parenchymal fibrosis and *not* to changes within the pleura.

Typically asbestosis causes more severe changes in the lower than the upper zones of the lungs. The progression of these changes has been studied both in animals and humans. As the microscopic grading scheme indicates, the lesions of asbestosis commence around the respiratory bronchioles followed by the terminal bronchioles, alveolar ducts and adjacent alveolar walls (Figs 10.4 and 10.5) and are first identified macroscopically in the subpleural regions of the postero-basal segments of the lower lobes. Fibrous bands then link up from adjacent lobules with loss of alveolar spaces and eventually coarse bands of collagen develop which may be accompanied by distortion of bronchiolar

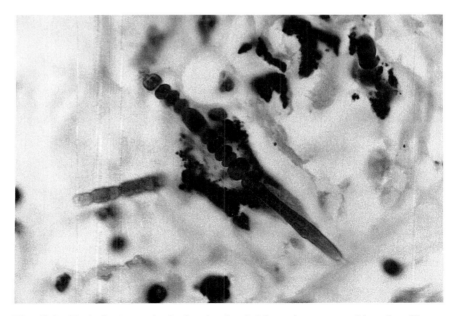

Fig. 10.2 Typical asbestos body showing beaded ferruginous coat with a clear fibrous core.

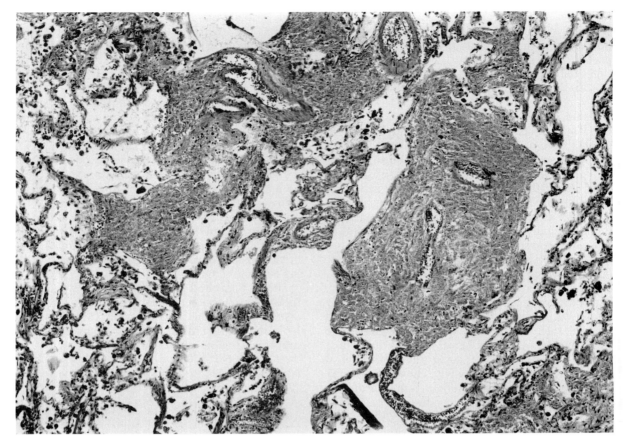

Fig. 10.3 Interstitial collections of dust-laden macrophages associated with little fibrosis from a man who had worked in the drying area of a china clay works.

structures to give honeycombing. Cuboidal, columnar and squamous metaplasia and dysplasia of the epithelium lining the bronchioles, alveolar ducts and alveoli around the scarred foci can occur.

Unfortunately scarring around respiratory bronchioles and membranous bronchioles and adjacent alveolar walls is frequent in the lungs of subjects not exposed to asbestos – therefore care must be taken not to over-interpret grade 1 and 2 lesions. It has been shown that these lesions tend to progress with age and are more prevalent in smokers compared with non-smokers.[30] It should also be realized that small irregular opacities on a chest X-ray, even in a subject with claimed asbestos exposure, may not equate with histological asbestosis on biopsy or at postmortem since there is a high prevalence of false-positive X-ray readings,[31] false-negative readings can also occur.

The presence of interstitial fibrosis on its own does not make a diagnosis of asbestosis but requires in addition the presence of asbestos bodies. These can be located within the alveolar spaces or interstitium. The number of asbestos bodies needed in paraffin sections is not established although Churg[32] has advocated a minimum of one per standard section. I think that there are several problems with this. First, it is often difficult to be certain that a ferruginous body is a true asbestos body on light microscopy when there is only one to look at. Second, the different types of asbestos have different propensities for forming asbestos bodies – amosite is greater than crocidolite and crocidolite is greater than chrysotile. When a pathologist observes an asbestos body it is almost always formed on an amphibole fibre. Third, when the number of fibres present is low a greater proportion will be coated;[33] the presence of other

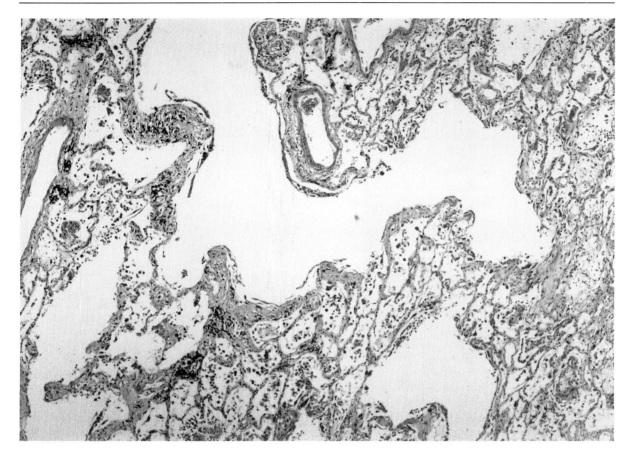

Fig. 10.4 Peribronchiolar fibrosis and fine fibrosis extending into alveolar ducts and adjacent alveolar walls (grade 2 fibrosis) in a man with substantial exposure to asbestos.

dusts will also influence the likelihood of coating. In these situations I consider an electron microscopic fibre count necessary. It is also advisable where substantial numbers of non-asbestos ferruginous bodies are also present. In cases of moderate to severe asbestosis, asbestos bodies are usually found quite easily in standard sections since there has been substantial exposure: it is uncommon to find high levels of asbestos fibres by electron microscopy of tissue digests in those cases where asbestos bodies are absent.

Asbestosis is often but not always associated with visceral pleural thickening and/or pleural plaques. However, there is no direct correlation between the extent of pleural lesions and the severity of asbestosis.

Centrilobular lesions

Many dusts first accumulate in the interstitial tissues surrounding the respiratory bronchioles and extend into the adjacent alveolar walls, so one could conceptualize the majority of pneumoconioses as being characterized by centrilobular lesions, at least in the early stages, for example asbestosis and silicatoses, but I shall restrict the group to those that are characterized by lesions that remain predominantly centrilobular rather than show diffuse spread even at advanced stages. These centrilobular lesions can show relatively little fibrosis, for example coal workers' pneumoconiosis, or severe fibrosis, for example silicosis. If they show little fibrosis they are impalpable and often re-

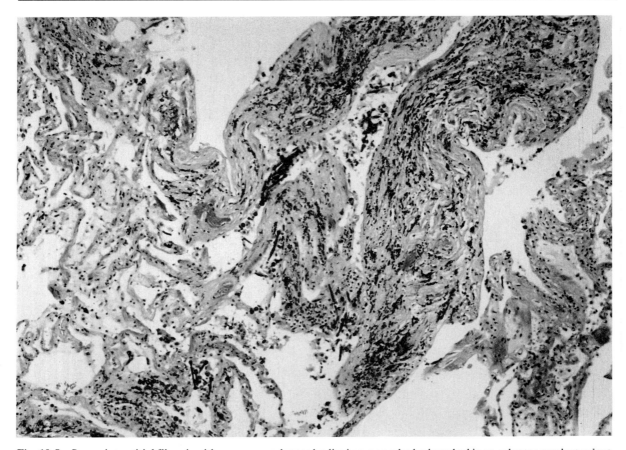

Fig. 10.5 Severe interstitial fibrosis with numerous asbestos bodies in a man who had worked in an asbestos products plant.

ferred to as macules. Macular lesions may occur in association with exposures to coal, iron, barium and tin. The diseases are usually divided into simple and complicated (PMF) and this is done arbitrarily on the basis of size of the largest dust lesion. If there is a dust lesion 1 cm or greater in size then it is termed complicated, if all lesions are less than 1 cm then it is termed simple. Usually clinical disability is only found in the complicated form of pneumoconiosis.

Coal workers' pneumoconiosis
The chances of developing simple or complicated coal workers' pneumoconiosis (CWP) increases with duration and severity of exposure but does not usually develop unless there has been more than a decade of exposure. The characteristic lesion consists of macrophages containing black dust located around the respiratory bronchioles and adjacent alveoli with little collagen formation.[34] As these

enlarge they become stellate and there is accompanying dilatation of air spaces ('focal' emphysema). Macroscopically there are multiple lesions which are stellate, black, impalpable (macules or primary dust foci), averaging 1–4 mm in diameter and most prevalent in the upper zones of the lung (Figs 10.6–10.8). A more variable feature is the presence of stellate or rounded palpable (nodules or secondary dust foci) lesions. Microscopically these lesions show more fibrosis. These nodules are more commonly observed in coal workers exposed to lower-ranked coals.[35]

In some cases of CWP ferruginous bodies may be seen which have rounded, fibrous or irregular platey black or brown cores, which to the uninitiated may cause confusion with asbestos bodies. As the dust lesions increase in profusion and size there is an increase in severity of emphysema[36] which appears similar to the centrilobular emphysema observed in individuals without occupational

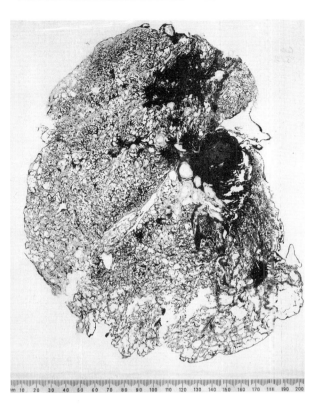

Fig. 10.6 Gough–Wentworth section from a coal worker's lung showing centrilobular emphysema, numerous stellate black macules and areas of PMF in the upper zones.

exposure. Usually there is little clinical disability or reduction in lung function in simple CWP.

Similar lesions can be observed in exposures to other carbonaceous materials such as carbon black and graphite.

In addition to the centrilobular lesions, interstitial fibrosis occurs in approximately 17% of coal workers.[37] Usually it is mild and not extensive but sometimes it can be severe and resemble cryptogenic fibrosing alveolitis, both clinically and pathologically (Fig. 10.9).

Silicosis

Chronic silicosis is characterized by the presence of discrete, well-demarcated nodules less than 1 cm in diameter in the simple form which are most prominent in the upper zones. It is unusual for silicosis to be detected radiologically with less than 20 years' exposure.

Macroscopically, the lesions are round or oval,

well demarcated, grey to black in colour, and hard. They may have a gritty sensation when cut with a knife due to calcification and there may be central necrosis. The earliest microscopic lesions consist of focal collections of dust-laden macrophage situated around respiratory bronchioles which are indistinguishable from lesions caused by silicate exposures. Fibrosis occurs relatively early in the centres of the lesions and with expansion and progression the classical silicotic nodule develops which is characteristic of silica exposure. In order for this lesion to occur dusts usually contain at least 5% or more silica. The classical silicotic nodule contains three zones: (1) the central zone consisting of fibrous tissue which is whorled, often hyalinized and shows focal calcification; (2) the midzone which contains circumferentially arranged layers of fibrous tissue; (3) the peripheral zone which consists of dust-laden macrophages, chronic inflammatory cells and irregularly arranged collagen (Fig. 10.10). The central zone may show necrosis; if so, acid-fast bacilli should be stained, even if no granulomas are evident, since there is a well-documented propensity to develop mycobacterial infection in silicosis. However, in my experience screening often proves negative since the necrosis is often ischaemic in origin. If one examines a silicotic nodule under polarized light the central zone shows relatively dull particles (mostly silica) whereas the central and peripheral zones show bright particles (mainly silicates).

This form is not usually associated with significant reductions of pulmonary function, whereas PMF silicotic lesions may be associated with clinical disability.

Mixed dust fibrosis

The mixed dust fibrotic lesion has an appearance halfway between a classical silicotic nodule and an interstitial silicate lesion and is considered to be caused by exposure to silica concomitantly with less fibrogenic minerals such as iron oxides, silicates and coal. They have been described in foundry workers, haematite miners, slate workers and pottery workers. In practice it is common to see mixed dust fibrotic lesions alongside classical silicotic nodules and stellate silicate-type lesions, for example in the lungs of slate workers. In general as the percentage of silica goes up, the greater the number of silicotic nodules. As the percentage of silica decreases one sees more mixed dust fibrotic nodules, and when the percentage becomes very low one sees only silicate-type lesions.

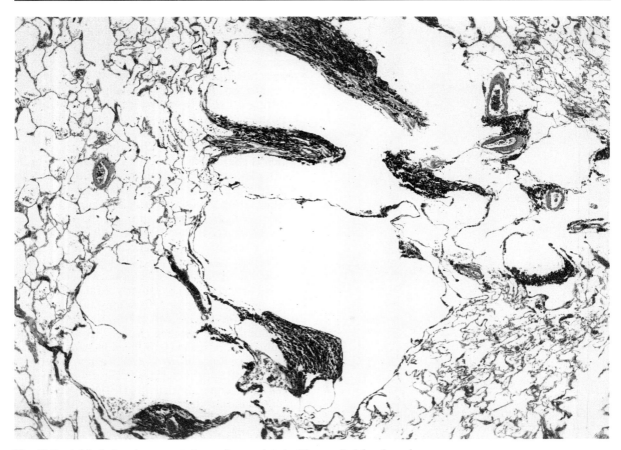

Fig. 10.7   A black dust-impregnated macule associated with so-called focal emphysema.

Macroscopically the mixed dust fibrotic lesion is stellate, firm and well demarcated. Microscopically, it has two main zones (Fig. 10.11): (1) a central zone of fibrous tissue which is frequently hyalinized; (2) a peripheral zone of radially and linearly arranged collagen fibres intermingled with dust-containing macrophages.

Massive lesions

Progressive massive fibrosis (PMF) or complicated pneumoconiosis is defined arbitrarily on size of the dust lesions. This has varied over periods of time and with different authorities. Radiologists who use the Labor Organisation International (ILO) classification of radiographs of the pneumoconioses require a minimum of 1 cm diameter whereas the Pneumoconiosis Committee of the College of American Pathologists recommend a minimum 2 cm diameter for a tissue dust lesion to be classified as a PMF lesion.[34] It is not very often that one encounters biopsies with this type of lesion but occasionally they are removed for medico-legal purposes or mistakenly for neoplasms. Usually PMF lesions are accompanied by a background of severe simple pneumoconiosis, since they are usually related to high cumulative dust exposures. Sometimes they develop some years *after* cessation of exposure. PMF lesions have been seen most commonly in CWP and silicosis but also have occurred with exposures to kaolin, asbestos and talc.

The PMF lesions may be unilateral or bilateral and are most frequently located in the upper lobes and situated more posteriorly than anteriorly. They can be very large and almost occupy a whole lobe and they may cross an interlobar fissure and occupy part of the adjacent lobe. The colour of the

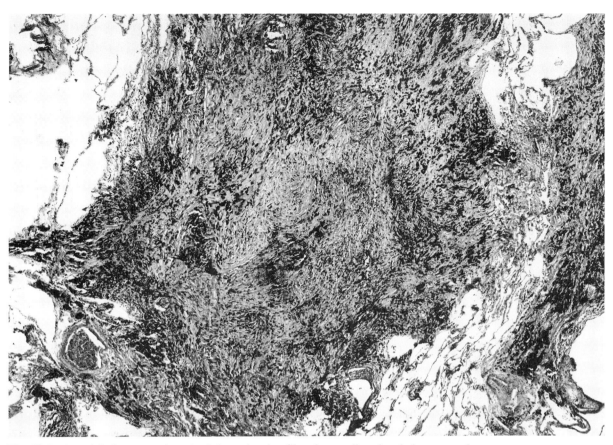

Fig. 10.8   A nodule from a coal worker's lung showing fibrosis as well as dust-laden macrophages.

lesion will depend on the predominant dust present – in coal workers it is black, in silicosis grey to black, in kaolinosis white. The lesions have been classified into two main types:[35] type 1, characterized by diffuse, solid, rubbery lesions which cut easily with a knife (Fig. 10.6); and type 2 formed by the fusion or conglomeration of oval whorled nodules which are hard and which cut with difficulty. Type 1 lesions are typically observed in coal workers exposed to higher-ranked coals and silicatoses such as kaolinosis. Type 2 lesions are typical of silicosis and are often seen in coal-workers exposed to the lower ranks of coal. Cavitation can occur in PMF lesions but is more common in type 1 lesions. Sometimes the PMF lesions do not fit neatly into either type and have features of both.

Microscopically, type 1 lesions consist of large quantities of dust mixed with randomly oriented reticulin and relatively low amounts of hyalinized collagen. In the case of coal there will be a relatively small amount of crystalline material visible under polarized light whereas in those lesions associated with silicate exposure there will be a considerable amount. In contrast, type 2 lesions consist of circumscribed nodules composed of a collagenous centre, a whorled intermediate zone and an outer zone of dust and inflammatory cells similar to the silicotic nodules. An additional feature which may be seen around either type of PMF lesion is the presence of a lymphocytic and plasma cell infiltrate within the walls of small vessels which simulates a vasculitis.

Rheumatoid pneumoconiosis (Caplan syndrome)
If individuals with a rheumatoid diathesis are exposed to dusts such as coal and silica, nodular lesions with an unusual appearance may develop relatively rapidly on a background of sparse dust lesions; this was first described by Caplan[38] in

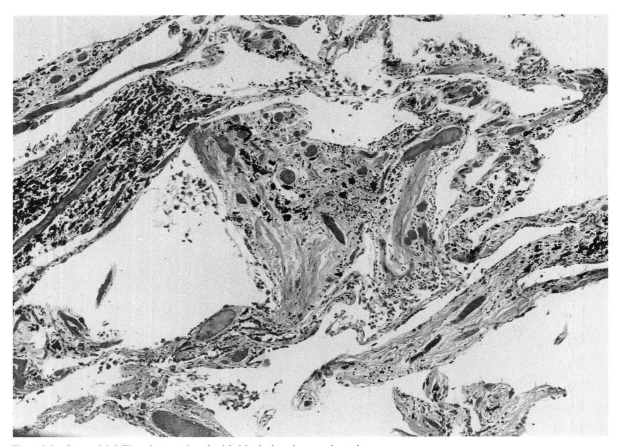

Fig. 10.9   Interstitial fibrosis associated with black dust in a coal worker.

coal miners from South Wales. The nodules are whorled, frequently subpleural but show no zonal predilection and vary in size from a few millimetres to several centimetres in size.[39] The nodules are laminated and may show central necrosis and calcification. The larger lesions often appear to be composed of conglomerate smaller nodules.

Microscopically, the nodules show a central zone of eosinophilic necrosis and nuclear debris surrounded by neutrophils, pallisaded macrophages and fibroblasts in the active lesions (Fig. 10.12) but in older lesions the circumference may be composed of fibrous tissue and chronic inflammatory cells only. Between the necrosis and the inflammatory zone a variable number of laminations composed of dust and dust-laden macrophages may be observed. The small vessels at the edge of the nodules frequently show infiltration of their walls by lymphocytes and plasma cells.

The major differential diagnoses are infective granulomatous lesions (especially tuberculosis) and silicotic nodules. The former usually show more giant cells, which are sparse in Caplan lesions, compact granulomas at the edge of and away from the nodules and acid-fast bacilli by direct staining. Silicotic nodules do not show pallisaded macrophages and fibroblasts and are much more fibrous; also there is a more uniform distribution of lesions.

Granulomatous disorders

Granulomatous lesions can be caused by a variety of particulate exposures including beryllium and organic dust. There are two main patterns of response: (1) a sarcoid-like reaction and (2) an extrinsic allergic alveolitis (hypersensitivity pneumonitis) reaction.

Chronic beryllium disease
Chronic beryllium disease is an insidious disease

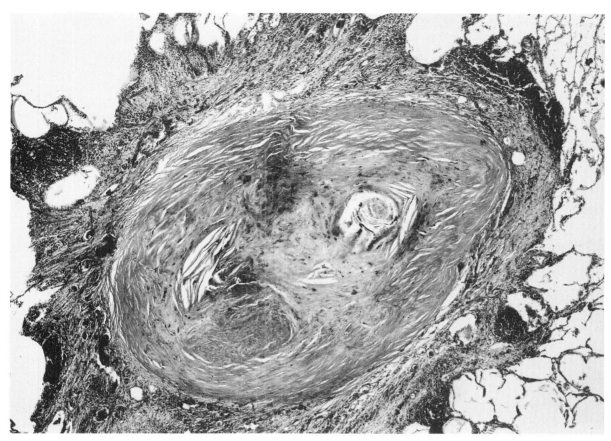

Fig. 10.10 A typical silicotic nodule from a North Wales slate worker.

which is very similar to sarcoidosis clinically and histopathologically. It can manifest from a month to 25 years after last exposure. Light microscopy of the lungs shows an interstitial chronic inflammatory infiltrate and compact well-delineated epitheliod granulomas located within the interstitium, bronchovascular bundles, septa and subpleural tissues.[40] The granulomas typically lack necrosis but occasionally they can show extensive necrosis. Fusion of granulomas can occur to give lesions 1–2 cm in diameter – these lesions often show marked hyalinized collagen. Schaumann and asteroid bodies are frequently present. Diffuse interstitial fibrosis can develop with the formation of honeycomb lung. Granulomas may be present in other organs.

After the exclusion of infectious agents the differential diagnosis includes extrinsic allergic alveolitis and sarcoidosis. In extrinsic allergic alveolitis the granulomas are usually less compact and evanescent and in sarcoidosis there is usually less interstitial inflammatory infiltrate. The diagnostic criteria for chronic beryllium disease are:

1. A history of exposure to beryllium.
2. Consistent clinical and radiological features.
3. The presence of sarcoid-like granulomas within the tissues.
4. Detection of beryllium within the lung tissues.
5. Evidence of beryllium hypersensitivity (beryllium skin patch test, beryllium lymphocyte transformation test).[41]

Ideally these should all be positive but this is not always practical or possible. There have been very rare reports of other metals causing disease similar to chronic berylliosis, namely aluminium[42,43] and titanium.[44]

Foreign body granulomas have been observed in subjects exposed to talc and mica;[19,20] sarcoid-type

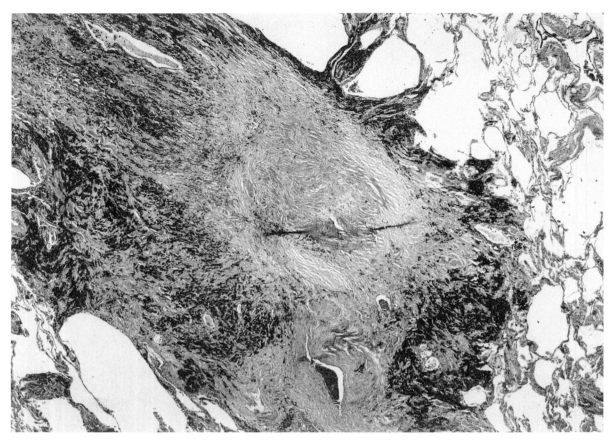

Fig. 10.11 A mixed dust fibrotic nodule from a coal worker exposed to substantial quantities of free silica.

granulomas have been reported rarely in association with talc.[20,21]

Extrinsic allergic alveolitis

Extrinsic allergic alveolitis is a general term which applies to disease caused by the inhalation of particulate (usually organic) matter through an allergic rather than toxic mechanism. Examples include farmer's lung, bird fancier's lung, air conditioner lung and mushroom worker's lung. The clinico-pathological features are similar in all of them. The disease can be acute or chronic. The acute episodes usually resolve but severe or repeated acute episodes or insidious chronic exposures may lead to diffuse interstitial fibrosis of the lungs with honeycomb formation. The acute form is characterized by an interstitial inflammatory infiltrate composed of lymphocytes, plasma cells and macrophages, non-caseating epithelioid gran-

ulomas within the interstitium and bronchioles, inflammation of the bronchiolar walls and oedema and inflammation of the pulmonary arteriolar walls.[45] The inflammatory process is maximal at the centre of the lobules. The granulomas are not usually as well developed as in sarcoidosis. Fibrosis within air spaces (bronchiolitis obliterans with organizing pneumonia (BOOP) or cryptogenic organizing pneumonia (COP) reaction) is observed in about half the cases. If biopsies are obtained several months after exposure to the antigen, the granulomas may have disappeared – contrast this with sarcoidosis where the granulomas persist throughout the disease. When the chronic stage is reached, if there has been no exposure to the antigen for several months, the microscopic picture will be similar to cryptogenic fibrosing alveolitis. However, one helpful feature is that the fibrotic changes are maximal in the upper lobes in chronic extrinsic allergic alveolitis whereas in cryptogenic

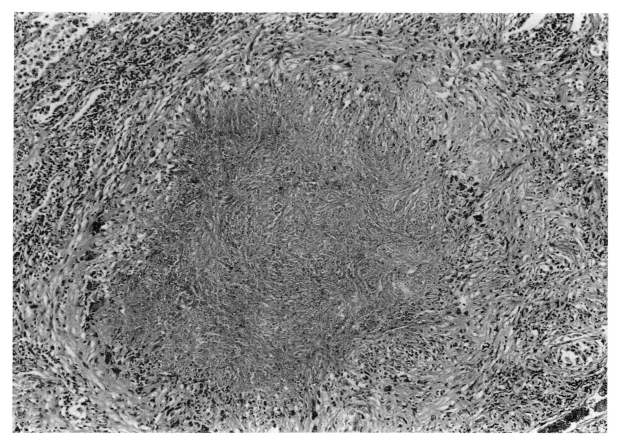

Fig. 10.12 A Caplan lesion showing central necrosis with neutrophilic debris surrounded by pallisaded histiocytes and occasional giant cells.

fibrosing alveolitis and asbestosis they are usually maximal in the lower.

Airway disorders

Some agents may cause acute or subacute pathological changes within small airways either alone or in combination with diffuse alveolar damage, particularly where there is exposure to high concentrations in confined spaces. These have included ammonia, oxides of nitrogen, fire smoke and sulphur dioxide.[14] The changes can include acute necrosis of the bronchiolar epithelium, obliterative bronchiolitis and combinations of bronchiolar and interstitial inflammatory changes – so-called BOOP or COP.[46,47]

Chronic small airway changes can result from exposure to mineral dusts such as coal dust, silica, aluminium oxide, iron oxide, sheet silicates and asbestos.[48] The changes include fibrosis of the walls of respiratory and non-respiratory bronchioles which can extend to alveolar ducts, often in association with pigment lying free and within macrophages.

Criteria for diagnosis

The main criteria for ascribing a particular pathology to a mineral exposure are shown in Table 10.4.

1. Preferably evidence of exposure should come from several disciplines but this is not always practical. There are often problems with the accuracy of occupational histories which can result in an exposure to a particular mineral being under- or over-estimated. Neoplasia and pulmonary fibrosis when caused occupationally

Table 10.4 Guidelines for diagnosis of a 'pneumoconiosis'

1. Evidence of exposure: history
   pathology
   mineral analysis
2. Appropriate clinical and radiological features
3. Exclusion of confounding factors
4. Biological plausibility

may be a result of exposures many years before. For example when mesotheliomas are related to asbestos exposure the date of first exposure to asbestos averages 30–40 years earlier. Silicosis, CWP and asbestosis are usually associated with mineral exposures in excess of 10 and often 20 years. Fibrosis can develop and progress after the person has ceased exposure to the particular mineral.

Sometimes the pathological appearances in themselves are enough for diagnosis, for example interstitial fibrosis associated with numerous asbestos bodies can be confidently diagnosed as asbestosis. However, the mere presence of a mineral within the lung tissues may not necessarily indicate that it was responsible for the pathological changes; mineral analysis of the lung tissues with quantitation may be very helpful. This is particularly relevant where the amount of material appears relatively low or where exposures are complex or poorly documented.

2. Appropriate clinical and/or radiological features should be present. For example on a biopsy a pathologist might observe some interstitial fibrosis associated with a dust. The chest X-ray should prove helpful in deciding whether this is a localized scar which has 'attracted' dust or is a diffuse process and therefore likely to be some form of pneumoconiosis.

The latent period between the exposure and the development of the pathology should be appropriate for the suspected agent. For example lung cancer rarely occurs in a period less than 20 years after first exposure to asbestos.

3. It should be remembered that natural disease can occur among particular groups of workers which might mimic an occupational disease; for example sarcoidosis and cryptogenic fibrosing alveolitis can occur among asbestos-exposed workers which might be misdiagnosed as

asbestosis. Therefore it is important to exclude confounding factors.

A particular problem is separating occupationally induced from naturally (or smoking) induced lung cancers in asbestos-exposed workers. In these cases the amount of interstitial fibrosis and asbestos fibre load are important in the ascription of the tumour to asbestos.

4. Biological plausability implies that there is epidemiological and experimental evidence that a particular mineral exposure can result in a particular pattern of lung pathology.

References

1. Craighead JE, Kleinerman J, Abraham JL *et al.* Diseases associated with exposure to silica and nonfibrous silicate minerals. *Arch Pathol Lab Med* 1988, **112**, 673–720.
2. Gibbs AR, Craighead JE, Pooley FD, Wagner JC. The pathology of slate workers' pneumoconiosis in North Wales and Vermont. *Ann Occup Hyg* 1988; **32**, 273–278 (supplement 1).
3. Gibbs AR, Jones JSP, Pooley FD, Griffiths DM, Wagner JC. Non-occupational malignant mesotheliomas. In: Bignon J, Peto J, Saraci R (eds) *Non-Occupational Exposure to Mineral Fibres*, vol. 90. Lyon: IARC Scientific Publications, 1989, pp. 219–228.
4. Sterner JH, Eisenbud M. Epidemiology of beryllium intoxication. *Arch Indust Hyg* 1951; **4**, 123–151.
5. Churg A, Green FHY. *Pathology of Occupational Lung Disease*. New York: Igaku-Shoin, 1988.
6. Morgan WKC, Seaton A. *Occupational Lung Diseases*, 3rd edn. WB Saunders, in press.
7. Abraham JL, Hertzberg MA. Inorganic particles associated with desquamative interstitial pneumonia. *Chest* 1981; **80**, 67S.
8. Corrin B, Price AB. Electron microscopic studies in desquamative interstitial pneumonia associated with asbestos. *Thorax* 1972; **27**, 324–331.
9. Herbert A, Sterling E, Abraham JL, Corrin B. Desquamative interstitial pneumonia in an aluminium welder. *Hum Pathol* 1982; **13**, 694–699.
10. Davison AG, Haslam PL, Corrin B *et al.* Interstitial lung disease and asthma in hard metal workers: bronchoalveolar lavage, ultrastructural and analytical findings and results of bronchial provocation tests. *Thorax* 1983; **38**, 119–128.
11. Prakash UB, Barham SS, Carpenter HA, Dines DE, Marsh HM. Pulmonary alveolar proteinosis: experience with 34 cases and a review. *Mayo Clinic Proc* 1987; **62**, 499–518.
12. McCuen DD, Abraham JL. Particulate concentrations in pulmonary alveolar proteinosis. *Environ Res* 1978; **17**, 334–339.

13. Miller RR, Churg AM, Hutcheon M, Lam S. Pulmonary alveolar proteinosis and aluminum dust exposure. *Am Rev Respir Dis* 1984; **130**, 312–315.

14. Wright JL, Churg A. Diseases caused by metals and related compounds, fumes and gases. In: Churg A, Green FHY (eds) *Pathology of Occupational Lung Disease*. New York: Igaku-Shoin, 1988, pp. 31–72.

15. Nemery B. Metal toxicity and the respiratory tract. *Eur Respir J* 1990; **3**, 202–219.

16. Begin R, Cantin A, Masse S. Recent advances in the pathogenesis and clinical assessment of mineral dust pneumoconioses: asbestosis, silicosis and coal pneumoconiosis. *Eur Respir J* 1989; **2**, 988–1001.

17. Gibbs AR, Wagner JC. Diseases due to silica. In: Churg A, Green FHY (eds) *Pathology of Occupational Lung Disease*. New York: Igaku-Shoin, 1988, pp. 155–175.

18. Wagner JC, Pooley FD, Gibbs AR, Lyons J, Sheers G, Moncrieff CB. Inhalation of china stone and china clay dusts: relationship between the mineralogy of the dust retained in the lungs and pathological changes. *Thorax* 1986; **41**, 190–196.

19. Gibbs AR. Human pathology of kaolin and mica pneumoconioses. In: Bignon J (ed.) *Health Related Effects of Phyllosilicates*, NATO ASI series G21. Berlin: Springer-Verlag, 1990, pp. 217–226.

20. Craighead JE. Pathological features of pulmonary disease due to silicate dust inhalation. In: Bignon J (ed.) *Health Related Effects of Phyllosilicates*, NATO ASI series G21. Berlin: Springer-Verlag, 1990, pp. 179–190.

21. Gibbs AR, Pooley FD, Griffiths DM, Mitha R, Craighead JE, Ruttner JR. 'Talc pneumoconiosis' – a pathological and mineralogical study. *Hum Pathol* 1992; **12**, 1344–1354.

22. Hinson KFW, Otto H, Webster I, Rossiter CE. Criteria for the diagnosis of and grading of asbestosis. In: Bogovski P (ed.) *Biological Effects of Asbestos*, vol. 8. Lyon: IARC Scientific Publications, 1973; pp. 54–57.

23. Craighead JE, Abraham JL, Churg A *et al*. The pathology of asbestos associated diseases of the lungs and pleural cavities: diagnostic criteria and proposed grading schema. *Arch Pathol Lab Med* 1982; **106**, 542–596.

24. Wagner JC, Newhouse ML, Corrin B, Rossiter CE, Griffiths DM. Correlation between lung fibre content and disease in East London asbestos factory workers. *Br J Industr Med* 1988; **45**, 305–308.

25. Ren H, Lee DR, Hruban RH *et al*. Pleural plaques do not predict asbestosis: high resolution computed tomography and pathology study. *Mod Pathol* 1991; **4**, 201–209.

26. Roggli VL. Asbestos bodies and nonasbestos ferruginous bodies. In: Roggli VL, Greenberg SD, Pratt PC (eds) *Pathology of Asbestos-associated Diseases*. Boston/Toronto/London: Little, Brown and Company, 1992, pp. 39–76.

27. Crouch E, Churg A. Ferruginous bodies and the histologic evaluation of dust exposure. *Am J Surg Pathol* 1984; **8**, 109–116.

28. Clarke GM. Phyllosilicates as industrial minerals. In: Bignon J (ed.) *Health Related Effects of Phyllosilicates*, NATO ASI Series G21. Berlin: Springer-Verlag, 1990, pp. 31–46.

29. Gibbs AR, Pooley FD, Griffiths DM, Mitha R. Silica and silicate pneumoconioses. A pathological and mineralogical study. *Annals Occup Hyg. Inhaled Particles VII* 1994, pp. 851–856.

30. Adesina AM, Vallyathan V, McQuillen EN, Weaver SO, Craighead JE. Bronchiolar inflammation and fibrosis associated with smoking. *Am Rev Respir Dis* 1991; **143**, 144–149.

31. Sluis-Kremer GK, Hessel PA, Hnizdo E. Factors influencing the reading of small irregular opacities in a radiological survey of asbestos miners in South Africa. *Arch Environ Hlth* 1989; **44**, 237–243.

32. Churg A. The diagnosis of asbestosis. *Hum Pathol* 1989; **20**, 97–99.

33. Pooley FD, Ransome DL. Comparison of the results of asbestos fibre dust counts in lung tissue obtained by analytical electron microscopy and light microscopy. *J Clin Pathol* 1986; **39**, 313–317.

34. Kleinerman J, Green F, Laqueur W *et al*. Pathology standards for coal workers' pneumoconiosis. *Arch Pathol Lab Med* 1979; **103**, 373–472.

35. Davis JMG, Chapman J, Collins P *et al*. Variations in the histological patterns of the lesions of coal workers' pneumoconiosis in Britain and their relationship to lung dust content. *Am Rev Respir Dis* 1983; **128**, 118–124.

36. Cockcroft A, Seal RME, Wagner JC, Lyons JP, Ryder R, Anderson N. Postmortem study of emphysema in coal workers and noncoal workers. *Lancet* 1982; **ii**, 600–603.

37. McConnachie K, Green FHY, Vallyathan V, Wagner JC, Seal RME, Lyons JE. Interstitial fibrosis in coal miners – experience in Wales and West Virginia. *Ann Occup Hyg* 1988; **32**, 553–560.

38. Caplan A. Certain unusual radiological appearances in the chest of coal miners suffering from rheumatoid arthritis. *Thorax* 1953; **8**, 29–37.

39. Gough J, Rivers E, Seal RME. Pathological studies of modified pneumoconiosis in coal miners with rheuamatois arthritis (Caplan's syndrome). *Thorax* 1955; **10**, 9–18.

40. Freiman DG, Hardy HL. Beryllium disease. *Hum Pathol* 1970; **1**, 25–44.

41. Jones Williams WD. Diagnostic criteria for chronic beryllium disease (CBD) based on the UK Registry 1945–91. *Sarcoidosis* 1993; **10**, 41–43.

42. Chen W, Monnat RJ, Chen M, Mottet NK. Aluminum induced pulmonary granulomatosis. *Hum Pathol* 1978; **9**, 705–711.

43. DeVuyst P, Dumortier P, Schandene L, Estenne M, Verhest A, Yernault C. Sarcoid-lung granuloma-

tosis induced by aluminium dusts. *Am Rev Repsir Dis* 1987; **135**, 493–497.

44. Redline S, Barna BP, Tomashefski J, Abraham JL. Granulomatous disease associated with pulmonary deposition of titanium. *Br J Industr Med* 1986; **43**, 652–656.

45. Seal RME. Pathology of extrinsic allergic bronchiolo-alveolitis. *Progr Respir Res* 1974; **8**, 66–73.

46. Gosink BB, Friedman PJ, Liebow AA. Bronchiolitis obliterans. *Am J Radiol* 1973; **117**, 816–832.

47. Geddes DM. BOOP and COP. *Thorax* 1991; **46**, 545–547.

48. Wright JL, Cagle P, Churg A, Colby TV, Myers J. State of the art. Diseases of the small airways. *Am Rev Respir Dis* 1992; **146**, 240–262.

11

Pulmonary tumours

B Mackay

Introduction

The classification of pulmonary tumours currently used by clinicians and pathologists is quite simple. At least 98% of the various neoplasms that arise within the lung fall into a small number of categories. Among these groups, however, many light microscopic appearances can be encountered. It is likely that lung carcinomas are not totally separate entities but rather are derived from a common endodermal precursor, and this shared origin may explain the morphological heterogeneity that is often seen in pulmonary tumours.

Analyses of large series of cases provide an indication of the relative incidence of the different categories. As an example, in a study of the characteristics of 1336 successive lung cancer patients,[1] the histological subtypes included 32% squamous cell carcinoma, 26% adenocarcinoma, 19% small cell carcinoma, 12% large cell carcinoma and 8% adenosquamous carcinoma. Of the patients 92% were smokers, 69% were men, and 68% were white. The figures vary among different series, and also from one country to another. In the Lung Cancer Study Group report of 1121 classified tumours, adenocarcinoma and squamous carcinoma were almost equal in incidence.[2] Adenocarcinoma is now the most frequent category in the USA and Japan, whereas squamous carcinoma continues to predominate in Europe.[3]

When studies are conducted over decades, changes in the pattern of lung carcinoma can be seen. In an examination of 505 cases spanning a period in which there was a significant alteration in tobacco-smoking habits, a shift in the location and histological type was observed.[4] Peripheral tumours made up 30.7% of the carcinomas before 1978 and 42% from 1986 to 1989, while the incidence of bronchioloalveolar adenocarcinoma more than doubled from 9.3% in the earlier period to 20.3% in the 1986 to 1989 period. Young patients are not spared the ravages of lung cancer: fewer than 5% of cases occur under the age of 40, and clinically the tumours are similar, but more females are affected than in older age groups.

In this chapter, comments on the techniques currently being used to diagnose and study pulmonary tumours are followed by a concise review of the pathology of the common primary lung tumours. A number of the uncommon lung tumours are also mentioned since even the rare entities must be considered when pondering a differential diagnosis. Pleural neoplasms, discussed in Chapter 13, and metastatic carcinomas must also be kept in mind.

Pathological assessment of a lung tumour specimen·

The first consideration of the pathologist who is handling, processing and interpreting a pulmonary tumour specimen is to provide information that will help in the management of the patient. Clinicians are primarily concerned with a separation of small cell from non-small cell tumours, but a detailed pathology report containing an accurate diagnosis and an account of the location and extent of the tumour is required for staging, and for retrospective clinico-pathological studies.

A number of factors can make the task of the surgical pathologist difficult. The most common reason for failure to reach a firm diagnosis is an inadequate specimen. Material provided from an intrathoracic tumour for microscopic study is often less than optimal because of technical problems. Bronchoscopic and mediastinoscopic biopsies are small and prone to crushing and drying artefact. Increasing reliance is being placed on fine needle aspiration biopsies for the initial or even the only diagnosis of a lung tumour, and while the material is frequently of good quality, it can be scanty and show distortion. The classification of lung tumours hinges on a recognition of architectural patterns as well as an assessment of the cytological features of the tumour cells, and there can be considerable overlap in both among the defined categories: lung tumours, and particularly the non-small cell carcinomas, are known for their heterogeneity. A small sample reflects only one area of the entire neoplasm. Loss of differentiation makes it even harder to reach a firm diagnosis.

Pathologists have at their disposal a number of techniques which complement conventional light microscopy in the diagnostic evaluation and prognostic assessment of lung tumours. They include immunocytochemistry, electron microscopy and flow cytometry, and there is currently considerable interest in oncogenes and proliferation markers. Each of these methods has its uses and limitations and they must be used selectively and the findings correlated with the light microscopy and clinical data.

Specimen handling

Every lung specimen should be processed with as little delay as possible to provide good preparations for microscopic study. A small bronchoscopic biopsy destined for paraffin embedding can be placed in formalin in the operating room. The common questions that are asked during intraoperative consultation on a resection specimen are the type of tumour, often already known from a fine needle aspiration biopsy, and adequacy of excision, which in the case of a lobectomy or pneumonectomy will require assessment of the bronchial resection margin and the relationship of the tumour to overlying pleura. This initial evaluation may provide an indication that special techniques are likely to be useful, and appropriate tissue specimens should be obtained at the time of accession of the specimen. Most relevant immunocytochemical procedures can be performed on paraffin sections, but it may be desirable to take a small piece of the tumour for flow cytometry, and possibly to place a thin slice in fixative for electron microscopy.

Light microscopy

Criteria for the classification of lung tumours must be based on features that can be seen in routine light microscopic sections, and they should be reproducible in order to allow comparison of results among different institutions and countries. The presence of mucin can be included since a mucin stain is available in most pathology laboratories. The importance of procuring well-preserved material can not be overemphasized. If sections show scanty and/or distorted tumour, it may be worth cutting at levels through the blocks in search of better foci. The interpretation of crushed cells can become a guessing game, although immunostaining for keratin is sometimes informative. The pathologist will understandably be reluctant to request rebiopsy, knowing the difficulties that are involved in obtaining a specimen from a lung tumour, but there are times when it is in the best interests of patient and clinician to obtain better material.

Histological grading

The level of differentiation in a non-small cell lung carcinoma can vary considerably within a small area of the tumour, so the assignment of a grade based on the histological appearance in a small specimen may be misleading with regard to the tumour as a whole. A separation of well and poorly differentiated tumours can be meaningful, but detailed grading has limited application in the management of patients with lung cancer. In a review of 2260 lung adenocarcinomas, it was found that survival data had little relevance to histological grade.[5]

Clinical staging

Data provided by the pathologist from the study of a resection specimen must include the size and location of the tumour, whether it is solitary or multiple, its distance from resection surfaces, bronchial margin, and pleural surface, and the status of any lymph nodes present within the specimen or separately submitted. The information is used to assign a pathological stage to the tumour, and the figures overall show a good correlation with survival. The tumour, node, metastases (TNM) nomenclature is used internationally. The prognostic implications of each category are a reflection of the relationship between extent of disease and the potential for cure of available therapies.[6]

The stage that is assigned after pathological study of a resection specimen is likely to be higher than the initial clinical stage. Among 1737 Japanese patients who underwent pulmonary resection for lung cancer,[7] 821 had stage 1, 248 stage 2, 465 stage 3a, 82 stage 3b, and 118 stage 4 disease. Based on postoperative staging, the patients were grouped differently. 536 had stage 1, 221 stage 2, 559 stage 3a, 159 stage 3b, and 258 stage 4 disease.

Most patients with small cell lung cancer have disseminated disease when the diagnosis is made, and the TNM staging system is less suitable than for non-small cell tumours, although it may be used to divide patients into prognostic subgroups.[8] The Veterans Administration Lung Cancer Study Group advocated subdivision of inoperable cases into limited disease, confined to one hemithorax including ipsilateral scalene nodes, and extensive disease, being any disease beyond the confines of the definition of limited disease.

Cytology

Increasing reliance is being placed on the use of cytological preparations to establish the initial and often the only diagnosis in a patient with a primary lung tumour. Fine needle aspiration biopsies can provide reliable results when interpreted by experienced cytopathologists. The accuracy approaches 100% for small cell lung cancer, but it is lower for non-small cell tumours. The procedure carries a low morbidity, but complications can occur, including pneumothorax. Disadvantages are the often small numbers of viable cells, difficulty in obtaining cells from a small lesion without several passes, and minimal sampling of an extensive lesion. Rapid evaluation of every smear by a cytologist is advisable.

Atypical cells and even an impression of malignancy is occasionally encountered in cytology specimens from patients who do not have cancer. Changes induced in epithelial cells by chemical and physical agents including chemotherapeutic drugs can simulate those seen in pulmonary tumours. They are more likely to be observed in sputum and in bronchial washings and brushings than in aspiration biopsies where clumps of cells are often present.

Immunocytochemistry

At the present time, immunoperoxidase techniques have limited application in the diagnostic evaluation and prognostic assessment of primary lung tumours. There have been many studies of possible markers for lung tumours, but most of the antibodies are not specific, and other than surfactant, a tumour marker which is unique for lung tumours does not exist.[9]

Electron microscopy

Ultrastructural study does not play a role in the routine diagnostic of lung neoplasms, but when a tumour cannot be confidently classified by routine light microscopy, it is often possible to obtain useful information with electron microscopy. It can clarify appearances seen by light microscopy, help define the relationships among different types of lung tumours, and answer questions of histogenesis. It can also reveal features of prognostic relevance. For instance, the number of cell junctions has been shown to be significantly tied to stage and survival time.[10]

Nucleolar organizer regions

A quantitative study of nucleolar organizer regions in 104 human lung carcinomas of different types led to the conclusion that the procedure will not reliably discriminate between various histological types of lung cancers. However, long-term follow-up of the patients must be awaited to assess the prognostic value.[11]

Flow cytometry

This technique is now widely used for the prognostic assessment of many types of tumours, and its value has been demonstrated in patients with lung cancer. In one study, patients with diploid squamous cell carcinomas were found to survive significantly longer than those with aneuploid tumours and the results were independent of stage. However, DNA ploidy was not a significant prognostic factor in non-small cell tumours, and ploidy did not correlate with other clinicopathological variables such as stage, nodal status, degree of differentiation, nuclear grade and mitotic rate.[12] Aneuploid tumours may show regional DNA heterogeneity.[13]

Proliferation markers

The proliferative activity of lung tumours can be assessed using the monoclonal antibody Ki-67. Survival in non-small cell carcinomas is better for tumours of lower proliferative rate up to approximately 2 years after operation, but by 5 years all tumours of a particular type show a similar survival curve. Small cell carcinomas have high labelling indices, in keeping with the aggressive course of the disease, while carcinoid tumours have low labelling rates.[14]

Molecular biology

Oncogenes and abnormalities of tumour suppressor gene expression hold promise as prognostic tools in lung cancer and they are presently the focus of many experimental studies. It may be that determination of their status in biopsy material will lead to the definition of new subgroups of lung cancer which tend to respond well or poorly to therapy. K-*ras* mutation has been shown to indicate poor prognosis in patients with adenocarcinomas, and amplification of c-*myc* or c-*ras* or allele loss at the H-*ras* locus predicts poor outcome in non-small cell carcinomas. Mutational activation of a ras oncogene, usually involving codon 12 of the K-*ras* gene, is detected in about 30% of lung adenocarcinomas, but K-*ras* mutations are uncommon in other carcinomas and rare in small cell lung cancer. K-*ras* mutations occur with significantly higher frequency in smokers than in non-smokers.[15]

Analysis of oncogene amplification, rearrangement and mutation requires the use of techniques such as the various blotting methods and the polymerase chain reaction, so use in the routine assessment of lung tumours is restricted. Only a small number of growth suppressor genes have been characterized in detail, but two members of this group which are frequently altered in lung cancer are the retinoblastoma and p53 genes.[16,17]

Part of the short arm of chromosome 3 appears to be important in the development of lung cancers of all histologies. By analogy with other tumours, the presence of a consistent deletion is believed to indicate that a tumour suppressor gene is encoded within the deleted region.[18]

Precursor lesions

Metaplastic transformation of the columnar cells of the bronchial passages to a squamous type of epithelium is more common in the lungs of smokers, and it is sometimes extensive within the respiratory passages, even extending into the terminal bronchioles.[19] Squamous metaplasia can probably revert to normal mucosa provided the inciting factor is removed, but it can also be a forerunner of the development of malignancy. In one study,[20] it was found in 83% of male lungs containing primary squamous cell or small cell carcinoma, but only in 43% of the lungs of both sexes with adenocarcinoma or metastatic tumours.

Squamous metaplasia is easily recognized histologically, but the transition from squamous metaplasia to mild dysplasia is subtle and more difficult to detect and assess. Dysplasia is generally graded as mild, moderate or severe, and to some extent the grading is subjective. The polarity of the cells is generally maintained in mild dysplasia but tends to be disturbed in moderate dysplasia. Greater variation in cell and nuclear size and shape is found in severe dysplasia and the cells have higher nuclear to cytoplasmic ratios and hyperchromatic nuclei with irregular nuclear profiles. In tissue sections, distinctions between moderate and severe dysplasia and between severe dysplasia and carcinoma *in situ* are often difficult and the interpretation will vary depending on the experience of the pathologist and the criteria that are considered important. The presence of dysplasia may be better detected in cytological preparations since they provide a more extensive sampling of the bronchial mucosa than a single small biopsy.

Patients with all degrees of dysplasia should be considered at risk for the development of carcinoma and should be carefully monitored with periodic sputa and radiological examinations. The significance of severe dysplasia in sputum samples was investigated by Risse *et al.*,[21] who found malignancy in 46% of their 46 patients on follow-up, and they recommended that a cytodiagnosis of severe dysplasia in sputum be followed by bronchoscopy with brushings of all suspicious areas and segmental bronchial washings, and if these should be negative, by sputum examinations at three-monthly intervals.

Atypical bronchioloalveolar cell hyperplasia is not related to smoking, but it appears to be a precancerous stage in the development of adenocarcinoma.[22] Papillary or bronchioloalveolar lesions suggesting the development of adenocarcinoma have been encountered in atypical adenomatous hyperplasia,[23] and atypical adenomatous lesions are not infrequently seen at the periphery of an adenocarcinoma.

In carcinoma *in situ* of the bronchial passages, the nuclear to cytoplasmic ratio is particularly high and the level of atypia may be greater than that seen in some well-differentiated or invasive squamous carcinomas. Spindle cell transformation can also occur in carcinoma *in situ* and it may be detected in brushing specimens.

Non-neoplastic lesions and benign tumours

A reactive process in the lung can simulate a true neoplasm histologically. The lesion formerly known by a number of names including plasma cell granuloma and now called inflammatory pseudo-tumour is composed of intersecting crude fascicles of fibroblasts, intermingled with histiocytes, plasma cells and lymphocytes. It is an inflamma-tory process but the nodule can be as large as 15 cm and can be mistaken for tumour.[24] The spindle cells look benign, the plasma cells are polyclonal, and epithelial markers will be negative.

In the older literature, most of the tumours that were called adenomas of the lung were carcinoids. True pulmonary adenomas are uncommon. They include the mucous gland adenoma and the poorly defined alveolar cell adenoma. Mucinous cyst-adenomas are mucus-containing, unilocular cysts lined by columnar mucinous epithelium with foci of stratification and papillary infoldings, and they appear to be distinct from mucous gland aden-omas.[25] Adenocarcinoma has been found arising in a mucinous cystadenoma.[26]

Bronchioloalveolar adenomas have been de-scribed. In a series of 247 consecutive resections for lung carcinoma, Miller[27] observed 23 patients who had a total of 41 incidentally found nodules which measured from 1 to 7 mm in diameter. Some were multiple, and in one patient they were bilateral. The cells displayed a lepidic growth pattern and variable cellular atypia, and the lesions were interpreted as an early, premalignant phase of glandular neoplasia with a potential for evolution to carcinoma.

Non-small cell carcinomas

Most primary lung tumours are carcinomas, and over 70% fall into the non-small cell group. For convenience, they are sometimes referred to collec-tively as large cell carcinomas of the lung or NSCLC to distinguish them from small cell carci-noma or SCLC. The non-small cell carcinomas are customarily subdivided into squamous cell carci-noma, adenocarcinoma, and undifferentiated large cell types. Many of the non-small cell tumours are not pure in their composition, and some fail to show convincing evidence of either squamous or glandular differentiation and are simply called poorly differentiated non-small cell carcinomas.

Squamous carcinoma

In Europe and many other countries, squamous cell carcinoma continues to be the most common subtype of lung carcinoma. There is a higher dose-response relationship between the number of cigarettes smoked and the duration of smoking with squamous carcinoma compared to adeno-carcinoma,[28] correlating with the fact that most squamous carcinomas are central tumours. Peri-pheral squamous tumours represented 16% of all resected squamous cancers in one study,[29] and compared with the central tumours, peripheral squamous carcinomas were associated with fewer symptoms at presentation and better survival. Symptoms in patients with centrally located tumours occur relatively early, but many patients with invasive carcinoma have no symptoms at the time of diagnosis and others have only a cough which may be ignored if the patient is a heavy smoker. Some squamous cancers form an exo-phytic endobronchial growth which is prone to cause obstruction with pneumonitis and atelec-tasis. Others become cavitary lesions, and contin-uity with a bronchus can result in haemoptysis.

A central squamous carcinoma is often diag-nosed by bronchoscopy or from cytological preparations. Squamous carcinoma of the lung has the same histological appearance as a squamous malignancy arising from other mucosal surfaces, and it will on occasion show as much keratinization as a well-differentiated cutaneous squamous carcinoma. The keratin may accumulate within the cells, giving the cytoplasm an eosinophilic appear-ance, or may accumulate extracellularly as con-centric whorls. When the cells are producing little or no keratin, the cytoplasm has a relatively clear appearance. A criterion that is used to distinguish a squamous carcinoma in the lung is the presence of so-called intercellular bridges. These sites of desmosomes are highlighted by retraction of the adjacent cytoplasm and they will not be visible if the cell surfaces are closely apposed. Marked variation in differentiation can be encountered in a squamous carcinoma within a small group of cells.

Necrosis within a solid mass of squamous carcinoma leaves a peripheral rim of cells with a stratified appearance (Fig. 11.1), and flattened cells on the surface of this layer are readily recognized in cytological preparations. The amount of fibrous stroma in a squamous cancer is usually less than in adenocarcinomas. Peripheral squamous tumours are smaller, have fewer mitoses, show less lymphatic invasion, and display a more intense stromal reaction than their central counterparts.[29]

In cytological preparations, squamous carcinoma cells usually have considerable amounts of cytoplasm and the nuclei are irregular. Major cytological indicators for squamous carcinoma are the presence of keratin and eosinophilic spindle cells with glassy or laminated cytoplasm.[30] The cytokeratin pattern in squamous carcinoma is of the stratified epithelium type.

In the Lung Cancer Study Group report,

patients with squamous cell carcinoma had an outcome superior to that of patients with adenocarcinoma in every TNM subset.[2]

Basaloid carcinoma

The term basaloid carcinoma has been applied to a variant of non-small cell lung carcinoma having an immunophenotype close to that of basal bronchial epithelial cells, with a low level of expression of low molecular weight cytokeratins.[31] The small tumour cells have sparse cytoplasm and moderately hyperchromatic nuclei, and they exhibit a lobular growth pattern with peripheral palisading. The pattern can be pure or mixed with the common types of non-small cell carcinoma. Neuroendocrine features do not appear to be present and the tumour may be derived from bronchial reserve cells. The prognosis is poor with a median survival rate of 22 months for stage 1 and 2 disease.[31]

Lymphoepithelioma-like carcinoma

A variant of non-small cell lung carcinoma which resembles the non-keratinizing carcinomas of the nasopharynx has been termed lymphoepithelioma-

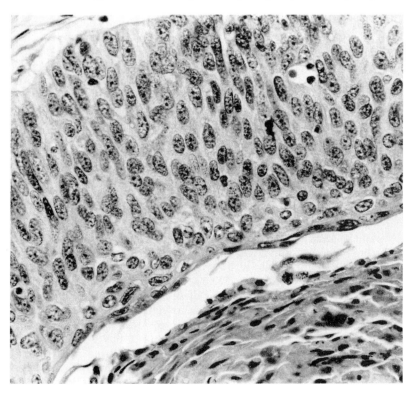

Fig. 11.1 Squamous cell carcinoma. The layer of cells presents a stratified appearance with flattened cells on the surface.

like carcinoma of the lung.[32] Only a small number of cases have so far been recorded.

Adenocarcinoma

Adenocarcinoma of the lung is now the most common lung cancer in Japan and the USA. It is less strongly associated with cigarette smoking than squamous cell carcinoma and small cell carcinoma. Adenocarcinoma is relatively more frequent in women and is uncommon in young patients.

Most lung adenocarcinomas form a discrete peripheral mass and the majority are under 5 cm when they are resected. Fewer than 3% of lung carcinomas are more than 10 cm in size, but 42% of these large tumours are adenocarcinomas.[33] A peripheral tumour is not likely to produce obstructive effects; adenocarcinomas are often asymptomatic and are found on routine chest X-ray. Chest pain, weight loss, cough and dyspnoea are symptoms of advanced disease. More than one adenocarcinoma may be present in the same patient, and even in the same lobe.

The cut surface of an adenocarcinoma is whitish or pale grey, and it glistens when mucin is present. There is often a desmoplastic reaction which confers a firm fibrous consistency on the tissue. The cells range from cuboidal to tall columnar, and they form a variety of architectural patterns. More than one is frequently seen in a single tumour. Large lumens can be bordered by a single layer of uniform columnar cells, or there may be sheets of cells containing many small acini. With loss of differentiation, progressive disruption of the pattern accompanies increasing degrees of cytological atypia.

Some solid pulmonary adenocarcinomas show evidence of mucin production in the presence of clear cytoplasm, signet ring cells or a cystic substructure.[34] A mucin stain is often useful to establish the diagnosis, since clear cytoplasm can also be produced by glycogen, lipid or degenerative changes.

By electron microscopy, acinar formation with microvilli is seen in lung adenocarcinomas and even small quantities of mucin are easily located. In solid adenocarcinomas, the acini are sometimes mere slits between cells, or intracytoplasmic lumens may rarely be present. A few dense granules of the size and appearance seen in non-ciliated bronchiolar (Clara) cells are common in cells of classic acinar and papillary pulmonary adenocarcinomas. Cytological preparations of

adenocarcinomas show uniform ovoid nuclei with smooth profiles. The cytokeratin pattern of adenocarcinomas of the lung is that of simple epithelium.

In the differential diagnosis of a lung adenocarcinoma, metastatic tumours must be considered, and it can be difficult or impossible to distinguish metastatic colonic adenocarcinoma from a primary lung tumour. Microfilament cores can be present within microvilli in tumours from both sites. A clear cell tumour in the lung may suggest metastatic renal cell carcinoma, particularly if the cells contain glycogen. It is sometimes necessary to rely on clinical and radiological studies to rule out an extrapulmonary primary site.

Attempts to subtype adenocarcinomas of the lung do not have great clinical value, in part because an architectural pattern is not always pure throughout a tumour. Histological types which can be recognized by light microscopy are acinar, papillary, bronchioloalveolar, and solid with mucus formation. Subtyping of 137 patients with radically resected stage 1 or 2 adenocarcinomas gave figures of 57% acinar, 12% papillary, 18% bronchioloalveolar and 14% solid with mucin production.[35] The bronchioloalveolar subtype is often recognized and designated by pathologists.

Peripheral adenocarcinomas frequently spread to regional lymph nodes and even a tumour smaller than 3 cm will show lymph node involvement in 50% of cases. There is a higher percentage of mediastinal lymph node metastases in adenocarcinomas compared with squamous carcinomas, and the lobar or hilar nodes in these patients may not be involved. Vascular invasion is also more common in adenocarcinomas and adenocarcinoma has a greater propensity for extrathoracic dissemination, especially to the brain. Among the histological subtypes, solid carcinoma with mucin production has the poorest prognosis.

Bronchioloalveolar adenocarcinoma

This variant of pulmonary adenocarcinoma has had many names since it was first described in 1876, but bronchioloalveolar adenocarcinoma has proved appropriate. The tumour is composed of columnar cells which spread using the distal air spaces for support. The architecture of the peripheral lung is thus preserved. Collagen deposition within the alveolar septae renders them rigid, and continued proliferation results in the formation of papillary fronds that protrude into alveoli or into larger cavities when the framework of the lung is disrupted. Many of the spaces become filled with mucinous secretion and it is typical to find small nests of tumour cells floating within these lakes of mucin, admixed with alveolar macrophages. Their presence in an aspirate or washing specimen will suggest the diagnosis.

The columnar cells vary in appearance. Some are uniform in height with clear cytoplasm which reflects the presence of mucin (Fig. 11.2), and the production of large amounts of mucin may obstruct airways and even cause electrolyte depletion from continued expectoration.[36] More commonly, the apical cytoplasm projects into the lumen above the level of the intercellular connec-

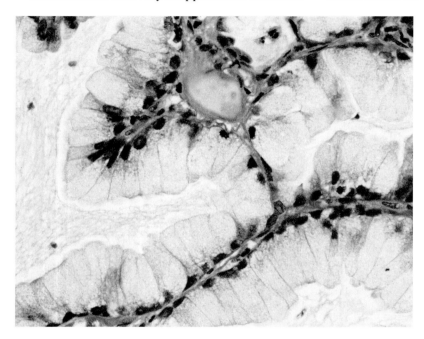

Fig. 11.2 Adenocarcinoma. Tall columnar cells with clear, mucin-filled cytoplasm rest on the alveolar framework.

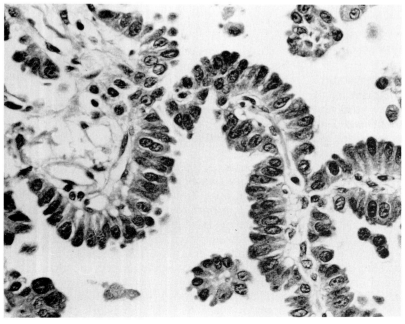

Fig. 11.3 Adenocarcinoma. The tumour cells coat thin septae and papillary projections and they protrude above the level of the intercellular attachments.

tions (Fig. 11.3), and these hobnail cells contain cytoplasmic granules similar to the protein secretory granules in non-ciliated bronchiolar (Clara) cells. Some mucin can often be found within the Clara-type cells. Infrequently, lamellar bodies like those in type 2 pneumocytes are present. The nuclear inclusions which are occasionally seen by light microscopy are aggregates of fine tubules formed by replication of the inner layer of the nuclear envelope. A bronchioloalveolar adenocarcinoma with a component of myoepithelial cells has been reported.[37]

Because of the admixture of bronchioloalveolar adenocarcinoma with classic adenocarcinoma in varying proportions, the consensus of diagnosis is relatively poor.[38] However, differentiation in bronchioloalveolar adenocarcinomas is predominantly towards cells of the distal airways, and the cells will generally mark for pulmonary surfactant apoproteins (Fig. 11.4). Both normal and proliferating type 2 pneumocytes and Clara cells synthesize, store and secrete similar or the same group of surfactant-associated proteins. In the normal lung, surfactant stabilizes the alveoli by reducing surface tension at the alveolar interface. Surfactant immunostaining or electron microscopy can be useful to distinguish a lung adenocarcinoma from mesothelioma or a metastatic adenocarcinoma.

Patients with bronchioloalveolar adenocarcinoma have a better survival rate than other adenocarcinoma subtypes,[1,35,39] but the mucin-rich tumours are more aggressive.[40]

Adenoid cystic carcinoma

Adenoid cystic carcinoma is more common in the trachea than within the lung. The histology is similar to this tumour in other locations, and tubular, cribriform and solid growth patterns may be seen. Perineurial invasion is often identified. Adenoid cystic carcinomas grow relatively slowly when compared with adenocarcinomas, but they are progressive tumours and the solid pattern is the most aggressive.[41]

Mucoepidermoid carcinoma

Mucoepidermoid carcinoma bronchial tumours are characterized by the copresence of epidermoid and mucus-secreting cells with an admixture of intermediate cells.[42] The histology is similar to the salivary tumours, and histological grading is relevant since low-grade tumours which are completely excised have the best prognosis.

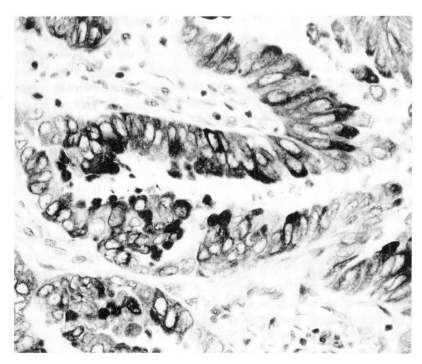

Fig. 11.4 Surfactant immunostaining in a bronchioloalveolar adenocarcinoma.

Adenosquamous carcinoma

In an adenosquamous carcinoma, zones of differentiated adenocarcinoma and squamous cell carcinoma are present. The two parts are often clearly separate, although they may impinge on one another. A small biopsy or a cytology preparation may only show one of the components. The frequency of this type of pulmonary carcinoma is less than 3% of all primary lung tumours.[43] Gender, race and age distributions are similar to other histological subtypes of lung carcinoma, and most patients are smokers.[44] The outcome of adenosquamous carcinoma is the same[45] or poorer[46] than that of pure adenocarcinomas and squamous cell carcinomas.

Sarcomatoid carcinoma

Spindle cell transformation occurs in a few squamous carcinomas and adenocarcinomas of the lung. A sarcomatoid tumour often presents at an advanced stage and lymph node metastasis is an important prognostic factor.[47] Immunostaining for keratin is useful to separate the tumour from a true sarcoma. Cagle *et al.*[48] reported two peripheral biphasic adenocarcinomas of the lung in which diffuse positive cytokeratin and negative vimentin immunostaining of the spindle-cell components supported an epithelial differentiation of the malignant spindle cells.

Undifferentiated large cell carcinoma

As pulmonary squamous carcinomas and adenocarcinomas dedifferentiate, their distinctive characteristics are progressively lost, and the cells approach a common nondescript morphology which constitutes undifferentiated large cell carcinoma. It is imprecise to refer to this tumour merely as large cell carcinoma since squamous and adenocarcinomas are also large cell carcinomas. Somewhere between 10 and 20% of pulmonary tumours are undifferentiated large cell carcinomas. The figure is closer to 10% when resection specimens are studied and mucin stains are performed and is even lower with electron microscopy. The tumours are often large, bulky, masses with extensive areas of necrosis or haemorrhage. They are typically aggressive and in many instances prove rapidly fatal.

Uniform round cells are closely packed in compact sheets devoid of any architectural pattern. A second form of undifferentiated large cell carcinoma is also referred to as giant cell carcinoma because it is composed of markedly pleomorphic cells with bizarre and multiple nuclei. The giant cell pattern may coexist with the compact type or form a component of adenocarcinoma or infrequently squamous carcinoma. There is probably no merit to recognizing a clear cell subtype: solid tumours with mucin fall within the adenocarcinoma category.

Neuroendocrine differentiation in non-small cell lung carcinomas

It has become recognized in the past few years that some non-small cell pulmonary carcinomas manifest neuroendocrine features (NE-NSCLC). They may be defined by immunoreactivity for neuroendocrine markers including L-dopa decarboxylase, chromogranin A, synaptophysin, neuron-specific enolase and Leu-7, and by their content of neurosecretory granules on electron microscopy.[49,50] There does not appear to be any morphological feature that will serve to identify these tumours in routine light microscopic sections, but more are adenocarcinomas than squamous carcinomas.[51] Both *in vitro* observations and preliminary clinical studies indicate a higher chemosensitivity for NE-NSCLC tumours than the non-neuroendocrine tumours, and NE-NSCLC may have a higher metastatic potential.[49] A statistically significant correlation between nodal status and neuroendocrine differentiation and between disease stage and neuroendocrine differentiation has been observed, although there was no correlation between neuroendocrine differentiation and survival.[52]

Small cell carcinoma

Approximately 20% of primary pulmonary tumours are small cell carcinomas. They are highly aggressive and have usually spread beyond the confines of the lung at the time of diagnosis. It is widely assumed that small cell lung carcinoma is a neuroendocrine tumour which forms a spectrum with carcinoid, but while neuroendocrine properties can be demonstrated in many small cell carcinomas, there are differences in epidemiology, morphology and behaviour with carcinoid tumours.

The majority of small cell carcinomas are

centrally located, correlating with the fact that cigarette smoking is a major aetiological factor. There used to be a high male predominance of as much as 10:1, but this is decreasing. Symptoms include cough, haemoptysis, neurological disorders, and endocrine hyperfunction which may be subclinical. Clinical staging is complicated by the fact that small foci of viable metastatic tumour cells may not be detectable radiologically.

In light microscopic sections, an architectural pattern usually cannot be seen in a small cell lung carcinoma, although the diffuse sheets of hyperchromatic cells with interspersed zones of necrosis are in their own way distinctive, and deposition of DNA from disrupted cells within vessel walls imparts an appearance which has been called haematoxyphil vascular change or basophilic encrustation. In some tumours the cells form circumscribed, interconnecting nests (Fig. 11.5), and rosettes and cords are seen in a few tumours. The tumour cells are about twice the size of small lymphocytes[53] and they are mostly ovoid and can be elongated. The shape of the nucleus is similar to that of the cell. The appearance of the nuclei is helpful. Their contour is smooth rather than jagged, and the chromatin is finely clumped but homogeneous in well-preserved material. Nucleoli are inconspicuous but there are exceptions and

some cells have prominent nucleoli. The size of the nuclei overlaps with some non-small cell carcinomas. A more useful feature is the high nuclear to cytoplasmic ratio;[54] the thin rim of cytoplasm is often barely perceptible in tissue sections. The puny cells are susceptible to artefactual distortion and crushing effect is common, but it must be remembered that a similar appearance can be produced by squashing in other tumours and in inflammatory disorders.

The diagnosis is made from cytological preparations in a high percentage of cases, provided adequate material is available. The tumour cells tend to be loosely cohesive and they may form small clusters or short cords. The cytoplasm is usually more visible in aspirate smears than in washing or brushing specimens or sputum. Nuclear features described for tissue sections are equally significant in the evaluation of cytological preparations.

The histology of a small cell carcinoma may change with therapy, altering towards a larger cell type, and comparable morphological alterations have been observed in cell lines. *In vitro* exposure of classic, non-adherent small cell carcinoma lines to bromodeoxyuridine results in a rapid cell-line dependent change to an adherent, non-small cell phenotype, and the change in morphology is

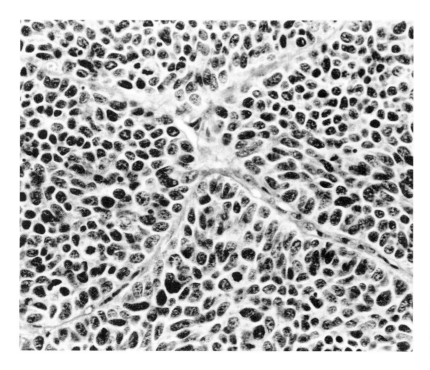

Fig. 11.5 Small cell carcinoma. The cells are forming well-defined nests separated by thin connective tissue partitions.

accompanied by a decreased expression of the amplified N-*myc* proto-oncogene.[55] These biological changes may in part be responsible for the acquisition of drug resistance.[56,57]

With immunoperoxidase methods, keratins can be demonstrated in most of the tumours. Small cell lung carcinomas can often be shown to possess neuroendocrine properties, but a battery of markers should be employed for this purpose. Neuron-specific enolase reactivity has been found in as many as 100% of cases, but this marker is also expressed in a variety of non-endocrine cells, so it should not be used as the sole evidence that a tumour is a small cell carcinoma. Chromogranin A is the most specific neuroendocrine marker but in small cell carcinoma of the lung immunoreactivity cannot always be detected. The reason might be the small size of the granules and their paucity or absence in many tumours. Localization of chromogranin A-mRNA could be more useful.[58] Leu-7 is inconsistently expressed and is present in only about 20% of small cell carcinomas, so its diagnostic value is limited. Staining for neurofilament protein is probably a rare occurrence. Synaptophysin is useful as a broad spectrum neuroendocrine marker though the results may be disappointing if they are performed on formalin-fixed, paraffin-embedded tissue. A variety of amines which can be used as endocrine markers are produced by small cell lung carcinoma. Bombesin/gastrin releasing polypeptide reactivity has been reported in 20 to 69% of small cell lung carcinomas, but it is also found in some non-small cell carcinomas. Other regulatory peptides such as ACTH and calcitonin are infrequently expressed.

Well-preserved material is desirable to evaluate the ultrastructural features of small cell lung carcinoma. The ovoid-to-elongated shape of the cells and nuclei can be appreciated with the electron microscope. The cells typically have scanty cytoplasm and sparse organelles. Cell surfaces are smooth and neighbouring cells are closely apposed (Fig. 11.6); dendritic processes are not common. Large numbers of secretory granules are rarely present and, more often, scattered granules are found in a few of the cells. The granules are round and uniformly small, of the order of 100–120 nm in diameter. Cell junctions range from tiny desmosomes to poorly defined densities.

Since variations in the size of the cells can be seen by light microscopy from one small cell carcinoma to another, there have been attempts to use these features to subtype the tumours. The results have been inconsistent and they have not been shown to have prognostic implications. In resection specimens, a spectrum of appearances is often seen, and the morphology of the cells can change with therapy. Earlier attempts to subtype the tumours as oat cell or lymphocyte-like and intermediate should therefore be abandoned. This was the recommendation of the pathology panel of the International Association for the Study of Lung Cancer.[59] The panel did, however, recommend that two variants be recognized. In mixed small cell/large cell carcinoma, small cells are intermingled with larger cells throughout the tumour. In the combined variant of small cell carcinoma, a component of differentiated non-small cell carcinoma is present. The biological behaviour of these variants is still being elucidated, but a study of central nervous system (CNS) metastases did not reveal significant differences among the subtypes.[60]

As many as two-thirds of patients with small cell lung carcinoma have clinically evident metastases at the time of diagnosis, and the frequency with which other metastatic foci are detected depends to some extent on the diligence with which staging is performed. Spread to bone is seen in about 35% of patients followed by liver (25%) lymph nodes (10%) and CNS (10%). The survival of patients has improved significantly with combination chemotherapy, though it is still dismal. Up to 90% of patients will experience an objective response to therapy, but only about 10% will be alive after 3 years without evidence of disease.

Carcinoid tumour

Approximately 2% of primary lung tumours are carcinoids. The average age at the time of diagnosis is around 50, or about 10 years younger than for patients with carcinoma, and they can occur in children. There is no predilection for either sex. The tumour is usually solitary and from 1 to 5 cm. The greatest concentration of pulmonary endocrine cells is in the lobar bronchi and this is where carcinoids have their highest incidence. The ratio of central to peripheral carcinoids is about 4:1, but bronchial carcinoids do not appear to be related to smoking, occupation or other environmental factors. One-quarter of the patients are asymptomatic,[61] especially when the lesion is small and peripheral, and endocrine symptoms are unusual. Although bronchial carcinoids are usually solitary

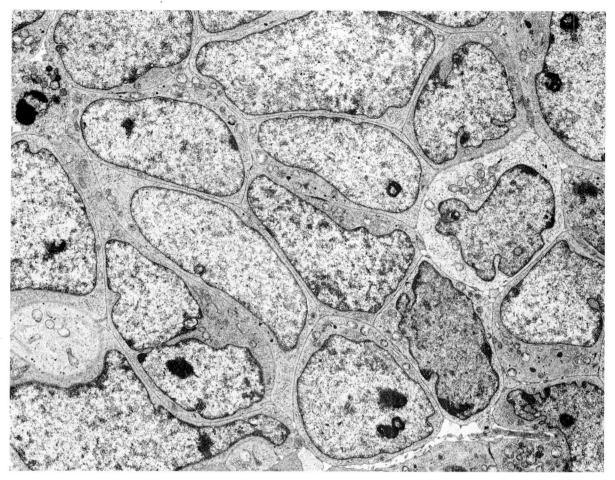

Fig. 11.6 Small cell carcinoma. Cells are ovoid rather than round and they are intimately apposed. The scanty cytoplasm contains very few organelles. The nuclear chromatin is finely and evenly clumped and nucleoli are inconspicuous. × 4800.

lesions, tumorlets may be multiple and must be distinguished from metastatic carcinoid tumour. Tumorlets are small groups of endocrine cells that lie adjacent to or surround bronchi, and they are observed with particular frequency in bronchiectasis.[62]

Many different patterns can be seen in lung carcinoids, but the commonest is nests of cells separated by delicate sheets of connective tissue containing slender vessels. Retraction of the cells from the stroma creates clefts which accentuate the architecture (Fig. 11.7). Other patterns are cords of cells which branch and curve, and the formation of small lumens. Even papillary projections occur and occasionally amyloid or mucin is found. The cells may be oncocytic, and melanin and psammoma

bodies are rare findings. A spindle cell pattern is common, particularly in peripheral tumours, and the cells may then attempt to form bundles (Fig. 11.8).

Cytological atypia in carcinoid tumours is of prognostic significance. From among 215 lung carcinoids, Arrigoni *et al*.[63] selected 23 atypical tumours which showed one or more of the following: nuclear pleomorphism, enlarged nucleoli, hyperchromatic nuclei, increased mitotic activity, abnormal nuclear to cytoplasmic ratio, increased cellularity, disorganized architecture and tumour necrosis. Sixteen of the 23 patients developed metastases whereas only 5.6% of typical tumours metastasized. Patients with atypical carcinoids are on average slightly older than those with typical

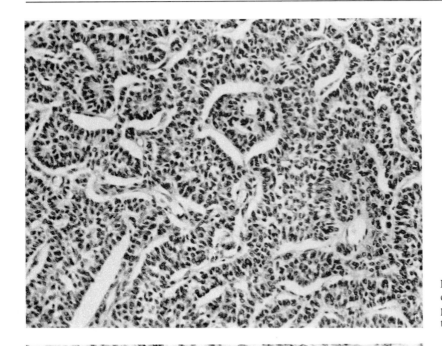

Fig. 11.7 Carcinoid. Retraction of cell nests from the connective tissue partitions accentuates the architectural pattern.

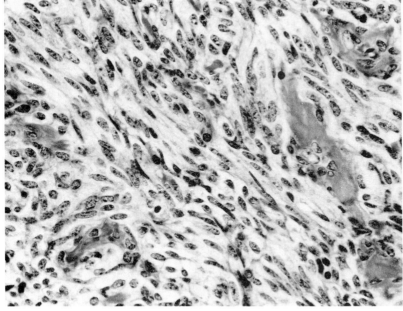

Fig. 11.8 Spindle cell carcinoid. The cells are forming crude fascicles.

tumours and the former are less likely to show endobronchial growth.[64] The spectrum of pulmonary neuroendocrine tumours does not readily fit into the traditional three-category classification scheme of typical carcinoid, atypical carcinoid and small cell lung carcinoma,[65] and the atypical tumours may also be called well-differentiated neuroendocrine carcinomas.[66]

Flow cytometry is of limited value in assessing the prognosis of lung carcinoids. Yousem et al.[67] did not find correlation of any histological variables with abnormal DNA content and they concluded that although abnormalities of DNA content, particularly aneuploidy, are common in atypical carcinoids, DNA ploidy cannot be used independently to assess malignant potential.

Cytology is not usually contributory since the tumour is as a rule covered by normal mucosa. Brushings and washings are often negative and a bronchoscopic biopsy may not reach the tumour. A fine needle aspiration biopsy can recover diagnostic material, and the frequently stripped cytoplasm (with resulting non-cohesive bare nuclei) coupled with the almost universal plexiform vascularity should allow an accurate cytological diagnosis.[68] The nuclear to cytoplasmic ratio is only slightly higher than in metaplastic cells. Nuclei are round and uniform, and the chromatin is more clumped than in a small cell carcinoma. Atypia in a carcinoid can often be recognized in cytological preparations. Compared with the typical tumours, the cell and nuclear size and shape are variable, there is less cytoplasm, nucleoli are prominent, the chromatin tends to be coarsely clumped, mitoses are frequent, and a background of necrosis may be seen.

Most carcinoids express epithelial and neuroendocrine markers as well as some specific endocrine cell products. The cells are often immunoreactive for chromogranin[69] (Fig. 11.9) and Leu-7. Bombesin/GRP reactivity was observed in 26 of 33 tumours (67%) and was found to be the most commonly expressed neuropeptide in lung carcinoids. Immunoreactivity for the *C*-terminal flanking peptide of human bombesin has also been reported in lung carcinoids, but its expression in these tumours is significantly lower than that of bombesin, being about 12% in one series.[9]

Every lung carcinoid contains secretory granules (Fig. 11.10) and they are usually round and uniform, though the size varies in different tumours. A range of 93 to 383 nm was found in a morphometric study.[54] Granules are occasionally angular rather than spherical, and they are sparse in a few tumours which have considerable numbers of intermediate filaments and in the mitochondrion-rich oncocytic tumours.

Survival figures for patients with carcinoid tumours of the lung are more meaningful when the typical and atypical tumours are separated. Patients with localized typical carcinoids have a good prognosis after resection, with a 95% five-year and 93% ten-year survival for pathological stage 1 disease.[70] In patients with the atypical form, five-year survival figures of from 25 to 41% have been recorded.[61,64] The most important factor influencing the prognosis is mediastinal lymph node involvement,[71] and atypical carcinoid neoplasms need more extended resection and lymphadenectomy.[72]

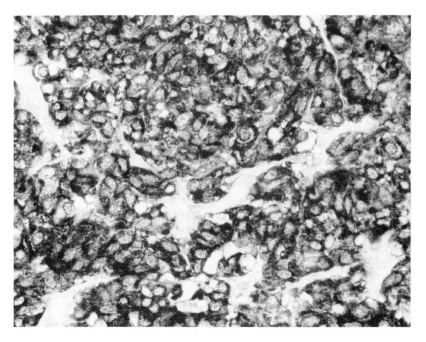

Fig. 11.9 Intense staining of carcinoid cells for chromogranin.

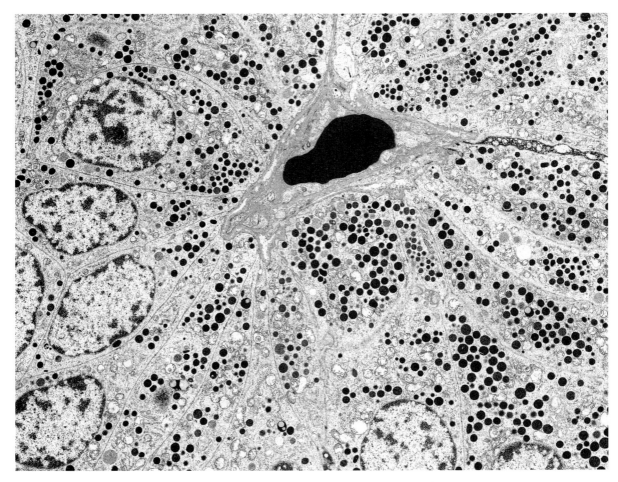

Fig. 11.10 Carcinoid. The cells contain large numbers of secretory granules. × 4800.

Uncommon pulmonary tumours

Only about 1% of primary pulmonary tumours fall within the uncommon bracket so they are not often encountered by the general surgical pathologist. Some are easily recognized from their histology but others may simulate a primary lung carcinoma. The uncommon lung tumours are described and illustrated in the comprehensive review by Dail.[73]

Hamartoma

A solitary hamartoma is the most common benign tumour of the lung. The size varies between 1 and 8 cm. Most are beyond the reach of the broncho-scope, but a needle aspiration biopsy will often provide the diagnosis. Only about one-third of patients have symptoms. Some hamartomas are multiple and may be accompanied by other developmental abnormalities and benign tumours.[74]

A hamartoma is composed of a mixture of cartilage, fat cells and undifferentiated mesenchymal cells, and there are usually entrapped slender clefts lined by distal respiratory cells, particularly type 2 pneumocytes. Some hamartomas are cystic, and calcification is sometimes present. Tumour growth may be detected if the patient is followed radiologically, but growth is slow and surgical resection is only necessary when expansion is recorded in young or middle-aged patients, or if symptoms are present.[75]

Sclerosing haemangioma

The sharply defined nodule is usually peripheral, solitary and under 5 cm. It is made up of sheets of round-to-ovoid epithelial cells which become cuboidal where they cover papillary projections (Fig. 11.11), and there are areas of fibrosis and vascular channels. The relative proportions of the components vary from one tumour to another and within the same tumour. Immunoperoxidase studies have shown that this unusual lesion is consistently negative for endothelial markers, but the cells stain for keratin and surfactant. Most patients are female, though they can be of any age. Once the diagnosis has been established, a limited resection is indicated.[76]

Epithelioid haemangioendothelioma

This neoplasm was initially named intravascular bronchioloalveolar tumour but it is now viewed as an epithelioid haemangioendothelioma, similar in its histology and immunoreactivity to the same tumour in the liver and soft tissues. Approximately 80% occur in women, including young adults. The nodules are usually smaller than 3 cm but they can be multiple. The cellular phase of the tumour is the advancing edge which pushes into neighbouring alveoli (Fig. 11.12). Deeper parts of the nodule are acellular and fibrosed. The cytoplasm of the tumour cells is eosinophilic and often vacuolated, and it stains for endothelial markers. Although the nodules are initially small and grow slowly, about half of the patients die from their tumour. Spread beyond the lungs is uncommon but the tumour can extend to the pleura and hilar lymph nodes may be involved.

Pulmonary blastoma

Pulmonary blastomas are mixed epithelial and mesenchymal tumours that recapitulate the structure of the developing lung at 10–16 weeks' gestation.[77] The sexes are equally affected and at least one-third of patients are asymptomatic. Microscopically, the tumours can be divided into two classes, those composed solely of malignant glands of embryonal appearance (well-differentiated fetal adenocarcinomas) and those with a biphasic appearance.[78] The malignant epithelium often contains chromogranin-positive cells, and vimentin, actin, and less frequently desmin and myoglobin, can be found in malignant stromal cells. The clinical course is not readily predicted from the histological appearance but the prognosis is on the whole poor and it is worse for the biphasic tumours.

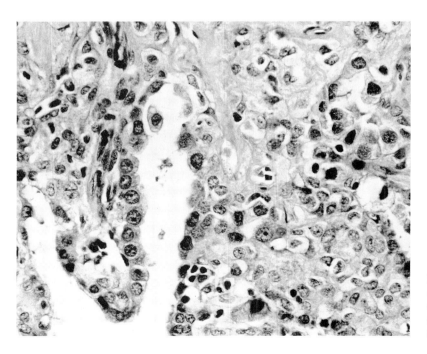

Fig. 11.11 Sclerosing haemangioma. A single layer of cuboidal cells covers a short papillary projection and similar cells cluster within the stroma.

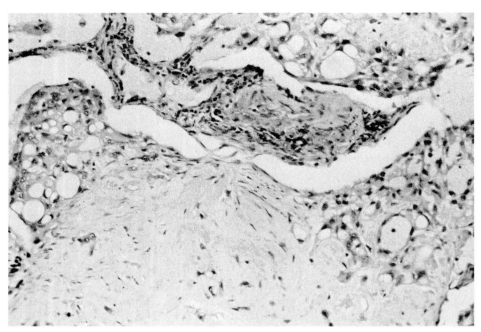

Fig. 11.12 Epithelioid haemangioendothelioma. Tumour cells, some with vacuolated cytoplasm, are present at the advancing edge of the tumour where it is pushing into alveolar lumens.

Benign clear cell tumour

This rare lesion is named descriptively because the cell of origin is not known. The large amount of cytoplasmic glycogen prompted the term 'sugar tumour'. It seems to be invariably benign and resection is curative, but it can be mistaken for a clear cell carcinoma of the lung or metastatic renal cell carcinoma. The bland cytological features should be sufficient to rule out a primary lung carcinoma, while atypia and mitotic activity are found in a metastatic renal cell carcinoma and the vascular pattern is coarser. If there is still doubt, electron microscopy will show glandular differentiation and cytoplasmic lipid in renal cell carcinoma.

Soft tissue tumours

A variety of types arise in the lung but they are all uncommon. Leiomyoma is the most frequent benign soft tissue tumour. A submucosal bronchial lipoma can produce obstructive effects by endobronchial growth. Haemangioma and benign peripheral nerve sheath tumours are not often found within the lung, but granular cell tumours are more common and they can be multiple and on occasion coexist with an underlying carcinoma. A pulmonary chondroma forms part of Carney's triad. The cellular areas of a localized fibrous tumour of the pleura are formed by fibroblasts in interlacing fascicles or a storiform arrangement, and the location will usually suggest the diagnosis.

Sarcomas arising in the lung are almost always solitary but they can be large and it may not be possible to distinguish one from a malignant spindle cell mesothelioma. In contrast, metastatic sarcomas within the lung generally form multiple small nodules. Pulmonary Kaposi's sarcoma is occasionally the presenting disorder in a patient with the acquired immunodeficiency syndrome (AIDS).

References

1. Sridhar KS, Raub W Jr, Duncan RC *et al*. Lung carcinoma in 1,336 patients. *Am J Clin Oncol* 1991; **14**, 496–508.
2. Mountain CF, Lukeman JM, Hammar SP *et al*. Lung cancer classification: the relationship of disease extent and cell type to survival in a clinical trials population. *J Surg Oncol* 1987; **35**, 147.

3. Mackay B, Lukeman JM, Ordonez NG. *Tumours of The Lung*. Philadelphia: WB Saunders, 1991, p. 100.

4. Auerbach O, Garfinkel L. The changing pattern of lung carcinoma. *Cancer* 1991; **68**, 1973–1977.

5. Mackay B, Lukeman JM, Ordonez NG: *Tumours of the Lung*. Philadelphia: WB Saunders, 1991, p. 105.

6. Mountain CF. Prognostic implications of the International Staging System for Lung Cancer. *Semin Oncol* 1988; **15**, 236.

7. Naruke T, Goya T, Tsuchiyar *et al*. Prognosis and survival in resected lung carcinoma based on the new International Staging System. *J Thorac Cardiovasc Surg* 1988; **96**, 440.

8. Byhardt RW, Hartz A, Libnoch JA *et al*. Prognostic influence of TNM staging and LDH levels in small cell carcinoma of the lung. *Int J Rad Oncol Biol Phys* 1986; **12**, 771.

9. Sheppard MN. Immunohistochemistry and *in situ* hybridization in the diagnosis and prognosis of lung cancer. *Lung Cancer* 1993; **9**, 119–134.

10. McDonagh D, Vollmer RT, Shelburne JD. Intercellular junctions and tumor behavior in lung cancer. *Mod Pathol* 1991; **4**, 436–440.

11. Soomro I, Patel, Whimster WF. Distribution and estimation of nucleolar organizer regions in various human lung tumours. *Pathol Res Practice* 1991; **187**, 68–72.

12. Sahin AA, Ro JY, el-Naggar AK *et al*. Flow cytometric analysis of the DNA content of non-small cell lung cancer. Ploidy as a significant prognostic indicator in squamous cell carcinoma of the lung. *Cancer* 1990; **65**, 530–537.

13. Sara A, el-Naggar AK. Intramoral DNA content variability. A study of non-small cell lung cancer. *Am J Clin Pathol* 1991; **96**, 311–317.

14. Tungekar MF, Gatter KC, Dunnill MS, Mason DY. Ki-67 immunostaining and survival in operable lung cancer. *Histopathol* 1991; **19**, 545–550.

15. Rodenhuis S, Groot P, Salomons G, Slebos R. Ras Oncogene activation and the expression of ras-related genes in human lung cancer. *Lung Cancer* 1993; **9**, 59–68.

16. Barbareschi M, Girlando S, Mauri FA. Tumour suppressor gene products, proliferation, and differentiation markers in lung neuroendocrine neoplasms. *J Pathol* 1992; **166**, 343–350.

17. Hirsch FR, Rygaard K. The role of the pathologist in the clinical management of lung cancer in the 90's. *Lung Cancer* 1993; **9**, 111–118.

18. Rabbitts P, Daly M, Douglas J, Ganly P, Heppell-Parton A, Sundarsen V. Deletion mapping of chromosome 3 in lung tumours. *Lung Cancer* 1993; **9**, 69–74.

19. Miyamoto T, Morita S, Tsunemoto H *et al*. Acute squamous metaplasia of the whole lung after combined radiation and chemotherapy in advanced lung cancer. *Gan No Rinsho* 1987; **33**, 89.

20. Tsuchiya E, Kitagawa T, Oh S *et al*. Incidence of squamous metaplasia in large bronchi of Japanese lungs: relation to pulmonary carcinomas of various subtypes. *Jpn J Cancer Res* 1987; **78**, 559.

21. Risse EK, Vooijs GP, van Hof MA. Diagnostic significance of 'severe dysplasia' in sputum cytology. *Acta Cytol* 1988; **32**, 629.

22. Weng SY, Tsuchiya E, Kasuga T, Sugano H. Incidence of atypical bronchioloalveolar cell hyperplasia of the lung: relation to histological subtypes of lung cancer. *Virchows Arch A: Pathol Anat Histopathol* 1992; **420**, 463–471.

23. Shimosato Y, Naguchi M, Matsumo Y. Adenocarcinoma of the lung: its development and malignant progression. *Lung Cancer* 1993; **9**, 99–108.

24. Kobzik L. Benign pulmonary lesions that may be misdiagnosed as malignant. *Semin Diagn Pathol* 1990; **7**, 129–138.

25. Kragel PJ, Devaney KO, Meth BM *et al*. Mucinous cystadenoma of the lung. A report of two cases with immunohistochemical and ultrastructural analysis. *Arch Pathol Lab Med* 1990; **114**, 1053–1056.

26. Davison AM, Lowe JW, Da Costa P. Adenocarcinoma arising in a mucinous cystadenoma of the lung. *Thorax* 1992; **47**, 129–130.

27. Miller RR. Bronchioloalveolar cell adenomas. *Am J Surg Pathol* 1990; **14**, 904–912.

28. Jedrychowski W, Becher H, Wahrendort J, Basa-Cierpialek Z, Gomola K. Effect of tobacco smoking on various histological types of lung cancer. *J Cancer Res Clin Oncol* 1992; **118**, 276–282.

29. Tomashefski JF Jr, Connors AF Jr, Rosenthal ES, Hsiue IL. Peripheral vs central squamous cell carcinoma of the lung. A comparison of clinical features, histopathology, and survival. *Arc Pathol Lab Med* 1990; **114**, 468–478.

30. Zusman-Harach SB, Harach HR, Gibbs AR. Cytological features of non-small cell carcinomas of the lung in fine needle aspirates. *J Clin Pathol* 1991; **44**, 997–1002.

31. Brambilla E, Moro D, Veale D *et al*. Basal cell (basaloid) carcinoma of the lung: a new morphologic and phenotypic entity with separate prognostic significance. *Hum Pathol* 1992; **23**, 993–1003.

32. Miller B, Montgomery C, Watne AL *et al*. Lymphoepithelioma-like carcinoma of the lung. *J Surg Oncol* 1991; **48**, 62–68.

33. Sano T, Naruke T, Kondo H *et al*. Prognosis for resected lung cancer patients with tumours greater than ten centimeters in diameter. *Jpn J Clin Oncol* 1990; **20**, 369–373.

34. Higashiyama M, Doi O, Kodama K, Yokouchi H, Tateishi R. Cystic mucinous adenocarcinoma of the lung. Two cases of cystic variant of mucus-producing lung adenocarcinoma. *Chest* 1992; **101**, 763–766.

35. Sorensen JB, Olsen JE. Prognostic implications of histopathologic subtyping in patients with surgically

treated stage 1 or 2 adenocarcinoma of the lung. *J Thorac Cardiovasc Surg* 1989; **97**, 245.

36. Homma H, Kiras Takashi Y *et al*. A case of alveolar cell carcinoma accompanied by fluid and electrolyte depletion through production of voluminous amounts of lung liquid. *Am Rev Respir Dis* 1975; **111**, 857.

37. Dekmezian R, Ordonez NG, Mackay B. Bronchioloalveolar adenocarcinoma with myoepithelial cells. *Cancer* 1991; **67**, 2356-2360.

38. Elson CE, Moore SP, Johnston WW *et al*. Morphologic and immunocytochemical studies of bronchioalveolar carcinoma at Duke University Medical Center, 1968-1986. *Ann Quant Cytol Histol* 1989; **11**, 261.

39. Sorensen JB, Hirsch FR, Olsen J. The prognostic implication of histopathologic subtyping of pulmonary adenocarcinoma according to the classification of the World Health Organization. *Cancer* 1988; **62**, 361.

40. Manning JT, Spjut HJ, Tschen JA. Bronchioalveolar carcinoma: the significance of two histopathologic types. *Cancer* 1984; **54**, 525.

41. Ishida T, Yano T, Sugimachi K. Clinical applications of the pathological properties of small cell carcinoma, large cell carcinoma, and adenoid cystic carcinoma of the lung. *Semin Surg Oncol* 1990; **6**, 53-63.

42. Lamotte F, Meurice JC, Dore P *et al*. Mucoepidermoid bronchial carcinoma. Review of the literature apropos of a new case. *Rev Pneumonol Clin* 1992; **48**, 29-32.

43. Naunheim KS, Taylor JR, Skosey C *et al*. Adenosquamous lung carcinoma: clinical characteristics, treatment, and prognosis. *Ann Thorac Surg* 1987; **44**, 462.

44. Sridhar KS, Raub WA Jr, Duncan RC, Hilsenbeck S. The increasing recognition of adenosquamous lung carcinoma (1977-1986). *Am J Clin Oncol* 1992; **15**, 356-362.

45. Ishida T, Kaneko S, Yokoyama H *et al*. Adenosquamous carcinoma of the lung. Clinicopathologic and immunohistochemical features. *Am J Clin Pathol* 1992; **97**, 678-685.

46. Takamori S, Noguchi M, Morinaga S *et al*. Clinicopathologic characteristics of adenosquamous carcinoma of the lung. *Cancer* 1991; **67**, 649-654.

47. Ro JY, Chen JL, Lee JS *et al*. Sarcomatoid carcinoma of the lung. Immunohistochemical and ultrastructural studies of 14 cases. *Cancer* 1992; **69**, 376-386.

48. Cagle PT, Alpert LC, Carmona PA. Peripheral biphasic adenocarcinoma of the lung: light microscopic and immunohistochemical findings. *Human Pathol* 1992; **23**, 197-200.

49. Hirsch FR, Skov BG. Neuroendocrine characteristics in bronchogenic adenocarcinoma and its clinical relevance. *Lung Cancer* 1993; **9**, 89-96.

50. Wick MR, Berg LC, Hertz MI. Large cell carcinoma of the lung with neuroendocrine differentiation. A comparison with large cell 'undifferentiated' pulmonary tumours. *Am J Clin Pathol* 1992; **97**, 796-805.

51. Vischer DW, Zarbo RJ, Trojanowski JQ *et al*. Neuroendocrine differentiation in poorly differentiated lung carcinomas: a light microscopic and immunohistologic study. *Mod Pathol* 1990; **3**, 508-512.

52. Sundaresan V, Reeve JG, Stenning S, Stewart S, Bleehen NM. Neuroendocrine differentiation and clinical behaviour in non-small cell lung tumours. *Br J Cancer* 1991; **64**, 333-338.

53. Lee TK, Esinhart JD, Blackburn LD, Silverman JF. The size of small cell lung carcinoma cells. Ratio to lymphocytes and correlation with specimen size and crush artifact. *Ann Quan Cytol Histol* 1992; **14**, 32-34.

54. Mackay B, Ordonez NG, Bennington JL *et al*. Ultrastructural and morphometric features of poorly differentiated and undifferentiated lung tumours. *Ultrastruct Pathol* 1989; **13**, 569.

55. McGarry RC, Feyles V, Tuff A *et al*. Induced morphological changes in human small cell lung carcinoma cells. *Cancer Lett* 1991; **61**, 67-74.

56. Mabry M, Nelkin BD, Falco JP. Transitions between lung cancer phenotypes – implications for tumour progression. *Cancer Cells* 1991; **3**, 53-58.

57. Stahel RA. Morphology, surface antigens, staging, and prognostic factors of small cell lung cancer. *Curr Opin Oncol* 1992; **4**, 308-312.

58. Hamid QA, Corrin B, Sheppard MN *et al*. Expression of chromogranin A mRNA in small cell carcinoma of the lung. *J Pathol* 1991; **163A**, 293-297.

59. Hirsch FR, Matthews MJ, Aisner S *et al*. Histopathologic classification of small cell lung cancer: changing concepts and terminology. *Cancer* 1988; **62**, 974.

60. Twijnstra A, Thunnissen FB, Lassouw G *et al*. The role of the histologic subclassification of tumour cells in patients with small cell carcinoma of the lung and central nervous system metastases. *Cancer* 1990; **65**, 1812-1815.

61. Stamatis G, Freitag L. Greschuchna D. Limited and radical resection for tracheal and bronchopulmonary carcinoid tumour. Report on 227 cases. *Eur J Cardiothorac Surg* 1990; **4**, 527-532.

62. Klinke F, Bosse U, Hofler H. The tumorlet carcinoid in bronchiectasis-changed lungs. An example of a multifocal endocrine tumour. *Pneumologie* 1990; **44**, 607-609.

63. Arrigoni MG, Woolner LB, Bernatz PE. Atypical carcinoid tumours of the lung. *J Thorac Cardiovasc Surg* 1972; **64**, 413.

64. Akiba T, Naruke T, Kondo H. Carcinoid tumour of the lung: clinicopathological study of 32 cases. *Jpn J Clin Oncol* 1992; **22**, 92-95.

65. Travis WD, Linnoila RI, Tsokos MG *et al.* Neuro-endocrine tumours of the lung with proposed criteria for large-cell neuroendocrine carcinoma. An ultrastructural, immunohistochemical, and flow cytometric study of 35 cases. *Am J Surg Pathol* 1991; **5**, 529–553.

66. Bonato M, Cerati M, Pagan A. Differential diagnostic patterns of lung neuroendocrine tumours. A clinico-pathological and immunohistochemical study of 122 cases. *Virchows Arch A: Pathol Anat Histopathol* 1992; **420**, 201–211.

67. Yousem SA, Wick MR, Randhawa P, Manivel JC. Pulmonary blastoma. An immunohistochemical analysis with comparison with fetal lung in its pseudoglandular stage. *Am J Clin Pathol* 1990; **93**, 167–175.

68. Anderson C, Ludwig ME, O'Donnell M, Garcia N. Fine needle aspiration cytology of pulmonary carcinoid tumors. *Acta Cytol* 1990; **34**, 505–510.

69. Totsch M, Muller LC, Hittmair A *et al.* Immuno-histochemical demonstration of chromogranins A and B in neuroendocrine tumours of the lung. *Hum Pathol* 1992; **23**, 312–316.

70. Harpole DH Jr, Feldman JM, Buchanan S *et al.* Bronchial carcinoid tumours: a retrospective analysis of 126 patients. *Ann Thorac Surg* 1992; **54**. 50–54.

71. Francioni F, Rendina EA, Venuta F. Low grade neuroendocrine tumours of the lung (bronchial carcinoids) – 25 years' experience. *Eur J Cardio-thorac Surg* 1990; **4**, 472–476.

72. DiGiorgio A, Tocchi A, Puntillo G *et al.* Tracheo-bronchial carcinoids: current therapeutic trends. *Ann Ital Chirurgica* 1990; **61**, 405–409.

73. Dail DH. Uncommon tumours. In: Dail DH, Hammar SP (eds) *Pulmonary Pathology*, ch. 29. New York: Springer-Verlag, 1988.

74. Gabrail NY, Sara BY. Pulmonary hamartoma syndrome. *Chest* 1990; 962–965.

75. Hansen CP, Holtveg H, Francis D *et al.* Pulmonary hamartoma. *J Thorac Cardiovasc Surg* 1992; **104**, 674–678.

76. Sugio K, Yokoyama H, Kaneko S *et al.* Sclerosing hemangioma of the lung: radiographic and pathological study. *Ann Thorac Surg* 1992; **53**, 295–300.

77. Yousem SA, Taylor SR. Typical and atypical carcinoid tumors of lung: a clinicopathologic and DNA analysis of 20 tumors. *Modern Pathol* 1990; **3**, 502–507.

78. Koss MN, Hochholzer L, O'Leary T. Pulmonary blastomas. *Cancer* 1991; **67**, 2368–2381.

12

Pulmonary lymphoma

P G Isaacson

Introduction

Primary lymphoma of the lung is a rare tumour. Most are low-grade B cell tumours and, as a result of their unusual histological appearance and excellent prognosis, in the past many cases have been mislabelled as 'pseudolymphomas'.[1] For these reasons their incidence is difficult to estimate. Where lymphoid infiltrates have been recognized as lymphomatous a number of terms have been used to describe them including 'lymphocytic lymphoma', 'lymphoplasmacytic lymphoma or immunocytoma', 'Waldenstrom's macroglobulinaemia', 'intermediate, mantle zone or centrocytic lymphoma' and 'follicle centre cell lymphoma'.[2] The formulation of the mucosa-associated lymphoid tissue (MALT) lymphoma concept,[3,4] together with the development of immunocytochemical and molecular techniques to detect monoclonality, has largely resolved the confusion surrounding primary low grade pulmonary lymphomas. Almost all are examples of MALT-type lymphoma.[5-8] Low-grade B cell pulmonary lymphomas of other types, including lymphocytic lymphoma (CLL), follicular (CB/CC) lymphoma and mantle cell (centrocytic) lymphoma, although occasionally described, are extremely uncommon and will not be discussed here. High-grade lymphomas of the lung are less common than low-grade MALT lymphomas;[8] a significant proportion of these result from transformation of low-grade MALT lymphoma while others are T cell or possibly NK cell lesions.[9,10] Primary pulmonary Hodgkin's disease, if it occurs at all, is extremely rare.

Pulmonary lymphoid tissue

In 1973, Bienenstock described MALT in the lung of rabbits which he termed bronchus-associated lymphoid tissue or BALT.[11] Subsequently others have not confirmed the consistent presence of BALT in either animals or humans.[12] The inconsistent presence of lymphoid aggregates in animal lungs seems to be the result of infection and none is seen in fetuses or immediately after birth.[12] Lymphoid tissue is extremely difficult to find in normal human lungs, but peribronchiolar lymphoid aggregates bearing a close similarity to Peyer's patches (i.e. MALT) have been observed in fetuses and neonates where there is evidence of infection.[13] In the condition known as follicular bronchiolitis, which occurs in the adult lung,[14] often in association with Sjögren's syndrome or other autoimmune diseases, well-developed Peyer's patch-like structures with a distinct lymphoepithelium are present (Figs 12.1 and 12.2). Under certain conditions, therefore, the lung can acquire MALT. Whether pulmonary MALT lymphomas arise in a background of conditions like follicular bronchiolitis is not known, but the association of some low-grade B cell pulmonary MALT lymphomas with Sjögren's syndrome[15] suggests that this may be the case.

Pulmonary lymphoma of MALT type

Clinical presentation

Most lymphomas of this type occur between the sixth and seventh decade with an equal sex inci-

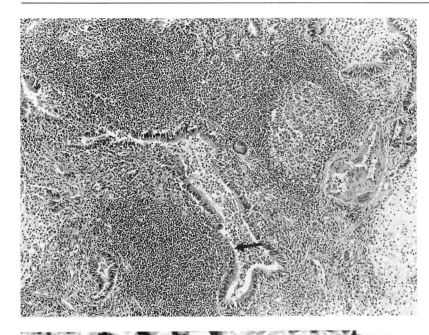

Fig. 12.1 An example of follicular bronchiolitis. The bronchiole is surrounded by a lymphoid infiltrate containing B cell follicles and aggregates of lymphocytes (B cells) are present within the eipthelium (arrow).

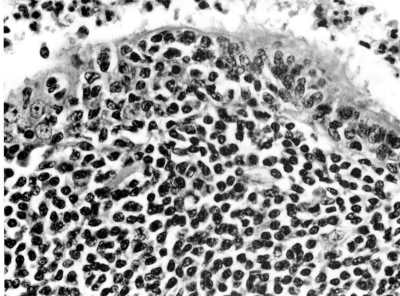

Fig. 12.2 Higher magnification of arrowed area in Fig. 12.1 showing the cluster of intraepithelial B cells.

dence. They commonly present as a solitary well-defined nodule in the lung found on routine X-ray examination (Fig. 12.3). Less commonly there may be multiple nodules and a pleural effusion is present. The majority of patients are in clinical stage 1 with fewer cases in stage 2 as shown by the presence of hilar lymphadenopathy. Most patients have few, if any, symptoms but cough, chest pain

and dyspnoea may all occur and a minority of patients complain of 'B' symptoms such as fever, weight loss and night sweats.[8]

In a small number of cases there is a history of Sjögren's syndrome[15] and in these cases it is possible that the lung lymphoma is secondary to spread from an undiagnosed MALT lymphoma of the salivary gland. Similarly, a past history of

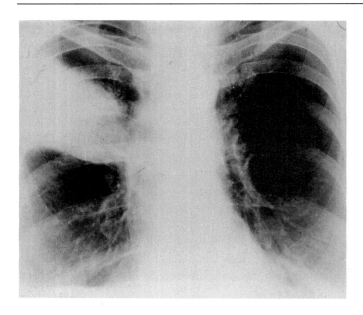

Fig. 12.3   X-Ray appearance of primary pulmonary lymphoma showing an infiltrate in the right upper zone.

conjunctival and lacrimal gland lymphoma has occasionally been documented.[5]

Histopathology

Low-grade MALT lymphoma
In most cases there is diffuse infiltration of the lung by small lymphoid cells containing a variable number of reactive B cell follicles (Fig. 12.4). The reactive follicles are surrounding by the lymphomatous infiltrate which invades bronchiolar epithelium to form lymphoepithelial lesions similar to those seen in other MALT lymphomas (Fig. 12.5).[16] At the edges of the lesion isolated follicles

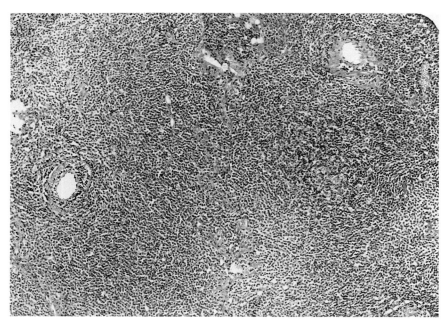

Fig. 12.4   Low-grade B cell pulmonary lymphoma of MALT type showing a diffuse infiltrate within which there are poorly defined follicles.

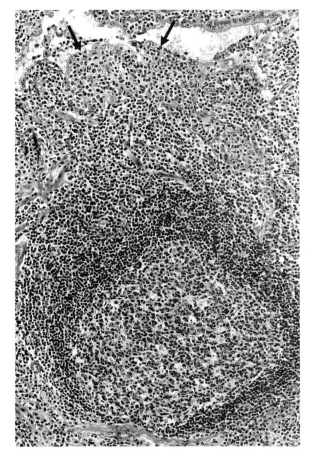

Fig. 12.5 Low-grade B cell pulmonary lymphoma of MALT type. A reactive B cell follicle, below, is surrounded by the lymphomatous infiltrate which invades bronchiolar epithelium to form lymphoepithelial lesions (arrows).

and an interstitial lymphoid infiltrate may be present. This can be mistaken for lymphocytic interstitial pneumonia (Fig. 12.6) but the typical characteristics of MALT lymphoma are present; the marginal zone distribution of the neoplastic cells around individual reactive B cell follicles is easily seen, as is their relationship to a 'dome' of bronchiolar epithelium with which they form lymphoepithelial lesions (Fig. 12.7). These small foci of lymphoma resemble the acquired MALT of follicular bronchiolitis (Fig. 12.2). In some cases the lymphoma has a markedly nodular pattern as if there has been incomplete coalescence of the small lymphomatous foci described above. This may result in an appearance that may be mistaken for follicular (CB/CC) lymphoma (Fig. 12.8).

Low-grade B cell MALT lymphomas of the lung show the same degree of cytological variation as other MALT lymphomas. The cells may resemble small lymphocytes (Fig. 12.9), show typical centrocyte-like morphology (Fig. 12.10) or resemble so-called monocytoid B cells (Fig. 12.11). Scattered transformed blasts are common and plasma cell differentiation is often present; this varies from the presence of small numbers of plasma cells or plasmacytoid cells with Dutcher bodies to large cohesive sheets of mature plasma cells (Fig. 12.12).

Follicular colonization[17] may occur in pulmonary MALT lymphomas. Reactive follicles are either partially or completely replaced by neoplastic cells which frequently show 'activation' (i.e. slight enlargement and increased mitotic activity or frank blast transformation (type 2 pattern)). Prominent follicular colonization results in a close resemblance to follicular (CB/CC) lymphoma.

High-grade B cell lymphoma
Transformation of low-grade MALT lymphoma to a high-grade lymphoma may occur and may be observed in either the same or subsequent biopsies.[8] The transformed blasts are not cytologically distinctive and may resemble centroblasts, immunoblasts or, more commonly, neither of these two classical cell types. Where no low-grade element is present it is impossible to be certain whether the lymphoma is of MALT type or not.

Immunohistochemistry

The immunophenotype of pulmonary MALT lymphoma is the same as that of other MALT lymphomas. The neoplastic CD20-positive B cells, which usually synthesize IgM, show light chain restriction which contrasts with the polytypic B cell follicles (Fig. 12.13). In cryostat sections the tumour is CD5-negative, CD10-negative[18] and positive with the marginal zone cell antibody UCL4D12.[19] Proliferation cell markers such as Ki-67 show that only scattered transformed blast cells are in cycle. There is usually a dense concentration of T cells.

Antibodies to cytokeratin enhance the demonstration of lymphoepithelial lesions (Fig. 12.14) while antibodies to follicular dendritic cells show evidence of numerous follicles, some of which have been overrun by the tumour and are not always apparent in routinely stained sections (Fig. 12.15). The phenomenon of follicular colonization is also more easily demonstrated using immunohistochemistry.

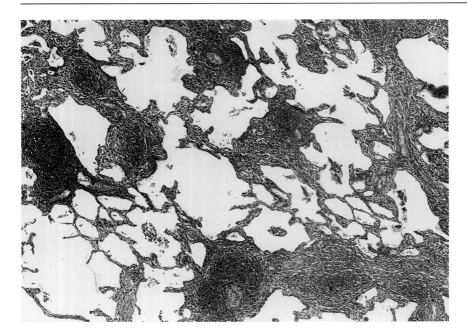

Fig. 12.6 The margins of a pulmonary MALT lymphoma showing an interstitial lymphoid infiltrate with reactive B cell follicles.

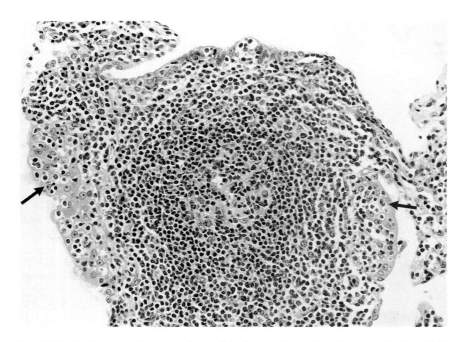

Fig. 12.7 Higher magnification of one of the lymphoid nodules illustrated in Fig. 12.6. A central reactive follicle is surrounded by lymphomatous cells which form lymphoepithelial lesions with bronchiolar epithelium (arrows).

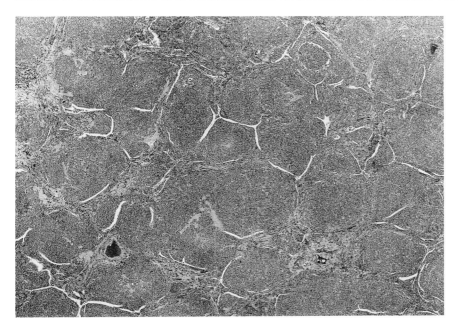

Fig. 12.8   Low-grade B cell pulmonary MALT lymphoma with a marked nodular pattern caused by the presence of numerous reactive follicles.

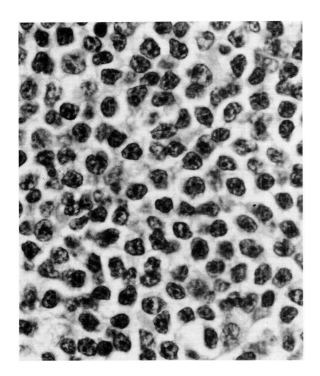

Fig. 12.9   The cells in this low-grade B cell pulmonary MALT lymphoma resemble small lymphocytes.

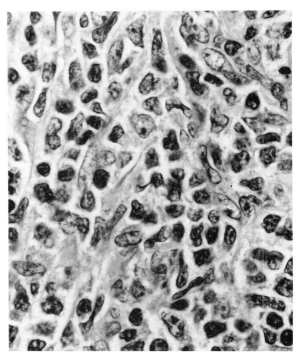

Fig. 12.10   The typical centrocyte-like appearance of the cells in low-grade B cell pulmonary MALT lymphoma.

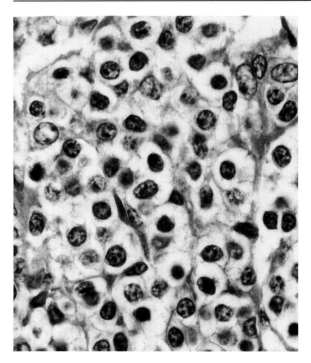

Fig. 12.11   Low-grade B cell pulmonary MALT lymphoma in which the cells resemble monocytoid B cells.

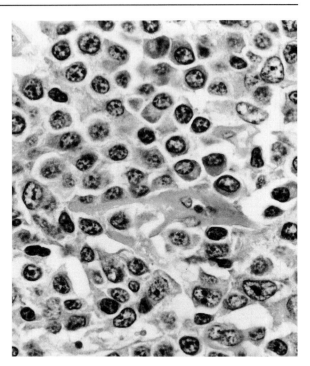

Fig. 12.12   Low-grade B cell pulmonary MALT lymphoma in which there is marked plasmacytic differentiation seen best at top of the illustration.

Molecular genetics

Demonstration of both immunoglobulin light chain restriction and gene rearrangement have been instrumental in defining low-grade MALT type lymphomas of the lung, many of which were previously called 'pseudolymphomas'. A search for characteristic cytogenetic abnormalities or oncogene rearrangements has been less successful and neither *bcl-2* nor *bcl-1* gene rearrangement has been detected in lymphomas of the lung. In three cases of primary pulmonary lymphoma, only one of which has been fully characterized as a MALT lymphoma,[18] a t(1;14) (p22;q32) translocation has been demonstrated. Attempts to clone the relevant breakpoint have so far been unsuccessful.

Differential diagnosis

For historical reasons, so-called 'pseudolymphoma' continues to be considered as a defined entity that is difficult to distinguish from lymphoma. The term 'pseudolymphoma' refers to any

lymphoid infiltrate that may be confused with lymphoma; the term arose as a device for the pathologist to offer a 'diagnosis' when in fact no diagnosis had been made; in using this term the pathologist is simply making a veiled confession of ignorance. The advent of immunohistochemistry and molecular genetics has resulted in the more certain criteria for a diagnosis of lymphoma which have been outlined above and as a result the term 'pseudolymphoma' should no longer be used. There are, nevertheless, a number of non-neoplastic reactive conditions that can be confused with lymphoma, especially in small closed needle biopsies. In this context it should be stressed that unless molecular techniques are readily available, a more generous open-lung biopsy should be performed if a diagnosis of lymphoma is under consideration. Conditions that may simulate lymphoma include the following:

Lymphoid interstitial pneumonia (LIP)
This condition is described as a diffuse, chronic interstitial pneumonia rich in small lymphocytes

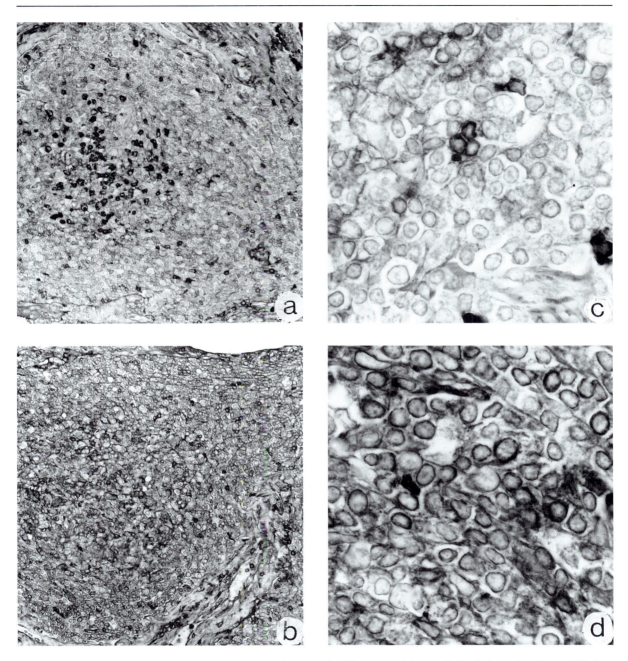

Fig. 12.13   Low-grade B cell pulmonary MALT lymphoma stained in (a) and (c) for κ light chain and in (b) and (d) for λ light chain. The follicle centre is polytypic but the tumour cells surrounding the follicle shown at higher magnification in (c) and (d) show λ light chain restriction (immunoperoxidase).

and plasma cells. There is slow and unremitting progression to diffuse interstitial fibrosis (honeycomb lung) in a third of the cases.[2,20] When defining this entity, Liebow and Carrington[20] already noted that there was a predilection for transformation to lymphoma. Herbert *et al.*[6] remarked on the close similarity between LIP and the peripheral regions of low-grade pulmonary MALT

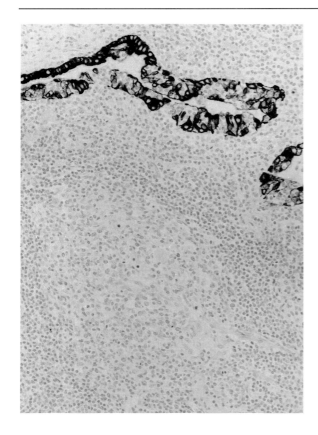

Fig. 12.14 Same section as Fig. 12.5 stained with anticytokeratin which highlights the lymphoepithelial lesions.

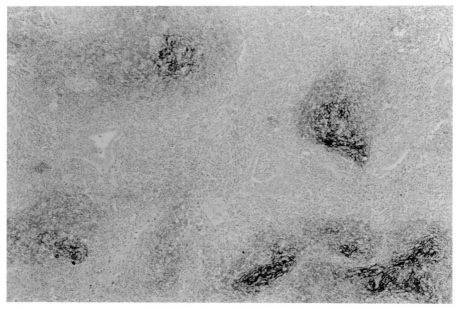

Fig. 12.15 Same case as Fig. 12.8 stained with CD21 to show the follicular dendritic cells of reactive B cell follicles.

lymphoma and Addis *et al.*[5] argued that most cases of LIP were in fact lymphomatous from the outset. Unlike LIP, MALT lymphomas form solid masses, but this may not be apparent in a biopsy and, indeed, is not always the case (Fig. 12.6). The presence of follicles and lymphoepithelial lesions favours a diagnosis of lymphoma. In doubtful cases evidence for monoclonality, either immunohistochemical or molecular, should be sought.

Acquired immunodeficiency syndrome (AIDS)
Patients with AIDS and other immunodeficiency states may develop both LIP and areas of confluent lymphoid infiltration which superficially resemble lymphoma.[21] These changes, which are commoner in children, may be caused by Epstein–Barr Virus (EBV). The infiltrate is, however, much more polymorphous and lacks any of the characteristic features of MALT lymphoma. Low-grade B cell lymphomas, although described in AIDS patients, are rare.

Follicular bronchiolitis
This condition has already been alluded to in the introduction to this chapter. It occurs in patients with autoimmune diseases, including Sjögren's syndrome, as part of a hypersensitivity state accompanied by blood eosinophilia and in patients with immunodeficiency including AIDS.[14] Follicular bronchiolitis effectively results in the formation of MALT around bronchioles and small bronchi. Not surprisingly, the appearances in a needle biopsy can resemble those of MALT lymphoma in which reactive follicles are a prominent feature. Extrafollicular infiltrates of centrocyte-like cells are not present in follicular bronchitis. In doubtful cases a diagnosis of lymphoma should only be made after demonstrating monoclonality using immunohistochemistry or molecular methods.

Miscellaneous conditions
A number of lung disorders with otherwise entirely characteristic histology may contain a dense lymphoid infiltrate. They are unlikely to be confused with lymphoma, especially if an adequate biopsy has been performed. Included in this group are plasma cell granuloma,[22] Castleman's disease,[23] intraparenchymal lymph nodes[24] and pulmonary hyalinizing granuloma.[25]

Clinical behaviour

Numerous reports attest to the remarkably indolent behaviour of low-grade B cell lymphomas of the lung.[1,5–8] While this is a feature of MALT lymphomas in general, the pulmonary MALT lymphoma seem to manifest the most favourable behaviour of the entire group. Long and meticulous follow-up is therefore necessary to evaluate its behaviour and few papers fulfil this requirement. The study of Li *et al.*,[8] which uses the data accumulated in the Kiel registry, is one such and has the advantage of having applied the MALT concept to the analysis of the cases. There were 43 cases of low-grade MALT lymphoma in this series; 33 were clinical stage 1, seven clinical stage 2 and three could not be staged. Treatment varied from simple surgical excision to excision followed by radio- or chemotherapy. Relapses occurred in 20 cases (46%), mostly in the lung but also in mediastinal, cervical, axillary and inguinal lymph nodes. Other sites of relapse included the stomach, skin, tonsil and, in only one case, the bone marrow. In three of eight cases in which histology of the relapse was obtained the tumour had transformed to a high-grade lesion. Despite this incidence of relapse, the survival of patients with low-grade pulmonary MALT lymphoma was no different from that of an aged matched group from the general population. This was also a finding of the large study of Koss *et al.*[1]

Insufficient numbers of high-grade B cell lymphomas of the lung have been studied to make definitive conclusions regarding their behaviour. The survival probability of the eight cases in the series of Li *et al.* was statistically no different from that of the low-grade lymphomas.

Angiocentric T cell lymphoma

In 1972 Liebow *et al.*[26] described a form of pulmonary angiitis and necrotizing granulomatosis which they called lymphomatoid granulomatosis. Later studies showed that frank malignant lymphoma occurred in some 12% of cases. With the evolution of more advanced diagnostic techniques including immunohistochemistry and molecular genetics it has become clear that most, if not all, cases of lymphomatoid granulomatosis of the lung are examples of T cell lymphoma.[27] The term angiocentric lymphoma[9,10,28] has evolved to describe this group of lymphomas seen not only in the lung, but also in the upper respiratory tract, where it was previously known as lethal midline granuloma, and in the skin.

Clinical presentation

Pulmonary angiocentric lymphoma is a disease of adults usually presenting with fever and haemoptysis. Dissemination of the lymphoma to other organs, including the brain, skin and kidney, may occur early in the course of the disease. Chest X-ray shows bilateral nodular lesions but hilar lymphadenopathy is uncommon.

Histopathology

Multiple, often well-defined nodular infiltrates are characteristic of pulmonary angiocentric lymphoma (Fig. 12.16). Confluent areas of necrosis are common and may surround a central surviving blood vessel. The cellular infiltrate is composed of macrophages, epithelioid cells and lymphocytes and intact alveolar septa walls may impose a pattern on this infiltrate resulting in a granulomatous appearance (Fig. 12.17). The lymphocytes in these diffuse areas usually show little or no atypia (Fig. 12.18). Typical Langhans' giant cells may accen-

tuate the granulomatous appearance. The lymphocyte infiltrate centres around blood vessels and invades their walls (Fig. 12.19) and it is here that the cytological features of these cells are best appreciated. An elastic van Gieson stain is useful in delineating the blood vessels and showing that the external and internal elastic lamina of pulmonary arteries are preserved in lymphoma, unlike in vasculitis where they are destroyed. Lipford *et al.*[28] have described three grades of atypia ranging from mild (Fig. 2.20) to moderate (Fig. 12.21) to severe (Fig. 12.22).

Immunohistochemistry

The majority of small cells in angiocentric lymphoma are CD3-positive, CD4-positive, CD8-negative T cells (Fig. 12.23), but characterization of the large atypical cells is more controversial. They are usually CD3-negative but may be CD3-positive and occasionally express CD30. Wong *et al.*[9] have found that in most cases the malignant cells are CD56-positive and have, therefore, concluded that this group, of lymphomas represent malignant transformation of natural killer (NK) cells. Macrophage markers reveal surprisingly large numbers of these cells.

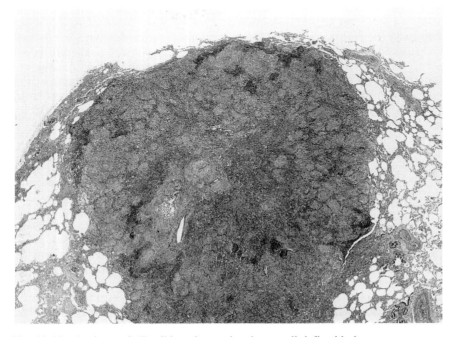

Fig. 12.16   Angiocentric T cell lymphoma showing a well-defined lesion.

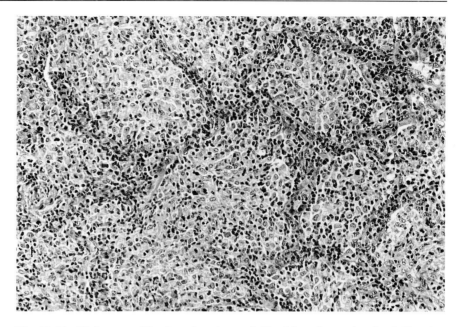

Fig. 12.17 Higher magnification of angiocentric T cell lymphoma showing infiltration of alveoli by epithelioid cells and lymphocytes. Surviving alveolar septa impose a granulomatous appearance on the infiltrate.

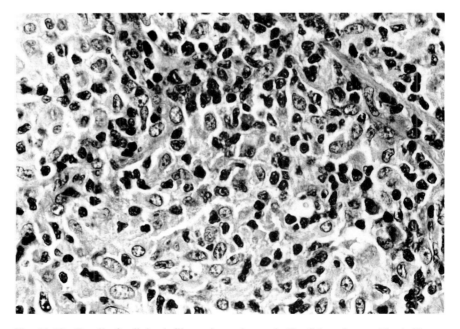

Fig. 12.18 Detail of cellular infiltrate in angiocentric T cell lymphoma. The infiltrate consists of macrophages, epithelioid cells and lymphocytes which show minimal atypia.

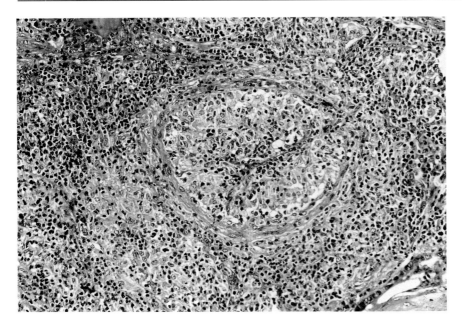

Fig. 12.19   Angiocentric T cell lymphoma showing the characteristic invasion of a blood vessel.

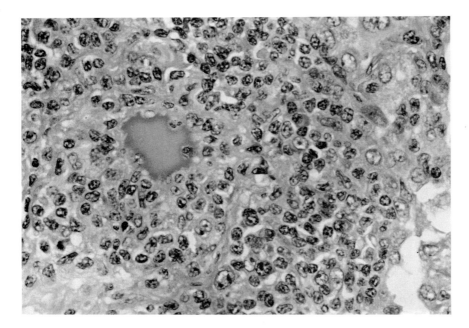

Fig. 12.20   Blood vessel invasion in angiocentric T cell lymphoma showing only mild atypia of the neoplastic T cells.

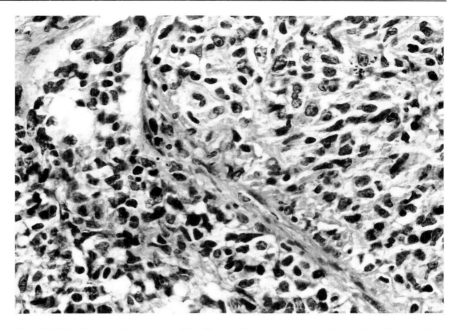

Fig. 12.21 The angio-invasive T cells in this angiocentric T cell lymphoma show moderate atypia.

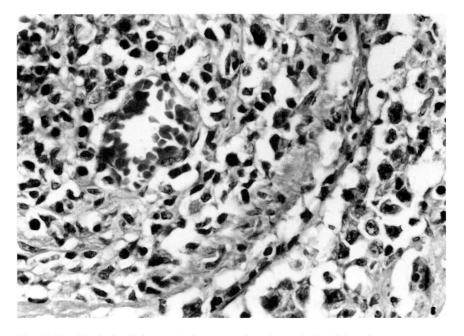

Fig. 12.22 Marked cellular atypia in a case of angiocentric T cell lymphoma.

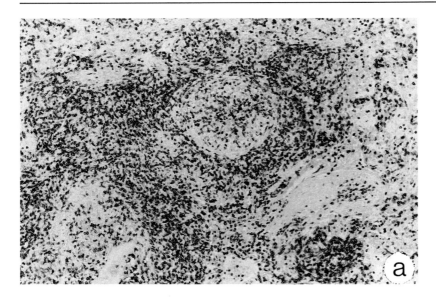

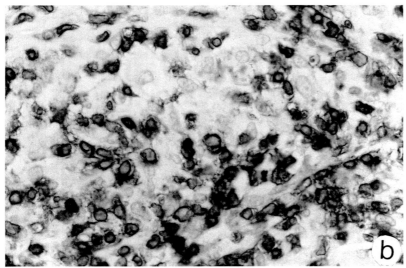

Fig. 12.23 (a) Angiocentric T cell lymphoma immunostained with CD3. Most cells, including those invading the central blood vessel, are CD3-positive T cells. (b) A higher magnification of blood vessel wall showing mild atypia of CD3-positive T cells.

Molecular genetics

T cell receptor gene rearrangement has been demonstrated in a minority of cases; most are negative for both T cell receptor and immunoglobulin gene rearrangement.[10] This would be in keeping with the suggestion that these angiocentric lymphomas are NK cell malignancies.

Clinical behaviour

Reliable information on the clinical behaviour of angiocentric lymphoma awaits the study of a large group of fully characterized cases. The current impression is that, despite intensive chemotherapy, 60% of patients are dead within 2 years.[10]

Other lymphomas

The lung may be the primary site of lymphomas and lymphoproliferative conditions associated with immunodeficiency. In this context a recent report of a series of cases with the features of lymphomatoid granulomatosis have been shown to represent an Epstein–Barr virus associated lymphoproliferative disorder.[29] A few reports of authenticated primary Hodgkin's disease of the lung exist.[30] The histology of these cases is similar to that of nodal disease.

References

1. Koss MN, Hochholzer L, Nichols PW, Wehunt WD, Lazarus AA. Primary non-Hodgkin's lymphoma and pseudolymphoma of lung: a study of 161 patients. *Hum Pathol* 1983; **14**, 1024–1038.
2. Yousem SA, Colby TV. Pulmonary lymphomas and lymphoid hyperplasias. In: Knowles DM (ed.) *Neoplastic Hematopathology*. Baltimore: Williams & Wilkins, 1991, ch. 32, pp. 979–1007.
3. Isaacson P, Wright DH. Extranodal malignant lymphoma arising from mucosa-associated lymphoid tissue. *Cancer* 1984; **53**, 2515–2524.
4. Isaacson PG, Spencer J. Malignant lymphoma of mucosa-associated lymphoid tissue. *Histopathol* 1987; **11**, 445–462.
5. Addis BJ, Hyjek E, Isaacson PG. Primary pulmonary lymphoma: a re-appraisal of its histogenesis and its relationship to pseudolymphoma and lymphoid interstitial pneumonia. *Histopathol* 1988; **13**: 1–17.
6. Herbert A, Wright DH, Isaacson PG, Smith JL. Primary malignant lymphoma of the lung: histopathologic and immunologic evaluation of nine cases. *Hum Pathol* 1984; **15**, 415–422.
7. Chetty R, Close PM, Timme AH, Willcox PA, Forder MD. Primary biphasic lymphoplasmacytic lymphoma of the lung. A mucosa-associated lymphoid tissue lymphoma with compartmentalization of plasma cells in the lung and lymph nodes. *Cancer* 1992; **69**, 1124–1129.
8. Li G, Hansmann ML, Zwingers T, Lennert K. Primary lymphomas of the lung: morphological, immunohistochemical and clinical features. *Histopathol* 1990; **16**, 531.
9. Wong KF, Chan JKC, Ng CS, Tsang WYW. CD56 (NKH1)-positive hematolymphoid malignancies. *Hum Pathol* 1992; **23**, 798–804.
10. Medeiros LJ, Peiper SC, Elwood L, Yano T, Raffeld M, Jaffe ES. Angiocentric immunoproliferative lesions; a molecular analysis of eight cases. *Hum Pathol* 1991; **22**, 1150–1157.
11. Bienenstock J. *Immunology of the Lung and Respiratory Tract*. New York: McGraw Hill, 1984.
12. Pabst R. Compartmentalization and kinetics of lymphoid cells in the lung. *Region Immunol* 1990; **3**, 62–71.
13. Gould S, Isaacson PG. Bronchus associated lymphoid tissue (BALT) in human fetal and infant lung. *J Pathol* 1993; **169**, 229–234.
14. Yousem SA, Colby TV, Carrington CB. Follicular bronchitis/bronchiolitis. *Hum Pathol* 1985; **16**, 700–706.
15. Hansen LA, Prakash UBS, Colby TV. Pulmonary lymphoma in Sjogren's syndrome. *Mayo Clin Proc* 1989; **64**, 920–931.
16. Isaacson PG, Spencer J. Malignant lymphoma of mucosa-associated lymphoid tissue. *Histopathol* 1987; **11**, 445–462.
17. Isaacson PG, Wotherspoon AC, Diss TC, Pan L. Follicular colonization ion B-cell lymphoma of MALT. *Am J Surg Pathol* 1991; **15**, 819–828.
18. Wotherspoon AC, Soosay GN, Diss TC, Isaacson PG. Low-grade primary B-cell lymphoma of the lung. An immunohistochemical, molecular and cytogenetic study of a single case. *Am J Clin Pathol* 1990; **94**, 655–660.
19. Smith-Ravin J, Spencer J, Beverley PCL, Isaacson PG. Characterization of two monoclonal antibodies (UCL4D12 and UCL3D3) that discriminate between human mantle zone and marginal zone B cells. *Clin Exp Immunol* 1990; **82**, 181–187.
20. Liebow AA, Carrington CB. Diffuse pulmonary lymphoreticular infiltrations associated with dysproteinemia. *Med Clin North Am* 1973; **57**, 809–843.
21. White DN, Mattay RA. Non-infectious pulmonary complications of infection with the human immunodeficiency virus. *Am Rev Respir Dis* 1989; **140**, 1763–1787.
22. Maier HC, Sommers SC. Recurrent and metastatic pulmonary fibrous histiocytoma/plasma cell granuloma in a child. *Cancer* 1987; **60**, 1073–1076.
23. Keller AR, Hochholzer L, Castleman B. Hyaline vascular and plasma cell types of giant lymph node hyperplasia of the mediastinum and other locations. *Cancer* 1972; **29**, 670–683.
24. Kradin RL, Spirn PW, Mark EJ. Intrapulmonary lymph nodes. Clinical, radiologic, and pathologic features. *Chest* 1985; **87**, 662–667.
25. Yousem SA, Hochholzer L. Pulmonary hyalinizing granuloma. *Am J Clin Pathol* 1987; **87**, 1–6.
26. Liebow AA, Carrington RB, Friedman PJ. Lymphomatoid granulomatosis. *Hum Pathol* 1972; **3**, 457–558.
27. Dunnill MS. Pulmonary granulomatosis and angiitis. *Histopathol* 1991; **19**, 297–301.
28. Lipford EH Jr, Margolick JB, Longo DL, Fauci AS, Jaffe ES. Angiocentric immunoproliferative lesions: a clinicopathologic spectrum of post-thymic T cell proliferations. *Blood* 1988; **72**, 1674–1681.
29. Guinee D, Jaffe E, Kingma D, Fisgback N, Walberg K, Krishnan J, Frizzera G, Travis W, Koss M. Pulmonary lymphomatoid granulomatosis. Evidence for a proliferation of Epstein–Barr viruses infected B-lymphocytes with a prominent T-cell component and vasculitis. *Am J Surg Pathol* 1994; **18**, 753–764.
30. Radin AI. Primary pulmonary Hodgkin's disease. *Cancer* 1990; **65**, 550–563.

13

Mesothelioma

Samuel P Hammar

Introduction

Mesotheliomas are malignant neoplasms that arise from the serosal lining of the pleural, pericardial and peritoneal cavities. In contrast to most carcinomas and soft tissue sarcomas which usually grow as nodular masses, mesotheliomas have, in most instances, a tendency to grow in a diffuse spreading manner, and in the case of pleural mesotheliomas, typically encase the lung in a rind of tumour. In decades past, malignant mesothelioma was a diagnosis of exclusion and was stated to be made only at postmortem examination by demonstrating the typical growth pattern of the tumour and by ruling out a metastasis from another primary site. However, with today's technology, diagnoses of mesothelioma are made on small biopsies including needle biopsies of pleura, fine needle aspiration biopsies, and malignant cells in fluids such as pleural or peritoneal fluids. The ability to make a specific diagnosis on small specimens is often based on a knowledge of the immunohistochemical and ultrastructural features of mesotheliomas.

In the general population, mesotheliomas are rare tumours, accounting for less than 1% of all cancer deaths in the world. As noted by Chahinian,[1] Joseph Lieutaud described two pleural neoplasms which may have represented mesothelioma in a study of 3000 autopsy cases in 1767. E. Wagner[2] is credited with recognizing mesothelioma as a pathological entity in 1870. Robertson,[3] in 1924, gave an excellent pathological description of pleural neoplasms, some of which probably represented mesotheliomas. Unfortunately, he concluded that only sarcomas could be classified as primary malignant tumours

of the pleura, and that all epithelial neoplasms represented metastases from an unrecognized or unknown primary site. Klemperer and Rabin,[4] in 1931, reported five pleural neoplasms. Four of these were localized and had mesenchymal features, while one of them was diffuse, encasing the lung, and had a mixed epithelial and mesenchymal histological appearance. Klemperer and Rabin[4] were the first to divide tumours of the pleura into the localized and diffuse forms, indicating that the localized tumours most likely originated from subpleural connective tissue, were low-grade malignancies, and were potentially curable by surgical removal. They concluded that the diffuse neoplasms of the pleura arose from mesothelial cells lining the serosal surfaces, and could exhibit either an epithelial or a mesenchymal histologic appearance. Wedler,[5] in his study of lung cancer associated with asbestosis, referred to some neoplasms occurring in persons with asbestosis as 'tumors of the pleura'; these may have been either mesotheliomas or pseudomesotheliomatous carcinomas. Likewise, Merewether,[6] in a 1947 report of lung cancer in asbestos factory employees, described neoplasms referred to as 'tumors of the pleura', which may have also represented mesotheliomas.

Incidence

The world's interest in mesotheliomas was aroused by the report of Wagner, Sleggs and Marchand[7] in 1960 concerning 33 patients in the Northwestern Cape Providence of South Africa, who developed mesotheliomas, of which 32 had a history of exposure to asbestos. Dr Wagner recently recounted his experience with the discovery of mesotheliomas in South Africa,[8] and a recently published photograph and description of the first patient he encountered with mesothelioma.[9] In 1965, Dr Irving Selikoff and his colleagues[10] presented further data associating the development of mesotheliomas with asbestos exposure. Selikoff and colleagues have an on-going study of 17 800 asbestos insulation workers in the USA and Canada, and to date, approximately 10% of the recorded deaths in this cohort have been caused by mesothelioma.[11]

The incidence of mesothelioma varies considerably, although in the non-asbestos-exposed population is extremely low. The highest incidence occurs in locations in which persons are occupationally exposed to asbestos such as shipyards, refineries, and asbestos mines, mills and factories. In Trieste, Italy, a shipbuilding city, Biava *et al.*[12] reported an incidence of 21.4 cases per million population per year. The highest incidence at the present time is in Australia, the majority of cases probably a direct result of exposure to crocidolite asbestos.[13] Several recent papers have described mesotheliomas in young (less than 40 years old) persons who were most likely occupationally exposed to asbestos through contact with a relative in their home, such as father, uncle, grandfather, etc. The apparent increased incidence of mesotheliomas may in part represent a greater awareness of this disease by clinicians, and more accurate diagnosis by pathologists.

Aetiology of mesothelioma

The majority of cases of diffuse malignant mesothelioma in the world are caused by asbestos. The exact percentage of cases of mesothelioma which are caused by asbestos will never been answered with certainty. The various percentages of cases stated to be caused by asbestos are often based on clinical histories of exposure to asbestos, which in this author's experience are often notoriously inaccurate in determining who has and has not been exposed. This author has performed asbestos digestion analysis for asbestos bodies and fibres on several cases of diffuse malignant mesothelioma in which no history of exposure to asbestos was elicited from the patient, and has found asbestos body/fibre counts to be significantly elevated in most cases. The occupations in which asbestos exposure can potentially occur are numerous,[14]

and many persons do not realize they have been occupationally exposed to asbestos. The threshold amount of asbestos necessary to induce mesothelioma is unknown, although most studies have suggested a dose–response relationship. The majority of patients who develop mesotheliomas, who are exposed occupationally to asbestos, do not have asbestosis and in many cases, the concentration of asbestos in the lungs of these people is only slightly elevated over that seen in the non-occupationally exposed group. Cases of mesothelioma have been reported which have been stated to be caused by asbestos, in which the concentration of asbestos in the lung tissue is within what is considered a non-occupational range.[15,16]

The exact mechanism by which asbestos causes mesothelioma is unknown, although at this time asbestos is thought to be a primary carcinogen capable of causing mutagenic changes in the DNA of cells leading to the development of a neoplastic cell.[17] An experimental report published in 1987[18] suggested that the asbestos fibres became trapped in the pleura, primarily in the stomata of the parietal pleura, which then led to inflammation, fibrosis and eventually the development of a mesothelioma in which the carcinogenic effect of asbestos was one factor and the various growth factors induced by the inflammatory cells were another factor. Stanton *et al.*[19] and Gross[20] have shown experimentally that it is the physical characteristics of asbestos that are important in causing mesothelioma, rather than its chemical composition. Fibres greater than 8 µm long and less than 0.25 µm wide have been identified to be more tumorigenic than the shorter, thicker fibres. However, this may not be uniformly true, in that Churg and Wiggs[21] have shown that, in some instances, short, wide amphibole fibres are capable of causing mesotheliomas.

Several studies have investigated the tumorigenic capacity of asbestos fibres with perhaps one exception, the table published by Hillerdal[22] several years ago is generally still pertinent (Table 13.1). In general, the amphibole asbestos crocidolite is the most tumorigenic of the commercial asbestos fibres, and serpentine asbestos chrysotile is the least tumorigenic; amosite and tremolite are in between. Evidence suggesting that chrysotile is significantly less tumorigenic has been reviewed.[23] A recent paper published by Begin *et al.*[24] reviewed all cases of pleural mesothelioma seen and accepted by the Quebec Workmen's Compensation

Table 13.1   Features and relative tumorigenicity of asbestos fibre types and eronite

Fibre	Risk of mesothelioma	Length to diameter ratio	Durability in biological tissue	Human exposure
Fibrous zeolite (eronite)	Very high	Very high	Very good	Only environmental
Amphiliboles				
Crocidolite	Very high	Very high	Very good	Occupational and environmental
Amosite	Fairly high	High, but lower than crocidolite	Good	Occupational
Tremolite	Probably high	As amosite	Good	Environmental
Anthophyllite	None	Fairly low	Good	Environmental (formerly occupational)
Serpentine Chrysotile	Low	Most fibres very short	Poorer than other asbestos	Occupational
Glass fibres	None	Low	Probably poor	Occupational

From ref. 22 with the permission of the publisher.[22]
More recent information suggests that chrysotile may be more tumorigenic than previously believed (see text).

Board for related compensation of industrial disease. They identified 120 cases of pleural mesothelioma, seven of which occurred in women. The cases were subdivided into three groups according to workplace asbestos exposure. Forty-nine cases originated in the mines and mills of the Quebec Eastern Township region, 50 cases from the manufacture and industrial application sector, and 21 cases from industries where asbestos was not a major work material but often an incidental material. In the mining towns of Thetford and Asbestos, the incidence of mesothelioma was found to be proportional to the work force, suggesting that tremolite air contamination, which was 7.5 times higher in Thetford, was not a significant determinant in causing mesothelioma. The incidence of mesothelioma in the chrysotile miners and millers was 62.5 cases per million per year for the period 1980–90, well above the rate found in the North American population which was estimated to be between 2.5 to 15 cases per year per million in adult males. Begin and colleagues concluded that the incidence of pleural mesothelioma in chrysotile miners and millers, although not as high as in crocidolite workers, was well above the average seen in the North American male. They also concluded that comparative analyses of incidence of the disease in the two mining towns suggested that tremolite contamination was not a determining factor in producing mesotheliomas in

the chrysotile workers. They also found that the incidence of work-related mesotheliomas in the tertiary sector appeared to be increasing more rapidly than in other sectors, and constituted 33% of all cases they identified in the last four years of the study. These observations are of great significance, and would suggest that in this cohort the tumorigenicity of chrysotile is significant and perhaps has been underestimated. This author has observed several cases of mesothelioma in persons whose only occupational exposure to asbestos was to dust created in various brake- and clutch-related repair work, in which chrysotile asbestos was the only type of asbestos identified in their lung tissue.

Compared to asbestos, other causes of mesothelioma are rare. Eronite, a type of fibrous zeolite, is almost identical to asbestos in physical characteristics, and has been associated with a substantial number of mesotheliomas occurring in central Turkey, where this material is used in building materials.[25,26] Eronite has been shown experimentally to induce mesotheliomas in rats inoculated intrapleurally with this substance.

Therapeutic radiation, usually given for other types of neoplasms, has also been reported to cause a handful of diffuse malignant mesotheliomas. The most convincing case was one in which the tumour developed in the site of radiation, 20+ years after a primary Wilms tumour had been irradiated.[27] The patient expired, and lung tissue contained no increased concentrations of asbestos.

Anecdotal reports of mesothelioma caused by a variety of other agents have been reviewed and tabulated.[28] This author recently performed an extensive computer search on most of these agents, and could find no epidemiological reports that they were responsible for causing mesotheliomas. Potentially, any agent that injures the pleura can cause mesothelioma, although as chronicled by Henderson *et al.*,[29] agents like asbestos which are fibrous-type minerals with Stanton-like characteristics (aspect ratios greater than 32), are the ones capable of causing mesothelioma in experimental situations. In reality, the majority of these agents do not appear to cause mesothelioma, at least in the setting in which humans are exposed to them.

Finally, genetic factors might be important in the development of mesotheliomas, just as they are in lung cancer, although there has been no large-scale review of this issue. We recently published a case of two families in which mesothelioma developed and reviewed the literature on familial mesothelioma.[30] In one family, three brothers who developed mesothelioma had been employed as asbestos insulators, and had very high concentrations of asbestos bodies (>100 000 per gram of wet tissue) in their lung tissue. In the second family the father was exposed to asbestos and developed mesothelioma; the son who developed mesothelioma was theoretically exposed to asbestos from his father's clothing, although analysis of his lung tissue revealed concentrations characteristic of those present in non-occupationally exposed persons. One could perhaps argue that he had a genetically programmed increased sensitivity to the tumorigenic effect of asbestos.

Pathological features of mesotheliomas

Location and classification

In our experience with approximately 3000 cases of mesotheliomas, 90% have had a pleural origin and the other 10% a peritoneal origin. We have not seen a primary pericardial mesothelioma, but have seen approximately four cases of malignant mesothelioma arising in the region of the tunica vaginalis, a rare but well-known location for malignant mesothelioma, some of which appear to be caused by asbestos.[31]

Mesotheliomas are usually classified histologi-

cally into four major groups: epithelial; sarcomatoid (sarcomatous–fibrous); biphasic (mixed epithelial and sarcomatous); and desmoplastic. This classification is somewhat artificial and is a marked oversimplification of the many histological patterns of malignant mesotheliomas (Table 13.2). There are a variety of different histological epithelial types of mesotheliomas, and the sarcomatoid variety also exhibits significant heterogeneity. There is a histological type of mesothelioma that we refer to as transitional, which has histological features that are between an epithelial and a sarcomatoid mesothelioma. A poorly differentiated epithelial form of mesothelioma occurs, as does a small cell form, and an anaplastic giant cell mesothelioma. Some pathologists have the mistaken notion that mesotheliomas are composed of a uniform cell type. In this author's experience, the exact opposite is true. Even in the same tumour, one can see highly variable differentiation with epithelial areas, sarcomatoid areas and transitional regions. This author strongly believes that the more sections that are taken of a mesothelioma, the more histological variability will be observed. This opinion is supported by van Gelder *et al.*[32] who investigated the relationship between the histopathological diagnosis of mesothelioma and the amount of tissue examined. They found an incidence of 'mixed' histology in 36% of cases of needle biopsy (small) specimens, and 63% of thoracoscopic or autopsy (large) specimens. This author would suggest a new method to classify mesotheliomas: (1) The primary diagnosis would be 'malignant mesothelioma'; (2) the types of differentiation the mesothelioma is showing (with

Table 13.2   Histological classification of malignant mesothelioma

Epithelial
    Tubulopapillary
    Glandular
    Epithelioid (histiocytoid)
    Solid (poorly differentiated)
    Adenoid cystic
    Signet ring
Sarcomatoid (sarcomatous, malignant fibrous)
Mixed–biphasic (epithelial–sarcomatous)
Anaplastic–pleomorphic
Transitional
Desmoplastic

This histological classification lists the most frequent histological types but fails to indicate that in most instances malignant mesotheliomas show significant histological heterogeneity.

the relative amounts of each type) would be
indicated; and (3) 'special' features of the tumour
would be mentioned, such as extensive necrosis,
high mitotic rate, lymphatic/vascular invasion.
This scheme would only be useful in cases in which
larger specimens had been thoroughly sampled.

Macroscopic features of malignant mesothelioma

Mesotheliomas of any histological type are similar
macroscopically. Most grow in a spreading manner
and have a tendency to encase the lung by forming
a rind of greyish-white tissue that surrounds the
lung tissue, and not infrequently invades into it
(Figs 13.1 and 13.2). As suggested by Boutin *et
al.*,[33] mesotheliomas probably begin most frequently on the parietal pleura, and in this author's
opinion might well be of multifocal origin. They
often appear as diffuse greyish-white elevated
nodules that stud the parietal pleura, measuring
anywhere from 1 mm up to 1 cm. At the time of
autopsy, most mesotheliomas do not have this
distinct nodular appearance, but appear as a single
growth of tumour that encases the lung and often
extends into it along the connective tissue that
extends from the visceral pleura to form the
secondary lobules of the lung (Fig. 13.3). The
epithelial mesotheliomas that produce excessive
hyaluronic acid often are extremely slimy, similar
to what one would see in an excess mucus-producing adenocarcinoma of the lung. Sarcomatoid
and desmoplastic mesotheliomas are usually
'harder' and 'denser' than epithelial mesotheliomas, and may be more difficult to remove from
the thoracic cavity at postmortem examination.

Histological features

As previously stated, mesotheliomas can assume a
variety of histological appearances. In larger specimens, the more sections that are taken, the more
histological patterns will be observed (Fig. 13.4).
The epithelial form is most common, and it can
have a tubulopapillary configuration with or
without psammoma bodies, a glandular configuration or a relatively solid configuration (Fig. 13.5).

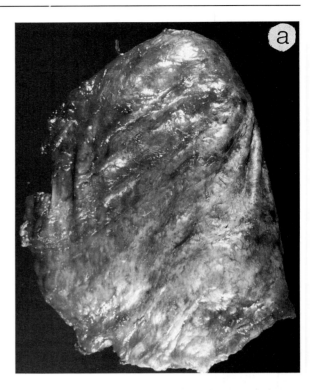

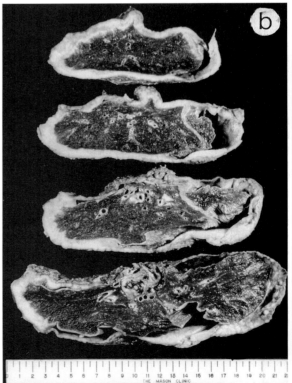

Fig. 13.1 Macroscopic appearance of pleuropneumonec-
tomy specimen. (a) Resected lung and parietal pleura. (b)
The specimen cross-sectioned from top to bottom. Note the
rind of greyish-white tumour encasing the lung at necropsy.

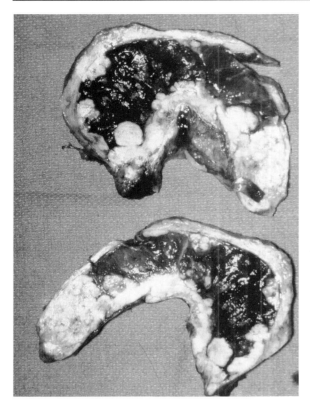

Fig. 13.2 In many cases of malignant mesothelioma, nodular masses of tumour directly invade the pulmonary parenchyma. In some instances this can cause the radiographic appearance of a parenchymal lung mass.

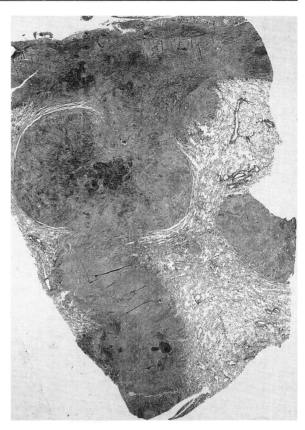

Fig. 13.3 This low-power light micrograph shows a pleural mesothelioma that is encasing the lung and growing into the lung parenchyma in the distribution of the connective tissue that forms the secondary lobules of the lung. × 4.

Some epithelial mesotheliomas have a histiocytic (epithelioid) appearance, being composed of relatively large histiocytic-appearing cells that form relatively large sheets, or an adenoid cystic pattern. Transitional mesotheliomas are composed of cells that are large, polygonal, round, and irregularly shaped, and which do not form glandular structures (Fig. 13.6). This pattern is infrequently present throughout the entire tumour, and regions of sarcomatoid differentiation or epithelial differentiation may be seen. The sarcomatoid type of mesothelioma is usually not a homogeneous neoplasm. It can range from an extremely uniform spindle cell neoplasm to an extremely pleomorphic tumour (Fig. 13.7). The mesothelioma shown in Fig. 13.8 occurred in a 67-year-old man who smoked cigarettes whose filter contained crocidolite asbestos. This tumour initially was thought to be some type of soft tissue sarcoma, probably a

malignant fibrous histiocytoma. As demonstrated by immunohistochemistry, the neoplastic cells had features suggesting mesothelial-type differentiation, expressing keratin and vimentin.

Of the four major histological types of mesothelioma, desmoplastic mesothelioma, which was described initially in 1980[34] and later expanded upon in 1982,[35] is the most deceptive and causes the most problems in diagnosis, especially when biopsy specimens are small. Desmoplastic mesotheliomas are composed predominantly of bland, hypocellular or moderately cellular tissue that often looks like benign reactive connective tissue (Fig. 13.9). The diagnosis of desmoplastic mesothelioma is made by identifying areas in which there is increased cellularity and cellular atypia, or the presence of necrosis, which often has a stellate

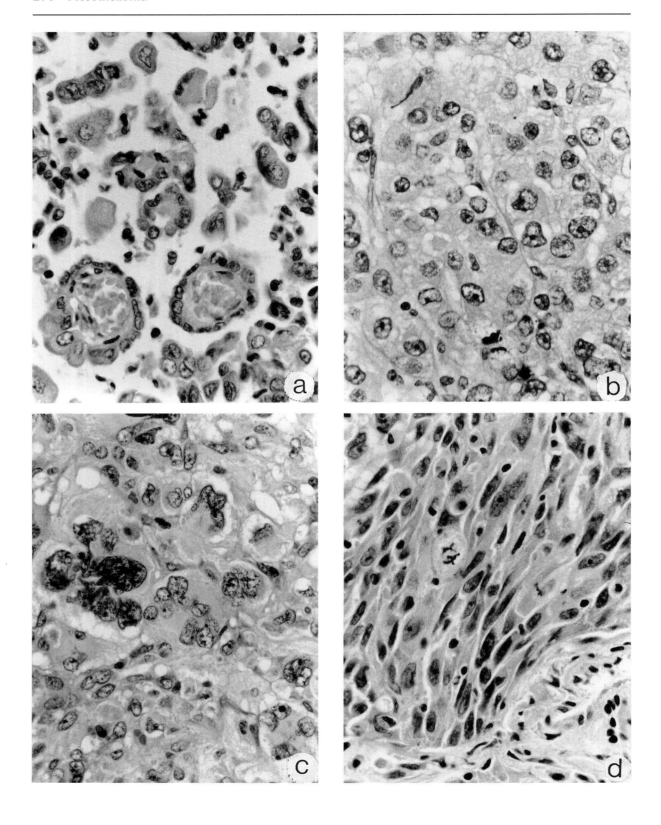

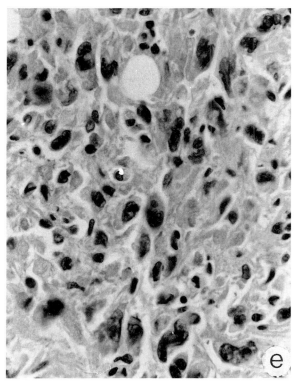

Fig. 13.4 (a–e) Representative regions of this mesothelioma showed a variety of histological patterns. This
histological heterogeneity is common in mesotheliomas if
multiple sections are taken. All × 550.

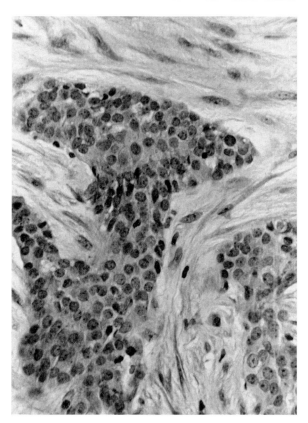

Fig. 13.5 This epithelial mesothelioma was composed
predominantly of solid sheets of relatively uniform
epithelial cells. × 300.

pattern. As desmoplastic mesotheliomas grow and
invade into the lung, they can produce a pattern
that looks identical to an organizing pneumonitis,
and taken out of context, can cause a problem in
diagnosis.

Another type of mesothelioma that is difficult to
put into a distinct category is what is referred to as
a small cell mesothelioma (Fig. 13.10).[23] We previously categorized this type of mesothelioma as an
epithelial mesothelioma, although when examined
by electron microscopy, the tumour does not have
distinct epithelial features, and by immunohistochemistry has a pattern more characteristic of a
sarcomatoid-type mesothelioma, expressing keratin and vimentin.

In 1988, Henderson et al.[36] reported an unusual
type of mesothelioma referred to as a lymphohistiocytoid mesothelioma, representing a variant of a
sarcomatoid mesothelioma, which was characterized by an infiltrate of lymphocytes and histiocytes
admixed with large, occasionally slightly spindle-

shaped cells (Fig. 13.11). As noted by these
authors, this tumour can be confused with a
lymphoma unless familiarity with this histological
pattern exists and appropriate ancillary studies,
namely electron microscopy and immunohistochemistry, are performed on suspected cases.

Histochemical features of mesotheliomas

Pathologists commonly use histochemical tests to
attempt to differentiate epithelial mesotheliomas,
especially those that are well or moderately well
differentiated, from pulmonary adenocarcinomas
or other types of metastatic carcinomas involving
the pleura. The tests that are most frequently done
include periodic acid Schiff (PAS), PAS–diastase
(PASD), mucicarmine and Alcian blue, or colloid-
al iron with and without hyaluronidase. Approximately 20% of epithelial mesotheliomas produce
hyaluronic acid in significant quantity, and the

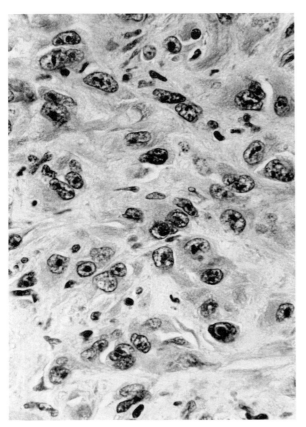

Fig. 13.6 This mesothelioma was composed of large variably-shaped cells and is what we refer to as a transitional mesothelioma. × 550.

demonstration of this substance and differentiation of it from neutral or slightly acidic mucin is supposedly helpful in differentiating a hyaluronic acid-producing epithelial mesothelioma from a neutral or slightly acidic mucin-producing pulmonary adenocarcinoma. While these tests are often done, and in some instances may be helpful, Benjamin and Ritchie[37] examined the staining results for glycogen and mucosubstances of 30 diffuse epithelial mesothelioma that had been fixed in formalin and found that some of the mesothelioma were positive for the mucicarmine and PASD, and that seven of 30 mesotheliomas failed to stain with any of the seven mucosubstance stains they used. They concluded that the histochemical staining reactions of epithelial mesotheliomas with mucopolysaccharide stains were too inconsistent to be of much value in diagnosing epithelial mesotheliomas. This author generally agrees with that

notion, although histochemical tests for mucosubstances continue to be used by many practising pathologists in differentiating epithelial mesotheliomas from pulmonary adenocarcinomas. Ernst and Atkinson[38] reported seven of 18 cases of epithelial mesothelioma that were mucicarmine-positive, and attributed this positive staining to hyaluronic acid. The review article on mesothelioma by the US–Canadian Mesothelioma Panel[39] also illustrated a mucicarmine-positive mesothelioma, and MacDougall et al.[40] recently reported a case that was mucicarmine, PASD and alcian blue hyaluronidase-positive that was confirmed by electron microscopy. We[41] compared the histochemical and immunohistochemical staining reactions of 10 epithelial mesotheliomas (diagnosis documented by ultrastructural examination) that were mucicarmine-positive, and compared them with 10 pulmonary adenocarcinomas. The adenocarcinomas were all primary nodular lung masses that were mucicarmine-positive. We concluded that the mucicarmine and PASD staining reactions in epithelial mesotheliomas were probably due to staining of the hyaluronic acid in these neoplasms. When the tissue sections were pretreated with hyaluronidase, there often was disappearance or reduction in the intensity of staining reactions with PASD and mucicarmine. In some instances, however, the staining was not completely eradicated. As will be discussed later, ultrastructural examination of such neoplasms can easily diagnose them as mesotheliomas and provide insight into the features responsible for the mucicarmine-positive staining.

Immunohistochemical features of mesotheliomas

Immunohistochemical techniques have made a contribution to our understanding of normal serosal tissues and mesotheliomas. Most of the immunohistochemical tests done to differentiate epithelial mesotheliomas from pulmonary adenocarcinomas have employed antibodies against keratin, vimentin, muscle-specific actin, carcinoembryonic antigen, LeuM1 (CD15), B72.3 and BerEP4. A few other antibodies not commercially available have also been used in the identification of epithelial mesotheliomas. As shown in Table 13.3, all mesotheliomas, including the sarcomatoid and desmoplastic form, express keratin, with most epithelial types of mesothelioma expressing both low- and high-molecular-weight keratin. Some

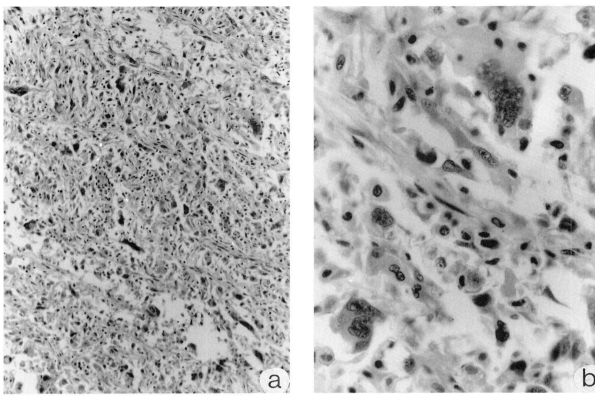

Fig. 13.7 (a,b) This sarcomatoid mesothelioma shows regions in which the neoplastic cells become large and anaplastic. (a) × 300, (b) × 550.

pathologists believe that the pattern of staining for keratin within a cell is helpful in differentiating an epithelial mesothelioma from a pulmonary adenocarcinoma, with epithelial mesotheliomas characteristically showing a perinuclear distribution of keratin staining, and adenocarcinomas showing a more diffuse staining pattern. In this author's experience using antibodies AE1/AE3 (Biogenics), and 35BH11 and 34BE12 (Enzo Biochem), we have not been able to differentiate epithelial mesotheliomas from pulmonary adenocarcinomas, and know by ultrastructural evaluation that some pulmonary adenocarcinomas show perinuclear aggregates of intermediate keratin filaments. Mullink et al.[42] have suggested that it is more common for epithelial mesotheliomas to coexpress keratin and vimentin than it is for pulmonary adenocarcinomas, but this has not been the experience of others[43] performing immunohistochemical tests on pulmonary adenocarcinomas. With respect to keratin immunostaining of epithelial mesotheli-

omas and pulmonary adenocarcinomas, Blobel et al.[44] reported that epithelial mesotheliomas contained keratin 5, as demonstrated by immunohistochemistry, whereas pulmonary adenocarcinomas did not. The study by Moll et al.[45] supported this finding in that 12 of 13 epithelial mesotheliomas expressed keratin 5 as demonstrated with monoclonal antibody AE14, whereas 21 pulmonary adenocarcinomas were negative, or in six cases, showed immunostaining of only a few cells.

Epithelial membrane antigen (EMA) and human milk fat globule protein-2 (HMFG-2) are positive in most well-differentiated to moderately differentiated epithelial mesotheliomas and pulmonary adenocarcinomas. However, as described by Leong et al.,[48] epithelial mesotheliomas have a tendency to show a cell membrane pattern of immunostaining for HMFG-2 and EMA, whereas pulmonary adenocarcinomas show cytoplasmic staining. Even this is not absolute, because some non-mucinous type of bronchioloalveolar cell

Table 13.3 Immunohistochemical features of malignant mesotheliomas

	LMWk	HMWk	Vimetin	EMA	HMFG-2	CEA	LeuM1	B72.3	BerEP4	HHF-35
Epithelial										
Tubulopapillary	+	+	±	±[a]	±[a]	−[b]	−[c]	−[d]	−[e]	−
Glandular	+	+	±	±[a]	±[a]	−[b]	−[c]	−[d]	−[e]	−
Epithelioid (histiocytoid)	+	+	±	±[a]	±[a]	−[b]	−[c]	−[d]	−[e]	−
Solid poorly differentiated	+	+	±[f]	±[g]	±[g]	−	−	−	−	−
Adenoid cystic	+	+	±	±[a]	±[a]	−[b]	−[c]	−[d]	−[e]	−
Signet ring	+	+	±	±[a]	±[a]	−[b]	−[c]	−[d]	−[e]	−
Sarcomatoid	+[h]	±[h]	+	−[i]	−[i]	−	−	−	−	±
Mixed-biphasic										
Epithelial	+	+	±	±[a]	±[a]	−[b]	−[c]	−[d]	−[e]	−
Sarcomatoid	+[h]	±[h]	+	−[i]	−[i]	−	−	−	−	±
Anaplastic (pleomorphic)	+[j]	±	+	−[k]	−[k]	−	−	−	−	−
Transitional	+	±	+	−[k]	−[k]	−	−	−	−	−
Desmoplastic	+	±	+	−	−	−	−	−	−	±[l]

[a] In the positive cases, the pattern of immunostaining is in a cell membrane distribution.
[b] Up to 15% of cases of epithelial mesotheliomas show focal, usually low-intensity immunostaining.
[c,d] Rare cases of epithelial mesothelioma show immunostaining, usually focal.
[e] In one report, up to 20% of epithelial mesotheliomas show focal immunostaining.
[f] Vimentin may be expressed in any type of epithelial mesothelioma, but is more frequently expressed in the epithelial mesotheliomas that are more poorly differentiated.
[g] Infrequently positive.
[h] Despite what is published, some sarcomatoid mesotheliomas do not immunostain for keratin with commercially available anti-keratin antibodies.
[i,j,k] Rare cases may be positive.
[l] Most cases are positive.
Abbreviations: LMWk, low-molecular-weight keratin; HMWk, high-molecular-weight keratin; EMA, epithelial membrane antigen; HMFG-2, human milk fat globule protein-2; CEA, carcinoembryonic antigen; LeuM1, CD15; B72.3, TAG-72; HHF-35, muscle-specific actin.

carcinomas of the lung will show a predominantly cell membrane staining for HMFG-2 and EMA. Antibodies against carcinoembryonic antigen (CEA), LeuM1, B72.3 and BerEP4 are the most useful tests in discriminating between an epithelial mesothelioma and pulmonary adenocarcinoma, but these tests are not absolute.[23,47-52] As a general rule, well-differentiated to moderately well-differentiated epithelial mesotheliomas show no immunostaining for CEA, LeuM1, B72.3 and BerEP4. However, up to about 15% of epithelial mesotheliomas can show focal areas of immunostaining for CEA,[53] which in this author's experience is seen most frequently in mesotheliomas that produce hyaluronic acid. Occasional mesotheliomas will also be positive for LeuM1 and B72.3.[54] The newest antibody used to differentiate epithelial mesotheliomas from pulmonary adenocarcinomas (anti-BerEP4), was initially stated to be almost exclusively found in pulmonary adenocarcinomas and negative in mesotheliomas.[55] However, a paper published about 6 months after this initial report[56] reported that up to 20% of epithelial

mesotheliomas showed focal areas of immunostaining for BerEP4, usually of low intensity, and one of the cases of malignant mesothelioma reported was diffusely positive. Thus, like most immunohistochemical tests, one has to be careful in interpreting the results; they should be interpreted in context with other findings.

Immunohistochemistry can be of general help in differentiating sarcomatoid (fibrous) mesotheliomas from soft tissue spindle cell sarcomas.[23] In the majority of cases, sarcomatoid mesotheliomas show immunostaining of the spindle cells for keratin and vimentin, and may show staining of the spindle cells for EMA and HMFG-2. In contrast, most sarcomas are negative for keratin and epithelial membrane antigen, although as recently reviewed by Miettinen,[57] almost all sarcomas can potentially produce keratin, but only two (epithelioid sarcoma, synovial sarcoma) express keratin with any frequency. One should be aware of the fact that many sarcomatoid mesotheliomas express muscle-specific actin, which usually correlates ultrastructurally with myofibroblastic differ-

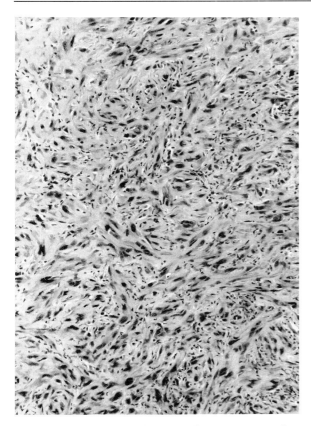

Fig. 13.8 This sarcomatoid mesothelioma was present in a man who smoked cigarettes containing crocidolite asbestos in the filter. × 300.

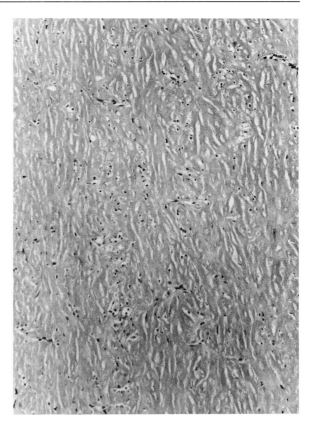

Fig. 13.9 Representative region of desmoplastic mesothelioma, showing the neoplasm to be composed of dense, relatively hypocellular connective tissue.

entiation by these neoplastic cells. Immunohistochemical studies are not particularly helpful in diagnosing desmoplastic mesotheliomas, since the primary differential diagnosis in such cases is fibrosing pleuritis (chronic pleural fibrosis), and in most instances the spindle-shaped cells in both conditions express keratin and vimentin, as well as muscle-specific actin.

Localized fibrous mesothelioma, which will be discussed later, is a neoplasm that probably arises from subpleural connective tissue cells and usually form distinct nodular masses within the peripheral portion of the lung or within the pleural cavity. By immunohistochemistry the spindle cells forming these neoplasms usually show immunostaining only for vimentin.

Ultrastructural features of mesotheliomas

Electron microscopy is most useful in evaluating

epithelial mesotheliomas, specifically in differentiating epithelial mesotheliomas from pulmonary adenocarcinomas or other adenocarcinomas. In fact, in this author's opinion, it is the single best test in differentiating an epithelial mesothelioma from a pulmonary adenocarcinoma, assuming that the tissue sample is adequate and has been properly fixed. Electron microscopy can be performed on small biopsy specimens, such as needle biopsies of pleura, fine needle aspiration biopsy of pleural masses, and on malignant cells in pleural effusions. The most important ultrastructural feature that allows the differentiation of a well- to moderately well-differentiated epithelial mesothelioma from another epithelial neoplasm is the presence of long thin sinuous microvilli that arise from the cell surfaces (Fig. 13.12). While there have been studies comparing the length to width ratios of mesothelioma microvilli to those seen in various types of adenocarcinomas,[23] this author feels that

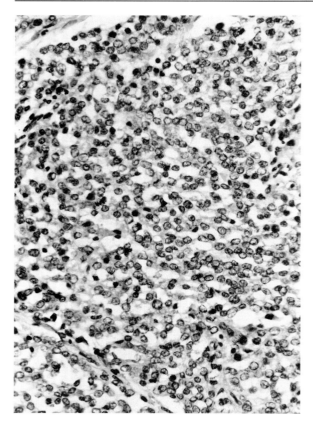

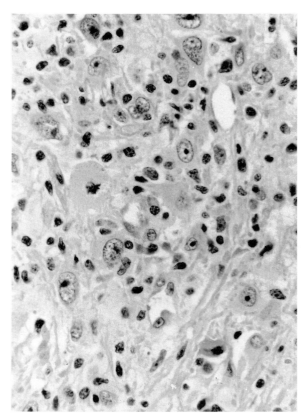

Fig. 13.10 This mesothelioma is one of the rarest histological types, and is composed of a uniform population of small round cells. × 125.

Fig. 13.11 This mesothelioma, referred to as a lympho-histiocytoid mesothelioma, is infiltrated by lymphocytes and histiocytes which obscure its features, and sometimes lead to an incorrect diagnosis of lymphoma. × 500.

measurement is not necessary, and in almost all cases, it is fairly obvious if the tumour has a microvillous pattern characteristic of a mesothelioma. Most pulmonary adenocarcinomas will have short microvilli, which in contrast to an epithelial mesothelioma, will be covered by a fuzzy glycocalyx (Fig. 13.13). The microvilli of an epithelial mesothelioma are smooth. Occasionally, we have encountered some pulmonary adenocarcinomas with long flowing microvilli, but these microvilli have always been covered by a fuzzy glycocalyx. Mesothelial cells also have other ultrastructural features that are not totally specific, but when taken in context with the microvillous pattern may be helpful in differentiating them from other types of neoplasms. Neoplastic epithelial mesothelial cells characteristically have abundant tonofilaments in their cytoplasm that are more frequently located in a perinuclear location than elsewhere,

often contain aggregates of cytoplasmic glycogen, and typically contain short profiles of rough endoplasmic reticulum. The desmosomes that connect the mesothelial cells to one another have been reported to be larger than the desmosomes connecting the cells of a pulmonary adenocarcinoma to each other.[58] Malignant mesothelial cells may form glandular structures, and where they form glands, the cells are characteristically connected to each other by junctional complexes. We[23,41] and Henderson et al.[59] have described certain ultrastructural features that we believe are relatively specific for mesotheliomas that produce large amounts of hyaluronic acid. Hyaluronic acid may occasionally 'coat' and obscure the surface of microvilli when this substance occurs in large amounts (Fig. 13.14). Likewise, the hyaluronic acid may crystallize, forming a fern-like pattern which represents aggregates of hyaluronic acid glycosaminoglycans, or crystallize into a tubular type of structure which we have seen only in hyaluronic acid-producing epithelial mesotheliomas (Fig. 13.15).

Recently it was thought that epithelial mesotheliomas had another unique ultrastructural feature; namely, the projection of microvilli into extracellular collagen.[60] This electron microscopic feature has been shown not to be specific since rare adenocarcinomas also display this feature. Most epithelial mesotheliomas show a fairly well-defined basal lamina separating the neoplastic cells from the connective tissue stroma, and in some instances there are a moderate number of cell membrane micropinocytotic vesicles at the cell membrane-basal lamina interface.

As mesotheliomas become more poorly differentiated (poorly differentiated epithelial mesothelioma–transitional mesothelioma), they lose most of their distinguishing ultrastructural characteristics. The ultrastructural features of such dedifferentiated mesotheliomas are relatively non-specific. Sarcomatoid mesotheliomas, in general, have non-specific ultrastructural features, although in the majority of them one can find evidence of 'epithelial' differentiation in the form of intracellular tonofilaments, occasionally well-formed desmosomes between the spindle-shaped tumour cells, and the presence of basal lamina surrounding a portion or all of the tumour cells. The combination of the electron microscopic findings and the coexpression of vimentin and keratin is highly suggestive of a sarcomatoid mesothelioma.

DNA Analysis, proliferative rate evaluation, and cytogenetic features of malignant mesothelioma

Two recent reports have evaluated cytogenetic abnormalities in malignant mesotheliomas. Tiainen *et al.*[61] performed successful cytogenetic analyses on cells obtained from tumours and/or pleural effusions in 34 of 38 patients. Clonal chromosomal abnormalities were detected in 25 patients, the majority being complex and heterogeneous, with no chromosome abnormality specific to mesothelioma. Nine patients had normal karyotypes and/or non-clonal chromosomal abnormalities. Translocations and deletions involving a breakpoint at 1p11–p22 were the most common structural aberrations. The number of copies of chromosome 7 short arms was inversely

correlated with survival, and a high concentration of asbestos fibres in the lung tissue was associated with partial or total losses of chromosomes 1 and 4, and a breakpoint at 1p11–p22.

Hagemeijer *et al.*[62] evaluated 40 confirmed cases of malignant mesothelioma, in 90% using cells from a pleural effusion. A normal karyotype was found in nine cases and complex karyotypic abnormalities were identified in 30 cases. The chromosomal changes were all complex and heterogeneous, with no consistent specific abnormality detected. Two main patterns of non-random abnormalities were found: (1) loss of chromosomes 4 and 22, 9p and 30p in the most abnormal cases, corresponding to a hypodiploid and/or hypo-tetraploid modal chromosome number; and (2) gain of chromosomes 7, 5 and 20 with deletion or rearrangement of 3p.

Several studies have evaluated DNA concentrations and proliferative indices in reactive mesothelial proliferations and in malignant mesothelioma. Croonen *et al.*[63] evaluated malignant cells in 11 pleural fluids and two ascitic fluids from six patients with malignant mesothelioma (five proven by biopsy) and found DNA-euploid cells in 10 of 13 effusions. They concluded that malignant mesotheliomas were usually DNA-euploid whereas adenocarcinomas were characteristically aneuploid.

Hafiz *et al.*[64] used cytophotometry to study the DNA content of cells in Feulgen-stained effusion specimens from 18 patients with mesothelioma and 14 patients with reactive mesothelial proliferations, and found that the mean DNA content of malignant mesotheliomas was significantly higher (30.5 ± 7.2) than reactive mesothelial cells (15.2 ± 2.9).

Frierson *et al.*[65] compared the DNA content of 28 fresh effusion specimens that contained abundant proliferating mesothelial cells with the ploidy profiles for 19 formal in-fixed paraffin-embedded malignant epithelial mesotheliomas. The cells in all cytologically benign fluid specimens were DNA diploid. Nine of 19 deparaffinized mesotheliomas were diploid, seven were aneuploid, and three had flow cytometric patterns suggesting hyperdiploidy. The authors concluded that while only 53% of epithelial mesotheliomas were aneuploid, the findings of DNA aneuploidy in an effusion specimen that contained atypical mesothelial cells supported a diagnosis of malignant mesothelioma.

Burmer *et al.*[66] studied the DNA content of 46

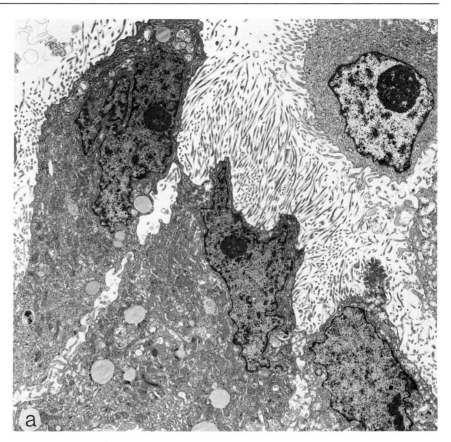

Fig. 13.12a Electron micrograph showing characteristic ultrastructural features of an epithelial mesothelioma. Note long thin sinuous microvilli arising from the cell surface of neoplastic mesothelial cells. × 5040.

cases of malignant pleural mesothelioma and found the majority (65%) to be DNA-diploid with low to intermediate proliferative rates, whereas 85% of primary lung carcinomas were DNA aneuploid and had high proliferative rates.

Dazzi *et al.*[67] evaluated tumour tissue from 168 formalin-fixed, paraffin-embedded specimens from 70 patients with malignant pleural mesothelioma (31 epithelial, 21 sarcomatous, 18 mixed). They found that 38.6% of mesotheliomas were diploid and 61.4% were aneuploid. A higher percentage of the epithelial mesotheliomas were diploid. The authors found no significant difference in survival in the patients whose mesotheliomas were aneuploid versus diploid. Patients whose tumour showed an S-phase percentage greater than the median of 6% had a significantly shorter survival than those whose tumours had a lower S-phase.

Tierney *et al.*[68] determined DNA concentration using DNA image analysis of Feulgen-stained slides in a series of 56 pleural lesions, including reactive proliferations, malignant mesotheliomas and metastatic tumours. Of the 36 cases of primary mesothelial lesions, data suggested that 11 were benign and 25 were malignant mesotheliomas. As determined by pathological diagnosis, there were eight false-negative benign interpretations and five false-positive malignant interpretations. The authors concluded that mesothelial lesions appeared to have a wide range of ploidy values regardless of their biological behaviour, and that ploidy could not be used as a reliable diagnostic index in diagnosing primary mesothelial tumours. With respect to the metastatic tumours, DNA analysis was associated with no false positives or false negatives.

El-Naggar *et al.*[69] analysed 23 immunohisto-

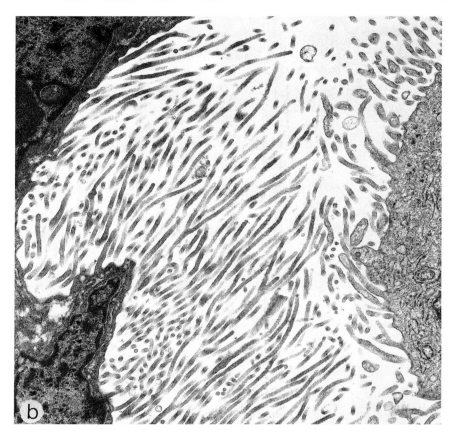

Fig. 13.12b Electromicrograph showing characteristic ultrastructural features of an epithelial mesothelioma. Note long thin sinuous microvilli arising from the cell surface of neoplastic mesothelial cells. × 12 800.

chemically and ultrastructurally proven pleural epithelial mesotheliomas by flow cytometry and compared them with 41 pulmonary adenocarcinomas. Multiple separate tissue blocks were evaluated from each neoplasm to assess DNA heterogeneity. They found that 80% of pulmonary adenocarcinomas and 100% of pleural mesotheliomas showed a homogeneous DNA ploidy. The majority (78%) of epithelial mesotheliomas were diploid, whereas 88% of pulmonary adenocarcinomas were aneuploid. The proliferative fraction (S-phase) of aneuploid adenocarcinomas was significantly greater than aneuploid epithelial mesotheliomas. These authors concluded that the DNA indices of epithelial mesotheliomas were significantly different than pulmonary adenocarcinomas.

Esteban and Sheibani[70] performed flow cytometric analysis of 45 malignant mesotheliomas and 41 pulmonary adenocarcinomas. Five cases of mesotheliomas were excluded due to high coefficients of variation. Five cases (14%) of malignant mesothelioma were aneuploid with DNA indices between 1.2 to 1.9 (mean 1.5), and three cases had increased $S + G_2M$ values. Of the aneuploid mesotheliomas, four were epithelial and one was sarcomatous. In contrast to malignant mesotheliomas, 31 (75%) of pulmonary adenocarcinomas were aneuploid. The authors concluded that in view of the marked differences between malignant mesotheliomas and pulmonary adenocarcinomas, ploidy analysis should be used in diagnostically difficult cases in which histochemistry, immunohistochemistry and electron microscopy could not provide an unequivocal diagnosis.

In reviewing the findings of DNA content in mesotheliomas, this author believes there is too much variability in the observations to draw any definite conclusions.

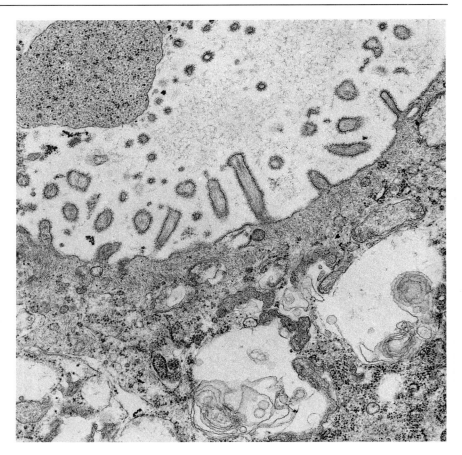

Fig. 13.13   This pulmonary adenocarcinoma cell shows short microvilli arising from the cell surface, which are covered by a fuzzy glycocalyx. × 16 000.

Histogenesis of mesotheliomas

In 1986, we published a paper[71] concerning reactive and neoplastic serosal tissues, trying to gain insight into the development and morphology of mesotheliomas. We found that injury to pleural tissue such as that caused by metastatic tumours or trauma often resulted in a proliferation of spindle-shaped cells which had a unique immunohisto-chemical profile, and a surprising ultrastructural appearance. These cells, which we have termed 'multipotential subserosal cells', coexpress vimentin, keratin, and muscle-specific actin, and by electron microscopy have the ultrastructural features of myofibroblasts. Experimental studies of rat serosal tissue, plus our own studies, suggested that it was the multipotential subserosal cell that was the cell that gave rise to the surface

epithelial mesothelial cell through a process of differentiation. There are many phenotypic similarities between reactive non-neoplastic serosal tissue and mesotheliomas. Desmoplastic and sarcomatoid mesotheliomas resemble the multi-potential subserosal cell with similar immunohisto-chemical features (positive for keratin, vimentin and muscle-specific actin) and in many cases ultrastructural features having a myofibroblastic appearance. Diffuse epithelial mesotheliomas resemble the surface mesothelium and it is probably likely that the surface mesothelium does give rise to epithelial mesotheliomas, since it is not an end-stage tissue incapable of turnover. Our interpretation of the origin of mesotheliomas using this model has been challenged by Henderson *et al.*,[29] who believe that the surface mesothelial cells dedifferentiate and become spindle-shaped, and

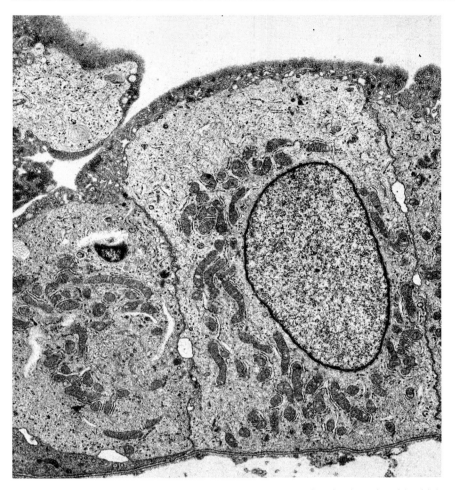

Fig. 13.14 This epithelial mesothelioma was producing abundant hyaluronic acid, which coated the surface of the neoplastic mesothelial cells, obscuring the microvilli. × 8000.

are responsible for forming sarcomatoid and desmoplastic mesotheliomas. From a practical point of view, it does not make much difference which theory is correct. What is important is that one is aware of the morphological features of sarcomatoid and desmoplastic mesotheliomas and how this profile is different from tumours that could be confused with mesotheliomas.

Other pleural neoplasms – differential pathological diagnosis

Other neoplasms involve the pleura and create major difficulties in differentiation from meso-theliomas. With respect to the epithelial mesothelioma, the most important differential pathological diagnosis is pulmonary adenocarcinoma. Histologically and cytologically, these tumours can look virtually identical. This is true for malignant cells in pleural fluid, as well as biopsies from the pleura and chest wall. An accurate diagnosis is possible, even on small tissue specimens, if there is an adequate sample of viable tumour upon which to perform additional tests; i.e. immunohisto-chemistry and electron microscopy.

There is one type of pulmonary carcinoma that can very closely simulate a malignant mesothelioma macroscopically (Fig. 13.16), referred to as a pseudomesotheliomatous carcinoma. This type of neoplasm was probably first described in 1940 by

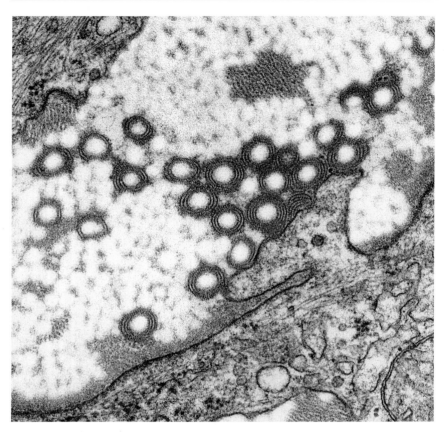

Fig. 13.15 Sometimes the hyaluronic acid formed by epithelial mesotheliomas crystallizes, producing a scroll-like pattern. × 40 500.

Babolini and Blasi[72] as a pleural form of primary lung cancer. The initial description of this tumour in Italy is of some interest, because Italy was the first country to mine and use asbestos commercially, and it is our opinion that most cases of pseudomesotheliomatous carcinoma of the lung occur in persons who are occupationally exposed to asbestos.[73] Using a combination of histochemistry, immunohistochemistry and electron microscopy, most cases of pseudomesotheliomatous carcinoma can be separated from mesotheliomas. The most frequent form of pseudomesotheliomatous carcinoma is an adenocarcinoma we have termed tubulodesmoplastic, in which small tubules of neoplastic epithelial cells are dispersed throughout the dense fibrous connective tissue pleura. For reasons that are unknown, these tumours grow like mesotheliomas, spreading over the pleural surface and encasing the lung. Unlike mesotheliomas, they are often CEA-positive and may be LeuM1-,

B72.3- and BerEP4-positive. By electron microscopy, they show evidence of non-mesothelial type differentiation such as glandular differentiation, squamous or small cell neuroendocrine differentiation. These tumours are thought to arise from small focal areas of cancer in the periphery of the lungs that do not form a dominant mass in the lung, but invade through the pleura and grow in a pleural distribution, like a malignant mesothelioma.

It should also be recognized that cases have been reported of localized malignant mesotheliomas that have the typical immunohistochemical and electron microscopic appearance of mesotheliomas. Three such cases were reported in abstract form in 1992,[74] and the abstract indicated that these had a significantly better prognosis with resection. This author has seen several such cases; one in particular was interesting pathologically and was perhaps unique. The tumour formed a large

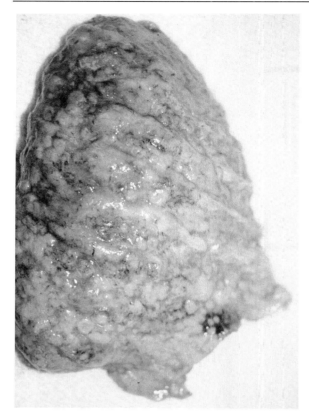

Fig. 13.16 This lung is encased by a rind of greyish-white tumour that has the macroscopic appearance of a malignant mesothelioma. Histologically this neoplasm showed squamous differentiation, and represents a pseudomesotheliomatous carcinoma.

distinct mass that was initially diagnosed as a localized fibrous tumour of the pleura, and was composed predominantly of spindle-shaped cells in which there were a few glandular-like structures. The spindle-shaped cells did not conform to the usual immunohistochemical pattern of a sarcomatoid mesothelioma in that they showed immunostaining only for vimentin and not for keratin. The epithelial component of the tumour showed immunostaining for keratin and was negative for CEA, LeuM1, B72.3 and BerEP4, and ultrastructurally had the characteristic features of an epithelial mesothelioma.

Certain metastatic tumours can produce localized masses involving the pleura, and one tumour, metastatic renal cell carcinoma, can cause a great deal of diagnostic confusion with a mesothelioma because the immunohistochemical pattern is often

the same, and some renal cell carcinomas can have fairly long, thin, smooth microvilli. We have also seen some localized masses that have involved both lung and pleura, and which were negative for CEA, LeuM1, B72.3 and BerEP4. These tumours were initially diagnosed as mesotheliomas, although when examined ultrastructurally they were found to have long microvilli which were covered by a glycocalyx; they were subsequently classified as pulmonary adenocarcinomas.

Most soft tissue sarcomas that involve the pleura or chest wall are localized masses that theoretically could be confused with a localized malignant mesothelioma. Immunohistochemically, these tumours usually show immunostaining only for vimentin, or possibly vimentin, actin and desmin, and do not coexpress keratin. Occasionally, a rare sarcoma of the chest wall and pleura can express keratin and can cause some diagnostic difficulty. Ultrastructural evaluation can occasionally be helpful in differentiating sarcomatoid mesotheliomas from spindle cell sarcomas in that sarcomatoid mesotheliomas usually show some evidence of

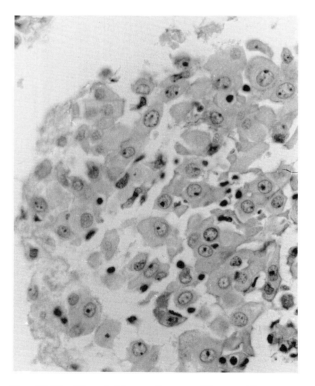

Fig. 13.17 Pleural biopsy from a 65-year-old woman showing aggregates of neoplastic round to polygonal cells. × 480.

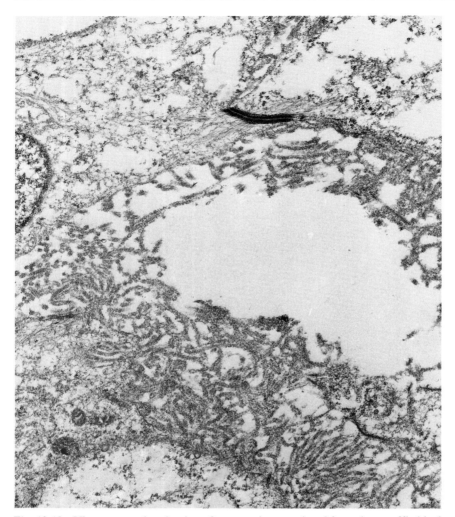

Fig. 13.18 Ultrastructural evaluation of tumour tissue retrieved from the paraffin block showed degenerative changes but typical features of an epithelial mesothelioma with long sinuous microvilli, and large desomsomes connecting the cells to each other. × 12 800.

epithelial differentiation such as basal lamina formation, tonofilament formation, and well-formed desmosomes, ultrastructural features that would generally not be seen in most soft tissue sarcomas. Other tumours involving the pleura that initially can be confused with sarcomatoid mesothelioma are angiosarcomas, which are extremely rare although they have been reported,[75] and other types of metastatic sarcomas.

Localized fibrous mesotheliomas were reviewed by Briselli, *et al.*[76] in 1981. In 1989, England *et al.*[77] evaluated 233 cases and suggested a name change to localized fibrous tumour of the pleura

based on evidence that these tumours were not of pleural origin, but were derived from submesothelial connective tissue cells. Macroscopically these tumours grow as well-demarcated (often ovoid) masses, and can be located within the pleural cavity and attached to the visceral pleura by a pedicle, or as a well-demarcated mass in the periphery of the lung. Histologically these neoplasms are composed of spindle cells of varying densities and are frequently associated with abundant extracellular collagen. The criteria used by England *et al.*[77] to diagnose malignancy were more than four mitoses per 10 high-power fields, pleomorphism, haemorrhage and necrosis. Most of the neoplasms evaluated by England *et al.*[77] were 5–10 cm in greatest dimension and weighed 100–400 g. Of the 169 localized fibrous tumours of the pleura available for follow-up, all of the benign and 45% of the malignant neoplasms were cured by surgical excision. Fifty-five per cent of patients with malignant tumours succumbed to their disease secondary to invasion, recurrence or metastasis.

Approach to the diagnosis

The diagnosis of malignant mesothelioma can be made in virtually any type of biopsy, including small biopsies, assuming there is a malignant neoplasm present, and the sample is not too degenerated to allow the performance of additional studies. As previously stated, we do not usually do histochemical studies, but go directly to immunohistochemistry and electron microscopy. Of these two ancillary techniques, electron microscopy is favoured in the epithelial tumours that are well- to moderately well-differentiated, but are often done in context with immunohistochemistry. Immunohistochemistry is a procedure that can be and is done in most laboratories, whereas electron microscopy is sometimes restricted; nevertheless, electron microscopy is more specific. If a small tumour biopsy is diagnosed as malignant, then one can perform the usual immunohistochemical profile on it, and decide if it has features more suggestive of mesothelioma or adenocarcinoma. Of the immunohistochemical tests, the most important for the determination of mesothelioma versus adenocarcinoma are CEA, LeuM1, B72.3 and BerEP4

From a practical point of view, it makes little difference what type of tumour is involving the pleura, since the overall prognosis is generally dismal. Nevertheless, a diagnosis is usually required, and in the USA many of the cases of asbestos-associated pleural neoplasms are litigated. It is sometimes for this reason that an accurate diagnosis is required.

Figure 13.17 shows a case we recently evaluated of a pleural biopsy composed of relatively large, polygonal to round cells that showed some cohesion but were not totally cohesive. By immunohistochemistry the tumour was positive for low- and high-molecular-weight keratin and focally positive for vimentin, being negative for CEA, LeuM1, B72.3 and BerEP4. We removed part of this tumour from the block for ultrastructural examination, and were able to identify features strongly suggestive of the diagnosis of an epithelial mesothelioma (Fig. 13.18). If one is certain that there are malignant cells in pleural fluid, a cell block can be studied by immunohistochemistry. Also, a sample of the pleural fluid can be spun down and the sediment either made into a cell block and treated as a tissue specimen for electron microscopy, or the sediment that contains the malignant cells can be processed directly for electron microscopic examination.

Clinico-pathological correlations

Antman[78] has adapted a clinical staging system from Butchart *et al.*[79] for patients with mesothelioma, according to the extent of the disease, similar to that done for primary lung cancer and other cancers. In general, there is a direct correlation between the stage of the disease and survival. Antman[78] reported survival rates of 16, 9 and 5 months in patients with stage 1, stage 2 and stage 3 disease, respectively. The average survival rate in most cases of mesotheliomas is between 9 and 12 months from the time of diagnosis to death. We have seen a few cases of well-documented invasive malignant mesotheliomas in which patients have lived as long as 10 years. Most of these long survivors have been women with epithelial mesotheliomas.

One other important factor to remember is that even though mesotheliomas tend to grow in a spreading manner, they frequently do metastasize.[23] They most frequently metastasize to regional lymph nodes and to the non-involved lung (usually pleural surface), and can metastasize to almost any organ in the body, including the brain.

References

1. Chahinian AP. Malignant mesothelioma. In: Holland JF, Frei E III (eds) *Cancer Medicine*. Philadelphia: Lea & Febiger, 1982, pp. 1744–1751.
2. Wagner E. Das tuberkelanliche lymphadenom. *Arch Heilk* 1870; **11**, 495–525.
3. Robertson HE. 'Endothelioma' of the pleura. *Am J Cancer* 1924; **8**, 317–375.
4. Klemperer P, Rabin CB. Primary neoplasms of pleura: report of 5 cases. *Arch Pathol* 1931; **11**, 385–412.
5. Wedler HW. Uber den Lungenkrebs bei Asbestose. *Deut Med Woch* 1943; **69**, 575–576.

6. Merewether ERA. *Annual Report of the Chief Inspector of Factories for the year 1947*. London: His Majesty's Stationery Office, 1949, pp. 78–81.

7. Wagner JC, Sleggs CA, Marchand P. Diffuse pleural mesothelioma and asbestos exposure in North Western Cape Province. *Br J Industr Med* 1960; **17**, 260–271.

8. Wagner JC. The discovery of the association between blue asbestos and mesotheliomas and the aftermath. *Br J Industr Med* 1991; **48**, 399–403.

9. Wagner JC. Asbestos and mesothelioma: a personal reminiscence. In: Henderson DW, Shilkin KB, Langlois SLP, Whitaker D (eds) *Malignant Mesothelioma*. New York: Hemisphere Publishing Corporation, 1992, pp. xvii–xxv.

10. Selikoff IJ, Churg J, Hammond EC. Relation between exposure to asbestos and mesothelioma. *N Engl J Med* 1965; **272**, 560–565.

11. Selikoff IJ, Seidman H. Asbestos-associated deaths among insulation workers in the United States and Canada, 1967–1987. In: Landrigan PI, Kazemi H, Selikoff IJ (eds). The third wave of asbestos disease: exposure to asbestos in place-public health control. *Ann NY Acad Sci* 1991; **643**, 1–14.

12. Biava PM, Ferri R, Spacal B *et al*. Cancro de lovora a Trieste: Il mesothelioma della pleura. *Sapere* 1976; **79**: 41–45.

13. Ferguson D. Malignant mesothelioma – the rising epidemic. *Med J Austral* 1989; **150**, 233–235.

14. Teta MJ, Lewinsohn HC, Meigs JW, Vidone RA, Mowad LZ, Flannery JT. Mesothelioma in Conneticut 1955–1977. *J Occup Med* 1983; **25**, 749–756.

15. Epler GR, Gerald MXF, Gaensler EA, Carrington CB. Asbestos-related disease from household exposure. *Respiration* 1980; **39**, 229–240.

16. Chen W, Mottet NK. Malignant mesothelioma with minimal asbestos exposure. *Hum Pathol* 1978; **9**, 253–258.

17. Walker C, Everett J, Barret JC. Possible cellular and molecular mechanisms for asbestos carcinogenicity. *Am J Industr Med* 1992; **21**, 253–273.

18. Moalli PA, Macdonald JL, Goodglick LA, Kane AB. Acute injury and regeneration of the mesothelium in response to asbestos fibers. *Am J Pathol* 1987; **128**, 426–445.

19. Stanton MF, Layord M, Tegeris A *et al*. Relation of particle dimension to carcinogenicity in amphibole asbestos and other fibrous minerals. *J Natl Cancer Inst* 1981; **67**, 695–975.

20. Gross P. Is short fibered asbestos a biological hazard? *Arch Environ Health* 1974; **29**, 115–117.

21. Churg A, Wiggs B. Fiber size and number in amphibole asbestos-induced mesothelioma. *Am J Pathol* 1984; **115**, 437–442.

22. Hillerdal G. Malignant mesothelioma 1982; review of 4710 published cases. *Br J Dis Chest* 1983; **77**, 321–343.

23. Hammar SP. Pleural neoplasms. In: Dail DH, Hammar SP (eds) *Pulmonary Pathology*. New York: Springer-Verlag, 1987, pp. 973–1028.

24. Begin R, Gauthier J, Desmeules M, Ostiguy G. Work related mesothelioma in Quebec, 1967–1990. *Am J Ind Med* 1992; **22**, 531–542.

25. Baris YI, Saracci R, Simonato L, Skidmore JW, Artvinli M. Malignant mesothelioma and radiological chest abnormalities in two villages in Central Turkey. *Lancet* 1981; **i**, 984–987.

26. Artvinli M, Baris YI. Environmental fiber-induced pleuro-pulmonary disease in an Anatolian village: an epidemiologic study. *J Natl Cancer Inst* 1979; **63**, 17–22.

27. Austin MB, Fechner RE, Roggli VL. Pleural malignant mesothelioma following Wilms' tumor. *Am J Clin Pathol* 1986; **86**, 227–230.

28. Peterson JT Jr, Greenberg SD, Buffler PA. Non-asbestos-related malignant mesothelioma: a review. *Cancer* 1984; **54**, 951–960.

29. Whitaker D, Manning LS, Robinson BWS, Shilkin KB. The pathobiology of the mesothelium. In: Henderson DW, Shilkin KB, Langlois SLP, Whitaker, D. (eds) *Malignant Mesothelioma*. New York: Hemisphere Publishing Corporation, 1992, pp. 25–68.

30. Hammar S. Familial mesothelioma: a report of two families. *Hum Pathol* 1989; **20**, 107–112.

31. Antman K, Cohen S, Dimitrov NV *et al*. Malignant mesothelioma of the tunica vaginalis testis. *J Clin Oncol* 1984; **2**, 447–451.

32. van Gelder T, Hoogsteden HC, Vandenbroucke JP, Van der Kwast TH, Planteydt HT. The influence of the diagnostic technique on the histopathological diagnosis in malignant mesothelioma. *Virch Archiv A Pathol Anat* 1991; **418**, 315–317.

33. Boutin C, Viallat JR, Rey F. Thoracoscopy in diagnosis, prognosis, and treatment of mesothelioma. In: Antman K, Aisner J (eds) *Asbestos Related Malignancy*. Orlando: Grune & Stratton, Inc, 1987, pp. 301–321.

34. Kannerstein M, Churg J. Desmoplastic diffuse malignant mesothelioma. In: Fenoglio CM, Wolff M (eds) *Progress in Surgical Pathology*. New York: Mason, 1980, pp. 19–29.

35. Cantin R, Al-Jabi M, McCaughey WTE Desmoplastic diffuse mesothelioma. *Am J Surg Pathol* 1982; **6**, 215–222.

36. Henderson DW, Attwood HD, Constance TI, Shilkin KB, Steele RH. Lymphohistiocytoid mesothelioma: a rare lymphomatoid variant of predominantly sarcomatoid mesothelioma. *Ultra Pathol* 1988; **12**, 367–384.

37. Benjamin CJ, Ritchie AC. Histological staining for the diagnosis of mesothelioma. *Am J Med Technol* 1982; **48**, 905–908.

38. Ernst CS, Atkinson BF. Mucicarmine positivity in malignant mesothelioma. *Lab Invest* 1980; **42**, 113–114 (abstract).

39. McCaughey WTE, Colby TV, Battifora H *et al.* Diagnosis of diffuse malignant mesothelioma: experience of a US/Canadian mesothelioma panel. *Mod Pathol* 1991; **4**, 342–353.

40. MacDougall D, Wang SE, Zidar BL. Mucin-positive epithelial mesothelioma. *Arch Pathol Lab Med* 1992; **116**, 874–880.

41. Hammar SP, Bockus DE, Remington FL, Rohrback KA. Mucicarmine positive epithelial mesothelioma: a histochemical, immunohistochemical and ultrastructural comparison with mucin-producing pulmonary adenocarcinoma. Submitted for publication.

42. Mullink H, Henzen-Logmans SC, Alons-van Kordelaar JM *et al.* Simultaneous immunoenzyme staining of vimentin and cytokeratins with monoclonal antibodies as an aid in the differential diagnosis of malignant mesothelioma from pulmonary adenocarcinoma. *Virch Arch B Pathol Anat* 1986; **52**, 55–65.

43. Upton MP, Hirohashi S, Tome Y, Miyazawa N, Suemasu K, Shimosato Y. Expression of vimentin in surgically resected adenocarcinomas and large cell carcinomas of lung. *Am J Surg Pathol* 1986; **10**, 560–567.

44. Blobel GA, Moll R, Franke WW, Kayser KW, Gould VE. The intermediate filament cytoskeleton of malignant mesotheliomas and its diagnostic significance. *Am J Pathol* 1985; **121**, 235–247.

45. Moll R, Dhovailly D, Sun T. Expression of keratin 5 as a distinctive feature of epithelial and biphasic mesotheliomas: an immunohistochemical study using monoclonal antibody AE 14. *Virch Arch B Cell Pathol* 1989; **58**, 129–145.

46. Leong AS-Y, Parkinson R, Milios J. 'Thick' cell membranes revealed by immunocytochemical staining: a clue to the diagnosis of mesothelioma. *Diag Cytopathol* 1990; **6**, 9–13.

47. Wick MR, Loy T, Mills SE, Legier JF, Manivel JC. Malignant epithelioid pleural mesothelioma versus peripheral pulmonary adenocarcinoma: a histochemical ultrastructural, and immunhistologic study of 103 cases. *Hum Pathol* 1990; **21**, 759–766.

48. Cibas ES, Corson JM, Pinkus GS. The distinction of adenocarcinoma from malignant mesothelioma in cell blocks of effusions: the role of routine immunohistochemical assessment of carcinoembryonic antigen, keratin proteins, epithelial membrane antigen, and milk fat globule-derived antigen. *Hum Pathol* 1987; **18**, 67–74.

49. Warnock ML, Stoloff A, Thor A. Differentiation of adenocarcinoma of the lung from mesothelioma: periodic acid-Schiff, monoclonal antibodies B72.3 and Leu M1. *Am J Pathol* 1988; **133**, 30–38.

50. Spagnolo DV, Whitaker D, Carrello S, Radosevich JA, Rosen ST, Gould VE. The use of monoclonal antibody 44-3A6 in cell blocks in the diagnosis of lung carcinoma, carcinomas metastic to lung and pleura, and pleural malignant mesothelioma. *Am J Clin Pathol* 1991; 322–329.

51. Kuhlmann L, Berghauser K-H, Schaffer R. Distinction of mesothelioma from carcinoma in pleural effusions. *Path Res Pract* 1991; **187**, 467–471.

52. Tickman RJ, Cohen C, Varma VA, Fekete PS, DeRose PB. Distinction between carcinoma cells and mesothelial cells in serous effusions; usefulness of immunohistochemistry. *Acta Cytol* 1990; **34**, 491–496.

53. Mezger J, Lamerz R, Permanetter W. Diagnostic significance of carcinoembryonic antigen in the differential diagnosis of malignant mesothelioma. *J Thorac Cardiovasc Surg* 1990; **100**: 860–866.

54. Otis CN, Carter D, Cole S, Battifora H. Immunohistochemical evaluation of pleural mesothelioma and pulmonary adenocarcinoma. *Am J Surg Pathol* 1987; **11**, 445–456.

55. Sheibani K, Shin SS, Kezirian J, Weiss LM. BER-EP4 antibody as a discriminant in the differential diagnosis of malignant mesothelioma versus adenocarcinoma. *Am J Surg Pathol* 1991; **15**, 779–784.

56. Gaffey MJ, Mills SE, Swanson PE, Zarbo RJ, Shah AR, Wick MR. Immunoreactivity for BER-EP4 in adenocarcinomas, adenomatoid tumours, and malignant mesotheliomas. *Am J Surg Pathol* 1992; **16**, 593–599.

57. Miettinen M. Keratin subsets in spindle cell sarcomas; keratins are widespread but synovial sarcoma contains a distinctive keratin polypeptide pattern and desmoplakins. *Am J Pathol* 1991; **138**, 505–513.

58. Burns TR1, Johnson EH, Cartwright T Jr, Greenberg SD. Desmosomes of epithelial malignant mesothelioma. *Ultra Pathol* 1988; **12**, 385–388.

59. Henderson DW, Shilkin KB, Whitaker D, Attwood HD, Constance TJ, Steele RH, Leppard PJ. The pathology of malignant mesothelioma, including immunohistology and ultrastructure. In: Henderson DW, Shilkin KB, Langlois SLP, Whitaker D (eds) *Malignant Mesotheliomas.* New York, Hemisphere Publishing Corporation, 1992, pp. 69–139.

60. Ghadially FN, McCaughey WTE, Perkins DG, Rippstein P. Diagnostic value of microvillus-matrix associations in tumours. *J Submicrosc Cytol Pathol* 1992; **24**, 103–108.

61. Tiainen M, Tammilehtol L, Rautonen J, Tuomi T, Mattson K, Knuutila S. Chromosomal abnormalities and their correlations with asbestos exposure and survival in patients with mesothelioma. *Br J Cancer* 1989; **60**, 618–626.

62. Hagemeijer A, Versnel MA, Van Drunen E, Moret M, Bouts MJ, van der Kwast Th H, Hoogsteden HC. Cytogenetic analysis of malignant mesothelioma. *Cancer Genet Cytogenet* 1990; **47**, 1–28.

63. Croonen AM, van der Valk P, Herman CJ, Lindeman J. Cytology, immunopathology and flow cytometry in the diagnosis of pleural and peritoneal effusions. *Lab Invest* 1988; **58**, 725–732.

64. Hafiz MA, Becker RL Jr, Mikel UV, Bahr GF. Cytophotometric determination of DNA in mesotheliomas and reactive mesothelial cells. *Anal Quant Cytol Histol* 1988; **10**, 120–126.

65. Frierson HF, Mills SE, Legier JF. Flow cytometric analysis of ploidy in immunohistochemically confirmed examples of malignant mesothelioma. *Am J Clin Pathol* 1988; **90**, 240–243.

66. Burmer GC, Rabinovitch PS, Kulander BG, Rusch V, McNutt MA. Flow cytometric analysis of malignant pleural mesothelimas. *Hum Pathol* 1989; **20**, 777–783.

67. Dazzi H, Thatcher N, Hasleton PS, Chatterjee AK, Lawson AM. DNA analysis by flow cytometry in malignant pleural mesothelioma: relationship to histology and survival. *J Pathol* 1990; **162**, 51–55.

68. Tierney G, Wilkinson MJ, Jones JSP. The malignancy grading method is not a reliable assessment of malignancy in mesothelioma. *J Pathol* 1990; **160**, 209–211.

69. El-Naggar AK, Ordonez NG, Garnsey L, Batsakis JG. Epithelioid pleural mesotheliomas and pulmonary adenocarcinomas: a comparative DNA flow cytometric study. *Hum Pathol* 1991; **22**, 972–978.

70. Esteban JM, Sheibani K. DNA ploidy analysis of pleural mesotheliomas: its usefulness for their distinction from lung adenocarcinomas. *Mod Pathol* 1992; **6**, 626–630.

71. Bolen JW, Hammar SP, McNutt MA. Reactive and neoplastic serosal tissue. A light microscopic, ultra-structural and immunohistochemical study. *Am J Surg Pathol* 1986; **10**, 34–47.

72. Babolini G, Blasi A. The pleural form of primary cancer of the lung. *Dis Chest* 1956; **29**, 314–323.

73. Robb JA, Hammar SP, Yokoo H. Psuedomesotheliomatous lung carcinoma: a rare asbestos-related malignancy readily separable from epithelial pleural mesothelioma. *Hum Pathol* (in press).

74. Myer J, Tazelaar H, Katzenstein L *et al*. Localized malignant epithelioid and biphasic mesothelioma of the pleura: clinicopathologic, immunohistochemical, and flow cytometric analysis of 3 cases. *Mod Pathol* 1992; **5**, abstract 679.

75. McCaughey WTE, Dardick I, Barr RJ. Angiosarcomas of serous membranes. *Arch Pathol Lab Med* 1983; **107**, 304–307.

76. Briselli M, Mark EJ, Dicersin GR. Solitary fibrous tumours of the pleura: eight new cases and review of 360 cases in the literature. *Cancer* 1981; **47**, 2678–2689.

77. England DM, Hochholzer L, McCarthy MJ. Localized benign and malignant fibrous tumours of the pleura. A clinicopathologic review of 223 cases. *Am J Surg Pathol* 1989; **13**, 640–658.

78. Antman KH. Current concepts: malignant mesothelioma. *New Engl J Med* 1980; **303**, 200–202.

79. Butchart EG, Ashcroft T, Barnsley WC, Holden M. The role of surgery in diffuse malignant mesothelioma of the pleura. *Semin Oncol* 1981; **8**, 321–328.

Index